Building Acoustics

Building or architectural acoustics is taken in this book to cover all aspects of sound and vibration in buildings. The book covers room acoustics but the main emphasis is on sound insulation and sound absorption and the basic aspects of noise and vibration problems connected to service equipment and external sources. Measuring techniques connected to these fields are also brought in. It is designed for advanced level engineering studies and is also valuable as a guide for practitioners and acoustic consultants who need to fulfil the demands of building regulations.

It gives emphasis to the acoustical performance of buildings as derived from the performance of the elements comprising various structures. Consequently, the physical aspects of sound transmission and absorption need to be understood, and the main focus is on the design of elements and structures to provide high sound insulation and high absorbing power. Examples are taken from all types of buildings. The book aims at giving an understanding of the physical principles involved and three chapters are therefore devoted to vibration phenomena and sound waves in fluids and solid media. Subjective aspects connected to sound and sound perception is sufficiently covered by other books; however, the chapter on room acoustics includes descriptions of measures that quantify the "acoustic quality" of rooms for speech and music.

Tor Erik Vigran is professor emeritus at the Norwegian University of Science and Technology, Head of the Acoustic Committee of Standards Norway, the Norwegian standardization organization, and member of several working groups within ISO/TC 43 and CEN/TC 126.

Building Acoustics

Tor Erik Vigran

CRC Press
Taylor & Francis Group
Boca Raton London New York

CRC Press is an imprint of the
Taylor & Francis Group, an **informa** business

A TAYLOR & FRANCIS BOOK

This translation has been published with the financial support of NORLA. Authorised Translation from the Norwegian language edition published by Tapir Academic Press of Nardoveien 14, NO-7005 Trondheim, Norway.

CRC Press
Taylor & Francis Group
6000 Broken Sound Parkway NW, Suite 300
Boca Raton, FL 33487-2742

First issued in paperback 2019

ISBN-13: 978-0-415-42853-8 (hbk)
ISBN-13: 978-0-367-86521-4 (pbk)

A catalogue record for this book is available from the British Library

Library of Congress Cataloging-in-Publication Data
Vigran, Tor Erik.
[Bygningsakustikk English]
Building acoustics / Tor Erik Vigran.
p. cm.
Includes bibliographical references and index.
1. Soundproofing. 2. Architectural acoustics. 3. Acoustical engineering. I. Title.
TH1725.V54 2008
729'.29–dc22
2007039258

Visit the Taylor & Francis Web site at
http://www.taylorandfrancis.com

and the CRC Press Web site at
http://www.crcpress.com

Contents

CHAPTER 3
Waves in fluid and solid media

CHAPTER 4
Room acoustics

CHAPTER 5
Sound absorbers

CHAPTER 6
Sound transmission. Characterization and properties of single walls and floors

CHAPTER 9
Sound transmission in buildings. Flanking sound transmission

List of symbols

SYMBOL	QUANTITY	UNIT
a	radius, length of edge	m
a	acceleration	m/s^2
a_i	equivalent absorption length	m
A	absorption area	m^2
b	slit width	m
B	permeability	m^2
B	bending stiffness per unit length	N·m
c	damping coefficient (mechanical)	N·s/m
c	sound speed (phase speed)	m/s
c_0	sound speed in air	m/s
c_B	bending wave speed	m/s
c_L	longitudinal wave speed	m/s
c_p	specific heat capacity at constant pressure	J/(kg·K)
c_S	shear wave speed	m/s
c_v	specific heat capacity at constant volume	J/(kg·K)
C	adaptation term	dB
C_{te}	early-to-late index	1
d	length	m
D	dipole moment	m^4/s
D	level difference	dB
$D_{v,ij}$	velocity level difference	dB
D_{50}	definition	1
D_θ	directivity factor	1
E	modulus of elasticity	N/m^2
E	energy	J
f	frequency	Hz
f_c	critical frequency	Hz
f_R	ring frequency	Hz
f_S	Schroeder limiting frequency	Hz
F	force	N
G	shear modulus	N/m^2
G	power spectral density function	[1]
h	impulse response	"
H	transfer function	1
I	sound intensity	watt/m^2
I	area moment of inertia per unit length	m^3
j	imaginary unit	
k, k_{mec}	stiffness (mechanical)	N/m
k	wave number	m^{-1}
k_B	wave number of bending wave	m^{-1}
k_s	tortuosity	1
K	bulk modulus	Pa

---

[1] The unit will depend on the actual physical quantity.

SYMBOL	QUANTITY	UNIT
K_{ij}	vibration reduction index	dB
L	edge length	m
L_n	impact sound pressure level	dB
L_p	sound pressure level	dB
L_w	sound power level	dB
LF	lateral energy fraction	1
m	mass	kg
m	mass per unit area	kg/m^2
m	power attenuation coefficient	m^{-1}
M	mobility (mechanical)	$m/(N \cdot s)$
p	sound pressure	Pa
p	probability density function	1
P	probability	1
P	pressure	Pa
P_0	atmospheric pressure	Pa
Pr	Prandtl number	1
q	scattering cross section	m^{-1}
Q	source strength	m^3/s
Q	Q factor	1
r	airflow resistivity	$Pa \cdot s/m^2$
r_0	radius	m
r_H	hall radius	m
$R(\tau)$	correlation function	2
R	room constant	m^2
R	sound reduction index	dB
R'	apparent sound reduction index	dB
R_w	weighted sound reduction index	dB
R_s	pressure reflection factor	1
s	stiffness per unit area	N/m^3
s_f	pore shape factor	1
S	area	m^2
T	time period, measuring time	s
T	reverberation time	s
T	transmissibility	1
u	velocity (of a surface)	m/s
U	circumference	m
v	sound particle velocity, surface velocity[3]	m/s
V	volume	m^3
w	sound energy density	J/m^3
W	power (acoustical, mechanical)	watt
Z_a	acoustic impedance	$Pa \cdot s/m^3$
Z_c	complex characteristic impedance	$Pa \cdot s/m$
Z_f	field impedance (normalized)	1
Z_g	surface impedance	$Pa \cdot s/m$

---

[2] The unit will depend on the actual physical quantity.

[3] In cases where u and v appear simultaneously, u is a surface velocity and v is the particle velocity in the wave.

SYMBOL	**QUANTITY**	**UNIT**
Z_m	mechanical impedance	N·s/m
Z_n	normalized impedance (to Z_0)	1
Z_r	radiation impedance	N·s/m
Z_s	specific acoustic impedance	Pa·s/m
Z_w	wall impedance (transmission impedance)	Pa·s/m
Z_0	characteristic impedance for air	Pa·s/m
α	attenuation coefficient	m^{-1}
α	absorption factor	1
α'	absorption exponent	1
β	phase coefficient	m^{-1}
δ	phase angle	radians
δ	viscous skin depth	m
ε	perforation ratio	1
γ_{xy}	coherence function	1
γ	adiabatic constant	1
μ	coefficient of viscosity	kg/(m·s)
η	loss factor	1
η_{ij}	coupling loss factor	1
φ	angle	radians
κ	thermal conductivity	watt/(m·K)
λ	wavelength	m
υ	Poisson's ratio	1
θ	angle	radians
ρ	density	kg/m^3
σ^2	variance	
σ	porosity, radiation factor	1
τ	transmission factor	1
ω	angular frequency	radians/s
ζ	damping ratio	1
Γ	propagation coefficient	m^{-1}
Λ	characteristic viscous length	m
Λ'	characteristic thermal length	m

NOTE For a quantity X, $\hat{X}$ signifies the amplitude value, $\tilde{X}$ the RMS-value and $|X|$ the modulus.

Preface

This book is mainly a translated version of a book that appeared in the Norwegian language in 2002, published by Tapir Academic Press, which originated from my many years of teaching at the Norwegian Institute of Technology (NTH) and the Norwegian University of Science and Technology (NTNU). This teaching included building acoustics, general noise abatement and acoustic measurement techniques. The book is therefore primarily intended for the engineering student but it should be possible to use it as a reference, partly because quite a number of references for further reading are included. This applies to books and journal articles as well as references to relevant standards, international as well as European. The author is painfully aware of the constant revisions of the latter group but as they normally retain the number these references also should have certain longevity. Although mainly a translated version of the Norwegian edition, quite a lot of new material is included, certainly in the chapter on sound absorbers, a field of particular interest for me in recent years.

The appearance of this English edition is wholly due to my friend and colleague Peter Lord, former professor at the University of Salford and long-time editor of the *Journal of Applied Acoustics*. Although getting many hints from colleagues abroad to translate the book, Peter really started the process by urging the present publisher to put some pressure on me.

The cooperation and inspiration offered by colleagues within the acoustics groups at NTNU and SINTEF is greatly acknowledged. I also take great pleasure in the many contacts with former students, coming back to me to discuss problems encountered in their professional life.

A special thanks to Arild Brekke, Sigurd Hveem, Ulf R. Kristiansen, Asbjørn Krokstad and Rolf Tore Randeberg for reading and commenting on the original Norwegian edition. For reading and commenting on some new material on room acoustics, I am indebted to Peter Svensson and to Arne Jensen, an expert on FEMLAB™, for providing additional FEM calculations.

NTNU, Trondheim
September 2007

Introduction

In recent years there has been an increased interest in office buildings, factory spaces and dwellings when the acoustics is of concern. It is acknowledged that reducing noise levels in the living environment of people does improve the quality of life and also contributes to an improvement in health. Legal requirements demanded by the authorities in various countries cover a wide range of characteristics — noise levels, airborne and impact sound insulation and reverberation time. In order to enforce such requirements, relevant measuring procedures must be provided, formulated in national or international standards. The international standards provided by ISO (International Standards Organization) have reflected the trend mentioned above, increasing both in number and covering broader aspects. On the European stage, the standard organization CEN has been very active in bringing out standards as a follow-up to the EU directives. The cooperation between ISO and CEN under the Vienna agreement has contributed substantially to the creation of standards of general acceptance.

However, measurement procedures applied in the laboratory or in the field is just one part of the story. Manufactures of building components and materials must have harmonized and practically oriented test methods and other guidelines to meet the regulatory requirements and consumer expectations in the quality of the products. This again is the task of the standards organizations.

Controlling the acoustical conditions, be it in the sound insulation, reverberation time or noise levels in a building or testing the acoustic properties of components in the laboratory may certainly be complicated tasks even for qualified personnel. There is, however, a problem area of another dimension than the above tasks — an accurate prediction of the acoustic conditions and properties. Nowadays, there certainly is a number of computer-based tools at the disposal of the building acoustics expert. However, without a thorough understanding of the physical principles one may easily go down the wrong track when new and novel constructions are needed. The author believes that insight into the physical phenomena and the ability to convert the knowledge into practical use is the mark of the expert, not a morass of lexicographical wisdom.

Furthermore, from the author's point of view the concept of building acoustics includes all types of acoustic and vibration phenomena related to buildings. Traditionally, one might envisage that this concept is limited to sound insulation problems in buildings whereas the design of rooms for proper conditions for music and speech, i.e. room acoustics, is something else. In the English language, the concept of architectural acoustics is often used to include all these aspects but in this book we shall use the former notion. In addition to the subjects of sound insulation and room acoustics, it would be natural to include all types of noise and vibration problems within the concept of building acoustics, whether the sources are internal, e.g. building service, or external, e.g. transport or industry. This book does not aim to do justice to all these topics but concentrate on the acoustic performance of building elements and constructions, in particular how they may be designed to obtain high sound insulation and absorption.

A chapter on room acoustics is also included but where large rooms are concerned the applications are generally directed towards industrial spaces, not performance spaces such as concert halls, theatres etc. This is a choice based on the experience that the reader will have fewer problems in finding an extensive literature on the acoustics performance of those rooms. As far as the noise and vibration aspect of building service equipment is concerned, these are not covered in any detail except for the important area of vibration isolation. The basis for this type of isolation is included as a part of the general description of mechanical oscillations.

Some readers will certainly miss a chapter on the subjective aspects of sound — the human hearing mechanism and the relationship between the objective, measurable quantities and the subjectively adjusted measurement quantities such as the frequency weighted sound pressure levels, using the A- or C-weighting curve, or the loudness or loudness level expressed in sone and phon, respectively. These aspects are, however, thoroughly treated in a number of books on noise and noise abatement allowing us to leave it out here. This said, it should be stressed that our ears are among the best instruments for acoustic analysis. A building acoustics "diagnosis" may often be put forward more easily by placing your ear to the wall than using an instrument.

The understanding of the performance of simple mechanical vibrating systems is basic for the understanding of the behaviour of complex systems such as the building elements and constructions one finds in buildings. Chapter 1 gives an overview of how oscillations of various types, periodic, transient as well as stochastic (random), are characterized and analysed in the time domain as well as in the frequency domain. In Chapter 2, the transfer of such oscillations through mechanical systems is treated, starting from systems made up of the concentrated (lumped) elements mass, spring and damper. This provides the base for a transition to continuous systems where wave phenomena become dominant.

In Chapter 3, we describe wave phenomena in fluid as well as in solid media, the sources of sound waves and their propagation in these media. Particular emphasis is placed on the subject of bending (flexural) waves to provide the background for treating sound transmission through building elements.

Chapter 4 is devoted to room acoustics with emphasis on the physical aspects. However, an overview of the room acoustic parameters for characterizing the acoustic quality with respect to transmission of music and speech is included. Important measuring quantities, which must be determined in practice, e.g. determining sound insulation, are also treated along with the expected measurement accuracy of these quantities.

Chapter 5 is wholly devoted to acoustic absorbing materials and constructions, modelling the absorption of sound in porous materials as well as the absorption offered by absorbers based on a resonator principle, membrane absorbers and absorbers of Helmholtz type. The last, based on microperforated panels, is given a broad treatment. Measuring methods for absorption and for the determination of material properties, important for modelling the absorption capability of absorbers, are thoroughly treated.

Chapter 6 introduces the measures used to characterize the sound isolating capability of building elements and constructions; i.e. the sound reduction index (also known as transmission loss) and impact sound pressure level, along with their frequency weighted counterparts. The treatment of sound transmission phenomena starts with a look at the ability of the elements to act as sound radiators, thereafter how these elements are vibrating when forced, either by point or distributed forces (pressure field). The treatment in this chapter is limited to single leaf partitions.

Statistical energy analysis (SEA) is a method for prediction of the dynamic behaviour of complex systems, containing both acoustic and structural elements. The method has gained wide acceptance for use in building acoustics and a short introduction is therefore given in Chapter 7, partly to give some background to the results presented in the remaining chapters.

Chapter 8 extends the treatment on sound transmission to composite elements such as double leaf constructions, sandwich elements etc. and the last chapter, Chapter 9, looks into methods for predicting the resulting sound insulation in a finished building. Here one takes account of all the different sound transmission paths between rooms putting emphasis on the flanking transmission, sound energy transmitted by way of the flanking members of the primary partition.

We have, as far as possible, tried to compare the various prediction models presented with relevant measurement results. A collection of results from sound insulation tests, compiled by Homb *et al.* (1983) (see references to Chapter 6), has been extremely valuable in preparing the Chapters 6 and 8. A number of data, both measured and predicted results, have also been reproduced from scientific journals and books. By reproduced we mean that the data are digitized from the original figures and plotted into new graphs. Permission from individuals and from journal publishers to do so is gratefully acknowledged.

Commercial program packages have been used for a couple of the examples but generally the calculations are performed by programs developed by the author and colleagues at NTNU and SINTEF. Important contributions have also resulted from cooperation with colleagues from the Katholieke Universiteit Leuven, Belgium, and l'Université de Maine, Le Mans, France. One of these programs, WinFLAG™, has for some years been commercially available.

Lastly, some comments must be made on the nomenclature used for the quantities. The quantities and units recommended by ISO are laid down in a new series of standards with the main number 80000 of which the acoustics field is covered by part 8. As a general rule, all quantities having the unit 1 (one) should be denoted factor, i.e. one should use absorption factor, transmission factor and so on, not absorption coefficient and transmission coefficient. We shall then reserve the notion "coefficient" for quantities having a unit different from 1. We have adhered to these recommendations in this book, which hopefully will not appear confusing for those accustomed to use the word coefficient whatever the unit might be.

T.E. Vigran

CHAPTER 1

Oscillating systems. Description and analysis

1.1 INTRODUCTION

Sound and vibration are dynamic phenomena, generally referred to as oscillatory motion. Audible sound is due to oscillations in the air pressure propagating as waves. The term vibration is used when talking about oscillations in mechanical systems. These may, depending on the dimensions of the system and the actual frequency, appear as wave motions of different kinds. It is then time to point out what is meant by the terms oscillation and wave. Exact definitions may be found in the international standard ISO 2041. To express it in a general way, we may say that an oscillation is the variation with time of the magnitude of a measurable quantity, the magnitude being alternately greater and smaller than some mean value. With a wave there is an oscillatory motion propagating through the actual medium and this motion is wholly dependent on the physical properties of the medium. There is a transport of energy but not of the medium itself.

The first step in measurement analysis of oscillations is usually a transformation of the motion into an electrical oscillation or signal, using some kind of transducer. This applies to all types of oscillatory motion with which we shall be concerned in building acoustics, whether we are measuring vibration quantities such as acceleration or velocity, or sound quantities such as pressure or particle velocity. In the current volume, the notion *signal analysis* is therefore synonymous with the analysis of the actual oscillatory motion.

In modern stand-alone analysing equipment the signals are digitized and one normally has at one's disposal a large menu of analysing options. Alternatively, with the common use of PCs in recent years, signals are input to sound cards in the PC and the analysing functions are implemented by software. This transforms the PC into a reasonably prized measuring system taking advantage of already existing resources, screen etc. No matter how these analysing functions are implemented, there is always a demand on the user for knowledge of modern signal analysis, in particular, on being aware of the possibilities and limitations of the actual "instrument" being used.

This chapter is aiming to give an elementary overview of the mathematical basis for some common types of signal analysis, in particular on the frequency analysis of sound and vibration quantities where the oscillations are described in the frequency domain. Some commonly used measurement quantities are introduced when it is natural to do so. There is a huge literature base on the subject, including textbooks. Some references can be found at the end of the chapter.

1.2 TYPES OF OSCILLATORY MOTION

Oscillatory motion may, as other kinds of physical phenomena, be characterized as either deterministic or stochastic (random). A process of some kind is denoted deterministic if it may be described by an explicit mathematical expression. Using knowledge of the actual physical phenomena and/or based on observations of the oscillatory motion we may in this case predict the future motion. As examples we may think of the motion of a pendulum, the motion of the pistons in a car engine etc.

For other processes we get data that never repeat themselves exactly. We have no possibility to predict the exact magnitude of a given measurement variable at a future time. Such processes are called stochastic and they may only be described by probability functions and statistical characteristics such as e.g. expected value (mean value) and standard deviation. The sound pressure (noise) from a jet engine, vibrations in a duct wall due to a turbulent flow inside the duct and the wind forces on a building during a storm are all examples of such processes. There are obvious problems using this simple classification scheme, whether it be deterministic or stochastic. As a practical guide one may decide the classification on the basis of how well one is able to reproduce the measured data in controlled experiments.

Normally, one will find various classification schemes for the two main types of process, an example being depicted in Figure 1.1. For a rough grouping one may denote a deterministic motion to be periodic or non-periodic, i.e. the oscillatory motion repeats itself after a period time T or it does not. A transient motion, a *pulse*, is the most important type of non-periodic motion.

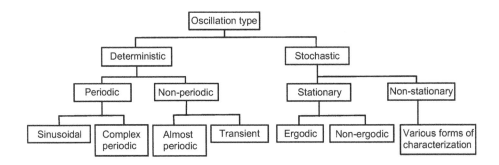

Figure 1.1 Classification of oscillatory motion.

Stochastic or random types of motion may roughly be divided into stationary and non-stationary types, i.e. the statistical properties are classified as invariant with respect to time or not. Again, there are of course problems connected to such a simple classification. In practical work, however, we will consider a process as being stationary if the statistical properties are constant over the time span in which we are interested in making observations. It is also important to note that these simple classification schemes do not exclude various combinations. A transient motion may, for example, also be stochastic.

It may already at this point be worth mentioning that stochastic motions, in practice stochastic signals, are commonly used when testing acoustical or mechanical systems. One may often tailor make the signal to cover just the frequency range needed. With digital systems, however, the test signals are not strictly stochastic but so-called pseudo-stochastic. This implies that they are periodic stochastic, i.e. they will eventually repeat themselves but the period will be very long, maybe several minutes. An important development in the measuring technique is by the use of signals where the periodicity of such stochastic or noise-like signals is turned to an advantage. This applies to the use of

MLS (maximum length sequences) which are binary sequences repeating themselves exactly. Measurements may then be performed where the useful signal, i.e. the MLS signal, has an amplitude not much higher or even lower than other disturbing signals (background noise). This is possible when performing a synchronous averaging over several sequences. As one may understand this presupposes that the tested system is time invariant.

A further development in the measuring technique is by using so-called chirp signals or swept sine signals. As the latter name implies, these are transient signals where the frequency sweeps from a starting frequency to a final frequency during the measurement period. The application of such signals in system testing is at least 30 years old but due to modern digital technology it is now in widespread use. We shall deal with such test signals in section 1.5. A recent international standard, ISO 18233, is devoted to the application of such new measurement technique in building acoustics.

1.3 METHODS FOR SIGNAL ANALYSIS

The methods or techniques that are used in practical signal analysis may be divided into three main groups as follows: 1) signal amplitude analysis, 2) time domain analysis and 3) frequency domain analysis (spectral analysis). Figure 1.2 gives an overview of some concepts associated with these main groups.

For many purposes the only information needed is the absolute value of the oscillation, normally specified by the RMS-value (root-mean-square value) or the peak value. The reason is that these values may be specified as a legal limit value, e.g. the maximum vibration amplitude for a certain type of machine, the maximum A- or C-weighted sound pressure level in a living environment or in a work space. These types of analysis may normally be carried out using simple and low-cost instrumentation.

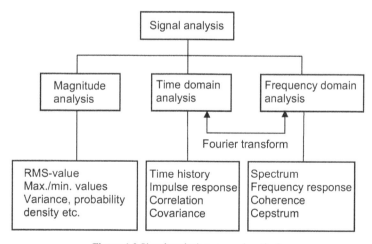

Figure 1.2 Signal analysis types and methods.

More advanced equipment is needed for accurate analysis in the time or frequency domain, in particular when information from many parts of a system is needed and/or when the task is to map the relationship between data from the various parts. Time domain analysis as cross correlation or the equivalent frequency analysis, i.e. transfer function analysis, requires two-channel instruments. When performing so-called modal analysis, where the purpose is to map the global vibration pattern of a mechanical system, a multi-channel instrument will be appropriate. As pointed out above these types of instruments are, owing to modern digital analysis methods, commonly available. However, the demand for knowledge to enable them to be used properly is by no means diminishing.

The main type of practical analysis in the fields of acoustics and vibration is frequency domain analysis. The emphasis will therefore mainly be placed on this type of analysis. The fundamental bases for this method are Fourier series and Fourier transforms. This chapter aims at giving an overview of the mathematical basis and furthermore on the modifications necessary for treating data in a digital form, i.e. performing digital signal analysis. Simulations and use of data in digital form are used to produce the illustrations below.

1.4 FOURIER ANALYSIS (SPECTRAL ANALYSIS)

Frequency domain analysis of acoustic and vibration signals is normally denoted by spectral analysis as we want to extract information, in more or less detail, of the frequency or spectral content. The analysis technique, based on Fourier series and Fourier transforms, will be demonstrated using a number of different types of signal: periodic, transient as well as stochastic.

1.4.1 Periodic signals. Fourier series

We now assume that we have a function $x(t)$ that varies periodically with time period T, i.e. we may write $x(t) = x(t + kT)$ where k is a whole number. According to Fourier's theorem such a function may be represented by the series

$$x(t) = a_0 + \sum_{k=1}^{\infty} \left(a_k \cos \frac{k 2\pi t}{T} + b_k \sin \frac{k 2\pi t}{T} \right). \tag{1.1}$$

The so-called fundamental frequency f_1 (in Hz) is given by the inverse of the period time T. Equation (1.1) therefore tells us that the function $x(t)$ may be expressed by the fundamental frequency and its harmonic components, hence

$$f_k = k f_1 = \frac{k}{T} = k \frac{\omega_1}{2\pi}, \tag{1.2}$$

where ω_1 is the fundamental frequency expressed by its angular frequency in radians per second. Using the latter leads to the equation in its simplest form but we shall, as far as possible use the most common measurement variable, the frequency in Hertz (Hz). Equation (1.1) may then be written

$$x(t) = a_0 + \sum_{k=1}^{\infty} \left(a_k \cos 2\pi f_k t + b_k \sin 2\pi f_k t \right), \tag{1.3}$$

where the coefficients a_k and b_k are given by the integrals

$$a_0 = \frac{1}{T} \int_0^T x(t) \, dt,$$

$$a_k = \frac{2}{T} \int_0^T x(t) \cos 2\pi f_k t \, dt \qquad (1.4)$$

$$\text{and} \quad b_k = \frac{2}{T} \int_0^T x(t) \sin 2\pi f_k t \, dt.$$

a_0 is the mean value of the function. For a periodic oscillation this value is by definition equal to zero but we shall for completeness retain it in the following derivations. In standard textbooks on mathematics one will find the coefficients readily calculated for a large number of functions. Before we give some examples, two alternative expressions for such Fourier series are worth looking into, expressions that are more common in signal analysis. The first alternative form of Equation (1.1) is

$$x(t) = c_0 + \sum_{k=1}^{\infty} c_k \cos\left(2\pi f_k t + \theta_k\right), \qquad (1.5)$$

where we may easily show that the Fourier coefficients (Fourier amplitudes) c_k and the Fourier phase angles θ_k are given by the following expressions:

$$c_0 = a_0,$$

$$c_k = \sqrt{a_k^2 + b_k^2},$$

$$\text{and} \quad \theta_k = \text{arctg}\,\frac{b_k}{a_k}.$$

Introducing complex numbers we may derive the second alternative form of Equation (1.1). Moivre's formula gives

$$\cos(n\theta) + j \cdot \sin(n\theta) = e^{jn\theta},$$

where j is the complex unit, and we may then write

$$x(t) = \sum_{k=-\infty}^{k=+\infty} X_k e^{j2\pi f_k t}, \qquad (1.6)$$

where $X_0 = a_0$

$$\text{and} \quad X_k\left(f_k\right) = |X_k| e^{-j\theta_k} = \frac{1}{T} \int_0^T x(t) e^{-j2\pi f_k t} dt \quad \text{with} \quad k = \pm 1, \pm 2, \pm 3,... \qquad (1.7)$$

The quantity $|X_k|$ is the modulus of the complex Fourier amplitude. Even if $x(t)$ is a real function we may mathematically represent it in complex form using both positive and negative frequencies. There is, however, no reason to ascribe any physical significance to these negative frequencies. They show up at an intermediate stage in the calculations and our physical measurement variables will always be functions of the positive frequencies. Note the symmetry in the Equations (1.6) and (1.7). The relationship to the coefficients used above will be

$$X_k = \frac{1}{2}(a_k - \mathrm{j}b_k),$$

$$X_{-k} = \frac{1}{2}(a_k + \mathrm{j}b_k).$$

$$(1.8)$$

Hence

$$|X_k| = \sqrt{X_k \cdot X_{-k}} = \frac{1}{2}\sqrt{a_k^2 + b_k^2} = \frac{1}{2}c_k$$

$$\text{and} \qquad \theta_k = \mathrm{arctg}\left(\frac{b_k}{a_k}\right).$$

$$(1.9)$$

1.4.1.1 Energy in a periodic oscillation. Mean square and RMS-values

In sound or vibration measurements the function $x(t)$ may, to mention a few examples, represent the pressure in a sound wave at a given position in a room; the displacement, the velocity or acceleration amplitude at a point on a wall or at a point on the surface of a machine. Using the first example we put $x(t) = p(t)$, where p is the pressure in the sound wave with the unit Pa (Pascal). The instantaneous energy transported per unit time and per unit area normal to the direction of propagation, i.e. the intensity of the wave, will be proportional to the square of the pressure. By using Equation (1.3) or (1.5), where we now put a_0 (or c_0) equal to zero, we obtain

$$\overline{p^2(t)} = \overline{x^2(t)} = \frac{1}{T}\int_0^T x^2(t)\,\mathrm{d}t = \frac{1}{2}\sum_{k=1}^{\infty}\left(a_k^2 + b_k^2\right) = \frac{1}{2}\sum_{k=1}^{\infty}c_k^2, \qquad (1.10)$$

where the line above the pressure squared denotes taking the mean value over time. The total energy flow is then given by the squared sum of the Fourier amplitudes. Considering the total sound pressure as being a sum of sound pressure components, each component associated with a given amplitude and frequency, it is reasonable to believe that each component carries its part of the energy, i.e. that the energy in the kth harmonic is proportional to c_k^2. However, we cannot claim this on the basis of Equation (1.10) only.

We may use a similar argument when $x(t)$ represents a velocity amplitude $u(t)$ of a part of a mechanical system, e.g. a vibrating plate. The mean square value of $u(t)$ will then be proportional to the kinetic energy of this part of the system. However, in practical measurement technique another quantity is more commonly used than the mean square value. This is obtained by taking the root of it and then we arrive at the *RMS-value* (root-mean-square-value) of the chosen variable. We then have

$$x_{\mathrm{rms}} = \sqrt{\overline{x^2(t)}} = \frac{1}{\sqrt{2}}\left[\sum_{k=1}^{\infty}c_k^2\right]^{\frac{1}{2}} = \left[\sum_{k=1}^{\infty}(c_{\mathrm{rms}})_k^2\right]^{\frac{1}{2}}, \qquad (1.11)$$

where c_{rms} denotes the RMS-value of the harmonic components. Using the complex form of Equation (1.6) we get

$$x_{rms} = \sqrt{\overline{x^2(t)}} = \sqrt{\sum_{k=-\infty}^{k=+\infty} |X_k|^2} = 2\sqrt{\sum_{k=0}^{k=+\infty} |X_k|^2} \qquad \text{with} \qquad X_0 = 0. \qquad (1.12)$$

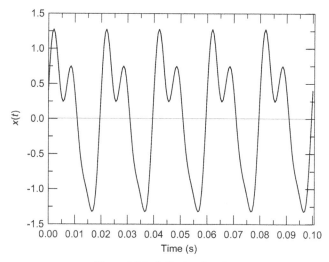

Figure 1.3 Periodic time function.

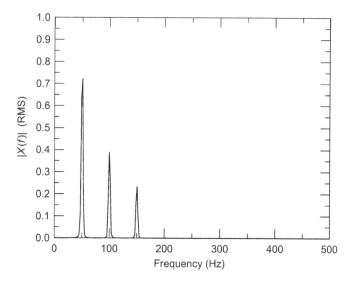

Figure 1.4 Frequency amplitude spectrum of the time function in Figure 1.3.

1.4.1.2 Frequency analysis of a periodic function (periodic signal)

An example of a periodic function is shown in Figure 1.3. Performing Fourier analysis we get the frequency amplitudes $|X_k|$ as shown in Figure 1.4. The calculations for this example are indeed performed by the so-called discrete Fourier transform (DFT) (see below), but this has no importance in this example. We have derived the time function by just summing three sinusoidal signals having amplitudes of 1.0, 0.5 and 0.3, respectively but also adjusting their relative phases. Performing the Fourier calculation, Figure 1.4 shows that these components will appear with the RMS-values reasonably correct.

1.4.2 Transient signals. Fourier integral

The above mathematical description may be adapted for non-periodic functions, e.g. transient time functions such as the sound pressure pulse from a gunshot or an explosion, the acceleration of a plate when hit by a hammer etc. Formally, we may say that the function is still periodic but the period T now goes to infinity. This makes the distance between frequency components infinitesimal, in the limiting case it goes to zero. We then get a *continuous* frequency spectrum. The Fourier series transforms into an integral and the Fourier coefficients will be a continuous function of the frequency, the so-called Fourier transform. Working from Equation (1.7), we may express the transform $X(f)$ as follows

$$X(f) = \int_{-\infty}^{+\infty} x(t)\, e^{-j2\pi f t} dt \qquad -\infty < f < +\infty, \qquad (1.13)$$

where $X(f)$ will exist if

$$\int_{-\infty}^{+\infty} |x(t)|\, dt < \infty.$$

$X(f)$ is called the *direct* Fourier transform or spectrum. If this is a known function we may use the *inverse* transform to find the corresponding time function $x(t)$. Using a similar modification of Equation (1.6) we may write

$$x(t) = \int_{-\infty}^{+\infty} X(f)\, e^{j2\pi f t} df \qquad -\infty < t < +\infty. \qquad (1.14)$$

Equations (1.13) and (1.14) are a Fourier transform pair. It should be noted that $X(f)$ again is a complex function with both positive and negative frequencies which applies even if the time function is real. It is also usual to express $X(f)$ using a polar notation as in Equation (1.7), i.e. we write

$$X(f) = |X(f)|\, e^{j\theta(f)}, \qquad (1.15)$$

where $|X(f)|$ and $\theta(f)$ are denoted amplitude spectrum and phase spectrum, respectively.

1.4.2.1 Energy in transient motion

Again letting $x(t)$ represent the time history of the sound pressure in air caused by a gun shot, an explosion or similar events, the total energy represented by the integral

$$\int x^2(t)\mathrm{d}t$$

must be finite. From Equations (1.13) and (1.14) it is easy to show that

$$\int_{-\infty}^{+\infty} x^2(t)\,\mathrm{d}t = \int_{-\infty}^{+\infty} |X(f)|^2 \mathrm{d}f = 2\int_{0}^{+\infty} |X(f)|^2\,\mathrm{d}f . \qquad (1.16)$$

In other words; we may find the total energy either by an integration of the time function or by an integration of the Fourier transform in the frequency domain. This is the reason why the squared modulus $|X(f)|^2 = X(f)\cdot X^*(f)$ is called the *energy spectral density*, where $X^*(f)$ is the complex conjugate of $X(f)$. The last form of the integral in Equation (1.16) is possible because the time function $x(t)$, representing all types of oscillatory motion, will be a real function and therefore $X(-f) = X^*(f)$.

1.4.2.2 Examples of Fourier transforms

In practice we certainly must, in the first place, put a finite limit on the time T when using our Fourier transform. Second, in measurement as well in calculations, the transform is used in a discrete form (DFT). In this section we shall, however, show some examples where the transform may readily be calculated analytically using Equation (1.13). In this way we may vary the parameters to illustrate some important relationships between the time and frequency domain representations.

A) A function describing a simple pulse of rectangular shape may be expressed as

$$x(t) = A \quad \text{for} \quad -T/2 \leq t \leq T/2$$
and $\quad x(t) = 0 \quad\quad$ otherwise.

Inserting this into the expression for $X(f)$ (see Equation (1.13)) we obtain

$$X(f) = AT\frac{\sin(\pi fT)}{\pi fT}.$$

Figures 1.5 and 1.6 show some examples of the time function and the resulting modulus $|X(f)|$ of the transform. The amplitude A is arbitrarily set equal to 100 and the time T is chosen 1, 5 and 10 ms, respectively. It should be noted that the spectrum broadens out with decreasing pulse duration. However, as the amplitude A is constant the spectral amplitudes must decrease. (Try to explain why.) An "infinitely" short pulse, represented by the so-called Dirac δ-function, gives an infinite broad spectrum of constant amplitude, a *white* spectrum. (Could you state the frequency amplitude of such a function?)

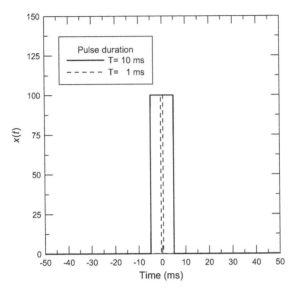

Figure 1.5 Time functions of rectangular pulses.

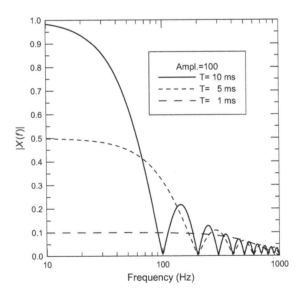

Figure 1.6 Modulus of the Fourier transforms of rectangular pulses.

B) The example above does not represent, for many reasons, an oscillatory motion you will find in real life. It does, however, illustrate important relationships concerning the form of a pulse and the corresponding frequency spectrum. The following example uses a more realistic type of motion; the function used represents the amplitude (displacement, velocity or acceleration) of a simple mass and spring system having a viscous damping small enough to make it oscillate. We assume that $x(t)$ is the displacement of the mass when we initially move it from its stable position and then release it. We may write

$$x(t) = Ae^{-at}\cos(2\pi f_0 t) \quad \text{for } t > 0$$

and $\quad x(t) = 0 \quad$ otherwise.

How quickly the motion "dies out" is determined by the constant a and Figure 1.7 shows a time section where a is 1 s^{-1} and 50 s^{-1}, respectively. The amplitude A is set equal to 1.0 and the frequency f_0 is 25 Hz for both curves[1]. The modulus of the Fourier transform will be

$$|X(f)| = A \cdot \left[\frac{a^2 + 4\pi^2 f^2}{\left[a^2 + 4\pi^2 \left(f_0^2 - f^2 \right) \right]^2 + 16\pi^2 a^2 f^2} \right]^{\frac{1}{2}} .$$

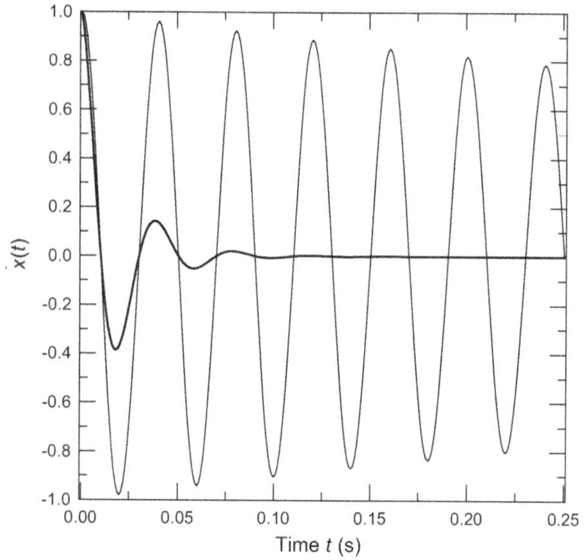

Figure 1.7 Time function of a damped oscillation (see the expression for *x(t)* above). The constant *a* is equal to 1 (thin line) and 50 (thick line), respectively.

This expression is shown in Figure 1.8 for the two values of a. It should be noted that the ordinate scale is logarithmic in contrast to that used in Figure 1.6. As expected we do get a very narrow spectrum around f_0 when the system has low damping. Setting a = 1 s^{-1} reduces the amplitude to 1/10 of the starting value after a time 2.3 seconds, i.e. after some 60 periods. When a is 50 s^{-1} the amplitude is down to 1/10 just after one period and the spectrum is then much broader.

---

[1] The frequency f_0 will not be independent of the damping of a real system. However, with the chosen variation in the damping coefficient a the variation in f_0 will be approximately 5%.

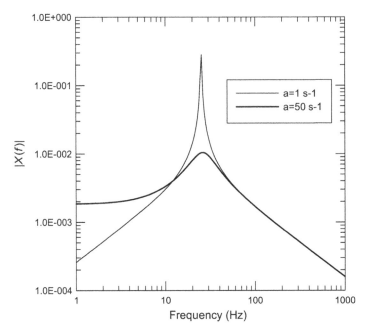

Figure 1.8 Fourier transform of time functions shown in Figure 1.7

1.4.3 Stochastic (random) motion. Fourier transform for a finite time *T*

As pointed out above, a stochastic process gives data, which in a strict sense never exactly repeat themselves. Figure 1.9 may be thought of as time samples of such processes. It could for example be the sound pressure measured at a given distance from a sound source of stochastic nature such as a jet engine, a nozzle for compressed air or a waterfall to mention a few examples. It is important to realize that such time histories or time functions, in practice, certainly being of finite length, are just samples of an infinite number of functions which could be attributed to the actual physical process. Collections of identical types of source will each give a different time function. All these possible time functions taken together make up what one in a strict sense calls a stochastic process. A collection of such time functions is what is called an *ensemble.*

In practice, however, just *one* such time history may be sufficient in our data analysis. The reason for this is that processes that represent physical phenomena often are *ergodic*. This means that we may extract all the necessary information from one single time function. This does not imply, however, that we will not experience non-linear phenomena when dealing with sound and vibration data analysis, analysis that demands ensemble averaging.

Lets make *x*(*t*) represent one such stationary time function, existing in a theoretical sense for all values of time *t*. Then we get

$$\int\limits_{-\infty}^{+\infty} |x(t)|\, dt = \infty,$$

which means that the Fourier transform according to Equation (1.13) does not exist.

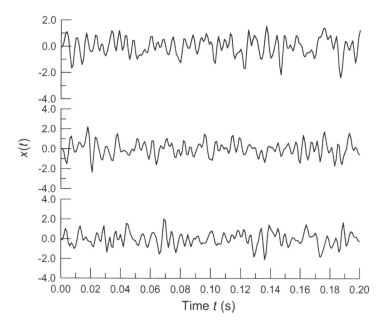

Figure 1.9 Examples of time functions of a stochastic signal. The signals (oscillations) are limited to a frequency interval of 10–250 Hz.

Obviously, we shall not be able to measure over an infinite time in any case. To perform an analysis the idea is to define a new time function $x_T(t)$, equal to the original function $x(t)$ in a time interval T but equal to zero otherwise. We then get an estimate of $X(f)$ by calculation of a Fourier transform over the time interval T

$$X_T(f) = X(f,T) = \int_0^T x_T(t)\, e^{-j2\pi ft}\, dt. \qquad (1.17)$$

A finite transform as shown here will always exist for a time segment of a stationary stochastic function. Taking Equation (1.6) into consideration we may see that for the discrete frequency components $f_k = k/T$ the transform in Equation (1.17) will give

$$X(f_k, T) = T\, X_k \quad \text{with} \quad k = \pm 1, \pm 2, \pm 3, \dots$$

This means that when performing the transform and letting the frequency f just to take on the discrete values f_k we get a Fourier series with period T. This is in fact the method used when processing data digitally. Before taking up that theme we shall introduce an important function used to characterize stochastic functions in the frequency domain, the *power spectral density function*, or in short, the *power spectrum*. Since $x_T(t)$ only exist in the time interval T we may calculate the mean square value as follows

$$\overline{x_T^2(t)} = \frac{1}{T}\int_0^T x_T^2(t)\,\mathrm{d}t = \frac{1}{T}\int_{-\infty}^{+\infty} x_T^2(t)\,\mathrm{d}t = \frac{2}{T}\int_0^\infty |X_T(f)|^2\,\mathrm{d}f\,, \qquad (1.18)$$

where the last expression is calculated using Equation (1.16). By assuming that the function is stationary (and ergodic) we may show that the mean square value of $x(t)$ is

$$\overline{x^2(t)} = \lim_{T \Rightarrow \infty}\overline{x_T^2(t)} = \int_0^\infty \lim_{T \Rightarrow \infty}\left[\frac{2}{T}|X_T(f)|^2\right]\mathrm{d}f = \int_0^\infty G(f)\,\mathrm{d}f\,, \qquad (1.19)$$

where $G(f)$ is the *power spectral density function* mentioned above and defined by the Fourier transform

$$G(f) = \lim_{T \Rightarrow \infty}\left[\frac{2}{T}|X_T(f)|^2\right]. \qquad (1.20)$$

Equation (1.20) gives the formal definition of the function but the reason for using such a term should be more obvious looking at Equation (1.19). Here one may see that $G(f)$ is the contribution per Hz to the mean square value, a quantity generally proportional to the power, hence the term *power spectrum* or more correctly *power spectral density*.

1.4.4 Discrete Fourier transform (DFT)

When processing data digitally one has, as pointed out above, to use a discrete set of frequency values. Calculating a finite Fourier transform using Equation (1.17) will give a Fourier series with period T. Here we shall just give a summary of this discrete type of transformation and point to some relationships important in measurement applications. When presenting the examples on spectral analysis (see section 1.4.5 below), we shall use this technique to simulate the results from a digital frequency analysis.

The first two steps in a digital analysis are 1) *sampling* and 2) *quantization*. The first step means that the signal $x(t)$ is substituted by a number N samples separated by a time step of Δt as illustrated in Figure 1.10. The length of the signal being analysed is then $T = N \cdot \Delta t$, and the calculations are performed treating the data as periodic with a time period of T. According to the *sampling theorem* this implies that the upper frequency limit in the analysis is $f_{co} = 1/(2\Delta t)$. This frequency is called the *Nyquist frequency* or *cut-off frequency*.

The quantization, performed by an analogue to digital converter (AD-converter), means an allocation of a finite set of numbers to the amplitudes of the sampled signal. This is illustrated in Figure 1.11, which shows the numbers or estimates allocated to the samples. The number of bits handled by the AD-converter gives the accuracy of these estimates. A 12-bit converter resolves the signal into $2^{12} = 4096$ steps, a 16-bit into 65536 steps etc.

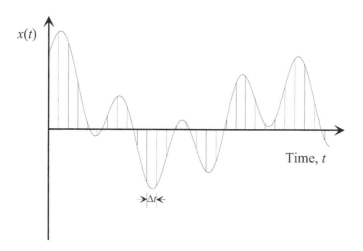

Figure 1.10 Sampling of a continuous signal by time intervals Δt.

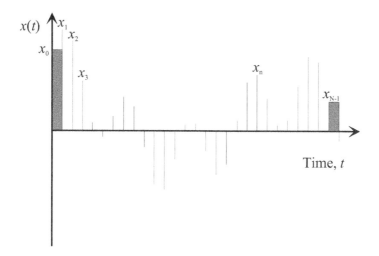

Figure 1.11 Allocating amplitude values to the samples. The marked bars indicate the values x_0 and x_{N-1} allocated to the first and the last sample.

The continuous time signal $x(t)$ is thereby substituted by a set of numbers $x_n = x(n\Delta t)$ with $n = 0, 1, 2, 3,..., N - 1$. The corresponding continuous transform $X(f)$ (see Equation (1.17)) will turn into a discrete sequence $X_k = X(k\Delta f)$ with $k = 0, 1, 2, 3,..., N - 1$. Per definition, the *discrete Fourier transform* (DFT) is given by

$$X_k = X(k\Delta f) = \frac{1}{N} \sum_{n=0}^{N-1} x_n e^{-j2\pi\frac{kn}{N}} \qquad k = 0, 1, 2, 3, ..., N-1, \qquad (1.21)$$

where Δf is $1/T$. The *discrete inverse Fourier transform* (IDFT) is accordingly given by

$$x_n = x(n\Delta t) = \sum_{k=0}^{N-1} X_k e^{j2\pi\frac{kn}{N}} \qquad n = 0, 1, 2, 3, ..., N-1. \qquad (1.22)$$

It should be noted that Equation (1.22) is an exact inverse function of Equation (1.21) and not an approximation. This is true in spite of our starting point being the continuous Fourier transform.

These equations form the basis of digital Fourier analysis. Looking at these equations one will see that a straightforward analysis using an N-points transform will require N^2 complex multiplications. Here the *fast Fourier transform* (FFT) algorithm is appearing as a saviour. Details on the procedure are outside the scope of this book. It is sufficient to point out the benefit of using this algorithm requiring $N \cdot \ln(N)$ operations only, as opposed to N^2 and the difference is quite formidable. Just for a "small" transform using $N = 2^{10} = 1024$, the ratio of these number of operations is 147. Using $N = 2^{16} = 65536$, the ratio will be 5900.

It may at this stage be appropriate to point out some further important phenomena appearing using DFT analysis. These are 1) *aliasing* and 2) *leakage*, where the first one is a unique problem in digital analysis. As indicated above, the upper frequency limit where reliable information can be extracted is $1/(2\Delta t) = \Delta f \cdot N/2 = f_s/2$, where f_s is the sampling frequency. Any signal components with higher frequencies present will not be detected but instead will be mistakenly assigned to lower frequency components. They are being "folded back" to the lower frequency components, which is the reason behind the term aliasing. There are two methods to escape this problem. One either uses a sampling frequency much higher than the expected highest frequency component in the signal or inserts sharp anti-aliasing filters which cut away or eliminate the components having frequencies above the Nyquist frequency.

The second phenomenon, leakage, means that one may get frequency components which in fact are not present in the actual signal. The main reason is that one is performing an analysis using a limited time segment. Starting or/and stopping the sampling at a time when the signal amplitude is not zero will create a discontinuity, i.e. we get a step in the value of the function giving components that are not present in the actual signal. The remedy is to apply various kinds of "window"; multiplying the time signal with a form function before taking the transform. These windows are symmetrical around $T/2$ and go to zero at each end. One then removes some of the energy but this may be compensated for after performing the analysis. There are a number of such windows in use, Hanning, Hamming and Kaiser-Bessel to mention a few.

1.4.5 Spectral analysis measurements

In practice, one will normally find that the sound or vibration phenomena to be analysed cannot be distinctly classified as periodic, transient or stochastic. They may contain elements of two or more types. Sound or vibration spectra from technical equipment such as a motor, a pump etc. will contain periodic components due to the rotation of blade wheels or gears together with stochastic components from turbulent flow of air or water in the cooling system. This will not cause any problem if the task is just to perform a simple amplitude analysis, i.e. to ascertain the RMS-value of the signal.

For a spectrum analysis, however, one has to consider exactly what sort of information is needed. Commercial frequency analysers may be divided roughly into two groups: the analysers with a fixed set of filters and the FFT analysers. One may find instruments where both types of analysis are implemented. The International Electrotechnical Commission (IEC) specifies the requirements for such instruments. Relevant examples are the standard for sound level meters, IEC 61672, Parts 1 and 2. The octave-band and fractional octave-band filters are specified in IEC 61260.

1.4.5.1 Spectral analysis using fixed filters

Analysers using fixed filters may be divided into two groups depending on the analysis being performed sequentially or in parallel. The latter type is called *real time analyser* because the entire spectrum is scanned at the same time. These analysers have a set of parallel filters, each filter covering a smaller or broader frequency band of the chosen frequency range. On the other hand, performing the analysis sequentially, which implies measuring one band at a time; one is dependent on the signal being stationary during the whole time of measurement. This is of course a serious limitation and the real time analysers now dominate the market.

In any case, we may for each frequency band determine a mean square value (or the RMS-value) for the signal passing through, which gives us an estimate of the spectral distribution. Measuring a stochastic signal this procedure will by definition give the power spectrum if we let the bandwidth Δf of the filters go to zero. We may write

$$G(f) = \lim_{\substack{\Delta f \Rightarrow 0 \\ T \Rightarrow \infty}} \frac{1}{\Delta f \, T} \int_0^\infty \xi^2(t, f, \Delta f) \, \mathrm{d}t. \qquad (1.23)$$

The symbol ξ describes the result after passing the signal $x(t)$ through the filter. In practice the averaging time T must be chosen giving consideration to the accuracy needed and the bandwidth must be adapted to the task. As for the latter, the question is whether detailed frequency information or more approximate estimates, i.e. mean values in octave or one-third-octave bands, are in demand.

The process of analysing a broadband signal using fixed filters with a given frequency bandwidth Δf is illustrated in Figure 1.12, using a sound pressure signal $p(t)$ as an example. After passing one of the filters indicated by its centre frequency f_0, we obtain a new time signal $p(t, f_0, \Delta f)$, a function of the selected band. Normally, we want to express the result in terms of the *sound pressure level* L_p in decibels (dB), as indicated by the lowermost sketch, for which we calculate the RMS-value and thereafter write the expression

$$L_p(f_0) = 10 \cdot \lg \left[\frac{p_{\mathrm{rms}}^2(f_0, \Delta f)}{p_0^2} \right] \qquad \text{(dB)}. \qquad (1.24)$$

The quantity p_0 is the reference value for sound pressure equal to $2 \cdot 10^{-5}$ Pa. For stochastic sound we observe from Equation (1.23) that $p^2_{\mathrm{rms}}(f_0, \Delta f)$ in the expression is an estimate of $G_p(f_0) \cdot \Delta f$. As for the frequency bands normally used, an octave band will have a bandwidth $\Delta f \approx 0.71 \cdot f_0$ and a one-third-octave band a bandwidth $\Delta f \approx 0.23 \cdot f_0$. As mentioned above, specifications for such filters are given in IEC 61260.

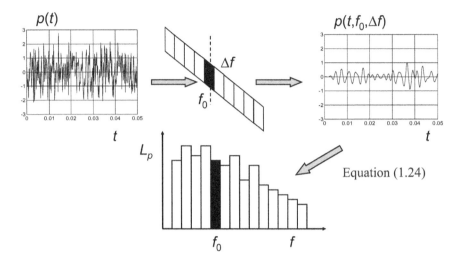

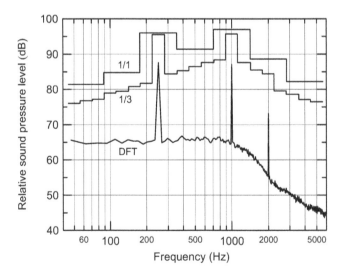

Figure 1.12 Frequency analysis of a sound pressure signal using fixed filters of bandwidth Δf. A filter having a centre frequency f_0 is indicated.

We shall also give a specific example of such analysis using these two types of band pass filter. Figure 1.13 shows the result of the analysis on a signal which could represent the sound pressure measured at a certain distance from a given source. In addition to the sound pressure levels using these filters, analysis is performed using a discrete Fourier transform as described in section 1.4.4.

Figure 1.13 Stochastic noise signal with added pure tone components. Analysis in octave bands (1/1) and one-third-octave bands (1/3) together with discrete Fourier transform analysis (DFT).

Applying DFT analysis however, we see that the sound (or noise), being mainly stochastic, also contain periodic components (pure tones). The measuring technique using fixed bandwidth filters will also give correct RMS-values for these pure tone components, c_{rms} in Equation (1.11), but in two conditions: 1) the bandwidth of the filters must be less than the distance between the pure tone components and 2) the rest of the signal inside the band must be negligible. The frequency resolution will however be poor and without the ability to repeat the analysis using another bandwidth it is difficult to decide whether or not a periodic component is present in the signal.

Without doubt, we may by looking at the one-third-octave band data in Figure 1.13 be reasonably sure that there are pure tones around 250 Hz and 1000 Hz but this is not so around 2000 Hz. Using DFT gives quite another opportunity to detect such components with certainty because one may choose the required frequency resolution[2].

1.4.5.2 FFT analysis

The breakthrough concerning the use of the discrete Fourier transform (DFT) came with the finding of an algorithm for a fast calculation, giving the fast Fourier transform (FFT). With modern FFT analysers the calculation time for equations such as (1.21) and (1.22) is of the order of milliseconds for many thousands of samples. The number of channels available in one instrument has also steadily increased. This offers the opportunity to map the global motion of a whole system in just one operation as opposed to using a single channel instrument capable of measuring at just one point at the time.

Many of the examples shown here are calculated using an FFT routine. It may be pertinent at this point to sum up the deliberations one must make before starting an analysis, this in spite of the "brain" the instrument maker has put into the instrument.

One may normally choose the number of samples N (1024, 2048, 4096, ...) or from a limited set. The next choice to decide on is the maximum frequency f_{max}. This normally results in setting the sampling frequency f_s to a minimum value of $2 \cdot f_{max}$ (a common choice is $2.56 \cdot f_{max}$). Furthermore, the anti-aliasing filter of the instrument is set to "cut away" all frequency components above f_{max}. What will the frequency resolution then be? From section 1.4.4 we know that the frequency lines will be

$$f_n = \frac{n}{T} = \frac{n}{N\Delta t} = \frac{nf_s}{N} \qquad \text{where} \quad n = 0, 1, 2, 3, ..., N-1.$$

As an example we may choose $N = 1024$ and f_{max} to be 5000 Hz. With $f_s = 2.56 \cdot f_{max} = 12800$ Hz, the total time of analysis T will be 80 milliseconds. The number of frequency lines below the Nyquist frequency will then be 512 with a frequency resolution $\Delta f = 12.5$ Hz. A commercial instrument will then present a total of 400 lines, i.e. all lines up to the chosen maximum frequency, $400 \cdot 12.5$ Hz = 5000 Hz. An alternative choice of $N = 2048$ will give 800 lines with a resolution of 6.25 Hz and so on.

This kind of analysis, called *base band analysis*, gives lines from 0 Hz to f_{max}. More often, one is interested in zooming in on a smaller frequency interval, which means that one would like to have all frequency lines f_n inside an interval given by $f_1 < f_n < f_2$. Most instruments have this option but imply that one must repeat the measurement. More details concerning this technique may be found in the rather extensive literature on the subject.

---

[2] The analysis is performed using a Hanning window, which gives good accuracy as for the frequency of the pure tone but less accuracy when it comes to amplitude.

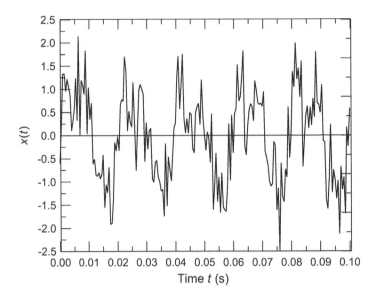

Figure 1.14 Example of a periodic signal with added Gaussian noise signal (normally distributed noise). The periodic signal is identical to the one shown in Figure 1.3.

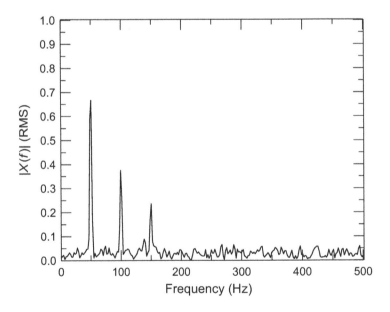

Figure 1.15 The Fourier transform of the signal given in Figure 1.14.

The next subject one should address is the accuracy of the amplitude values for the frequency lines. When the power spectrum is the goal (see Equation (1.20)), using the process outlined in the example above we have got an estimate based on the analysing time T. In the same way as when applying the fixed filter analysis, outlined in section 1.4.5.1, we have to select a total measuring time giving the desired accuracy. Measuring on a stationary signal we just have to repeat the process; performing an averaging over n records and thus obtaining an estimate for a total measuring time of $n \cdot T$.

We shall show a couple of examples of signal analysis using FFT both being simulations based on data in a digital format. Comparing with the processing performed in a commercial instruments this means that we are entering the process following the AD-converter.

In the first example, given in Figure 1.14, we are using the same periodic signal as shown in Figure 1.3 but now we have added a stochastic signal, a Gaussian distributed noise signal with zero mean value and a standard deviation equal to 0.5. Performing the Fourier transform we may see from Figure 1.15 that the periodic components show up again but now together with a contribution due to the noise.

The second example shows an analysis of a Gaussian distributed noise signal where the power spectrum is constant, i.e. $G(f)$ has a constant value G_0 for "all" frequencies. Such a signal is called *white noise*. This is of course an ideal concept, which is the reason for the quotation marks on the word *all*. There must in practice certainly be an upper limit in frequency. An example of the time signal of such noise is shown in Figure 1.16 whereas Figure 1.17 gives the corresponding power spectrum averaged using different number of records n.

The spectra are shown using a relative scale to show clearly the improvements in the accuracy by increasing the number n. Each curve is shifted by a factor of 10 from the previous one. Without doing so the curves will lie on top of one another. (Are you able to estimate the correct value of G_0 from Figure 1.17?)

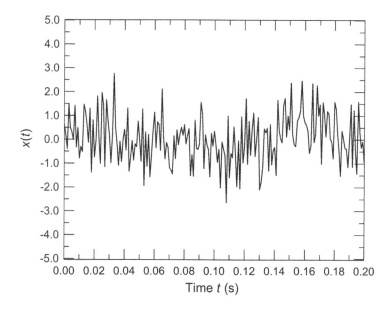

Figure 1.16 Noise signal in the time domain. The noise signal is "white" in the frequency range 0–500 Hz.

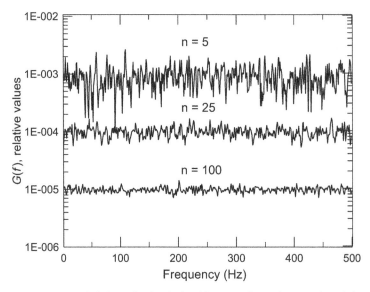

Figure 1.17 Power spectrum (relative values) calculated by averaging *n* time samples of the type shown in Figure 1.16, using $N = 1024$ and $f_s = 1000$ Hz. Note: The curves are shifted by a factor of 10.

1.5 ANALYSIS IN THE TIME DOMAIN. TEST SIGNALS

We have up to now concentrated on the frequency domain description of various types of oscillation (represented by signals). We have, starting out from a description in the time domain, made a transformation into the frequency domain, which is the most common way of describing sound and vibration phenomena. In some situations there are, however, other types of information that is required such as the statistical amplitude distribution or the *autocorrelation function*, the latter giving information on the dependence of data obtained at different times. It is reasonable in this context also to give some examples of this matter.

We have treated rather thoroughly the subject of stochastic noise, assuming a Gaussian distribution. The reason for this is partly due to the common use of such signals as test signals when measuring various types of transfer function. These are measurement situations where we want to map the relationship between a physical quantity representing the input to the system and another physical quantity representing the output or the response. This could be the velocity at a point in mechanical system excited by a force at the same point or at a more distant point. As another example, it could be the sound pressure level at a given position in a room when a given voltage is applied to the input terminals of a loudspeaker. The topic of transfer functions and applications is treated in the next chapter. Here we shall use a stochastic noise signal as an illustration of time signal analysis and further present a couple of deterministic signal types, also popular as test signals.

1.5.1 Probability density function. Autocorrelation

In Figure 1.2, the probability distribution was listed as one type of amplitude analysis. A more precise term is the *probability density function p(x)*, which gives us the probability of finding the amplitude of a signal $x(t)$ within a certain interval Δx. It is given by a limiting value as

$$p(x) = \lim_{\Delta x \Rightarrow 0} \frac{P(x) - P(x + \Delta x)}{\Delta x}, \qquad (1.25)$$

where P is the probability, a positive number between zero and one. In general, $p(x)$ will also be a function of time but for stationary and ergodic signals there will be no time dependence. In the literature, one will find a huge number of mathematical density functions with descriptions of their properties. The numbers of density functions associated with physical phenomena are infinite. As for the stochastic noise signal we have used up to now we have assumed it to be Gaussian distributed, which means that the density function is given by

$$p(x) = \frac{1}{\sigma\sqrt{2\pi}} e^{-\frac{x^2}{2\sigma^2}}, \qquad (1.26)$$

where the standard deviation σ is a measure of the width of the distribution. The square of this quantity is called *variance* and is given by

$$\sigma^2 = \int_{-\infty}^{+\infty} x^2 p(x)\, \mathrm{d}x. \qquad (1.27)$$

In the case of oscillations, the mean value is zero and this equation gives us the *mean square value*, i.e. the square of the *RMS-value*.

The density function, Equation (1.26), gives us the well-known bell-shaped curve as shown in Figure 1.18. It may, however, be interesting to see if this type of diagram could give information on other types of signal. As an example we may calculate $p(x)$ for a sinusoidal signal superimposed on Gaussian noise signals. The sinusoidal signal is given by $x(t) = \hat{x}\sin(\omega t + \theta)$, where the phase angle θ is considered to be a random variable. It may be shown that the probability density function (Bendat and Piersol (1980)),[3] in this case will be

$$p(x) = \frac{1}{\sigma_{\text{noise}}\pi\sqrt{2\pi}} \int_0^\pi e^{-\left(\frac{x - \hat{x}\cdot\cos\theta}{\sqrt{2}\sigma_{\text{noise}}}\right)^2} \mathrm{d}\theta. \qquad (1.28)$$

This function is shown in Figure 1.18 when the amplitude $\hat{x}$ is equal to one, i.e. the RMS-value $\tilde{x}$ or σ_{sinus} is approximately 0.71, using the RMS-value σ_{noise} as a parameter. When the latter is small one may see that the curve is approaching the density curve for a sine or cosine function, a curve resembling a hyperbolic function with high values near

---

[3] It is a misprint in Equation (2.35) in the reference.

to the maximum values. This is easily seen looking at a sinusoidal function. The function "spends more time" around the maximum values than around zero. For increasing σ_{noise} the function will, however, approach the common Gaussian curve again.

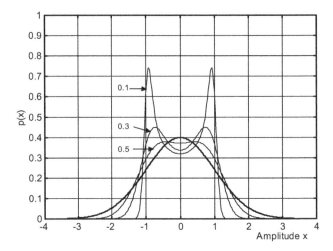

Figure 1.18 Probability density functions of a sum of a sinusoidal signal and a Gaussian noise signal. Thick curve – Gaussian noise only with $\sigma = 1$. Thin curves – sum of sinusoidal, having amplitude equal 1 and noise with σ equal 0.1, 0.3 and 0.5, respectively.

A probability density function thus gives information on how the amplitude is distributed. It does, in other words, tell us the amount of time it is expected to be within a certain range and furthermore, in which part of the time it is larger or smaller than some given value. It cannot, however, tell us anything about the time coherence in the signal. As an example it may tell us that the sound pressure is above 1 Pa during 1% of the time but it will not tell us whether this occurs due to pulses having a duration of 10 ms or 100 ms. We may obtain information of the latter type by looking at the signal at certain intervals τ in time.

We define the *autocorrelation function $R(\tau)$* as

$$R(\tau) = \lim_{T \Rightarrow \infty} \frac{1}{T} \int_0^T x(t) \cdot x(t + \tau) \, d\tau \, . \tag{1.29}$$

As is apparent from Equation (1.29), we take the value of the function $x(t)$ at a given time t and multiply it with the value at a later time $t + \tau$, thereafter taking the mean value of all such products. Before showing some examples we shall point to some important properties. First, this function gives a description of the signal in the time domain as equivalent to the *spectral density function* in the frequency domain. These functions comprise a Fourier transformation pair.

Assuming, as before, a stationary signal and performing the averaging process over a sufficiently long time, the function R will be independent of time and a function of τ only. Furthermore, the function is symmetric, $R(\tau) = R(-\tau)$ and it has a maximum for $\tau = 0$. In the latter case, we see from Equation (1.29) that this maximum is nothing more than the mean square value of the signal. Lastly, the value of the function will, for stochastic signals, approach zero with increasing τ.

Two examples of autocorrelation are shown in Figure 1.19, a pure sinusoidal signal and a Gaussian noise signal having variance equal to one. Mathematically speaking, the first one should turn out as a sinusoidal function but in practice the signal must be of finite length, therefore also in the numerical calculation used here. Both functions will therefore approach zero when increasing τ. (You may show that $R(0)$ is correct by applying the information above.)

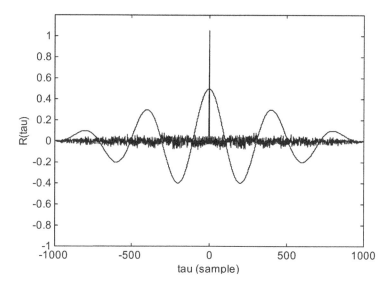

Figure 1.19 Autocorrelation function of stochastic noise and sinusoidal signal of length 5·π.

1.5.2 Test signals

As pointed out in the introduction to this section, the reason for the relatively broad coverage of Gaussian noise signal is their use as test signals. With their broadband character it may cover any frequency interval and the spectrum may be shaped if necessary. One will find an example of the latter in so-called *pink noise*, shaping the signal such that the energy in the test signal is constant when using filters having a constant relative bandwidth; octave or fractional-octave filters. Real test signals must of course have a finite length; they are *pseudo stochastic* repeating themselves after a given time. This property is normally not exploited when using this common type of noise signal generator. Using pseudo stochastic signals that exactly repeat themselves were first implemented applying MLS signals, a type that became very popular after being introduced in measuring equipment well over a decade ago.

We have called MLS signals as pseudo stochastic due to their noise-like properties but it must be stressed that they are wholly deterministic. There are, however, a multitude of deterministic signals that have been or are still being used for testing. These may be periodic (purely sinusoidal, amplitude modulated as well as frequency modulated signals etc.) or transient. As an example here we shall use the *swept sine* (SS) signal, a type also called *chirp* or *sinusoidal sweep* to give some names found in the literature. This is a signal where the frequency varies continually, linearly or nonlinearly, from a selected starting frequency to a selected stop frequency. Figure 1.20 gives an example where the frequency varies linearly with time. There are several reasons for the relatively recent popularity of these test signals.

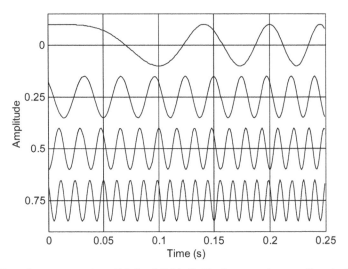

Figure 1.20 Example on a swept sinusoidal signal ("chirp"). The frequency increases linearly from 0–100 Hz in 1 second.

One may shape the spectrum by varying the sweep in different ways. A linear sweep with constant amplitude will give a white spectrum while increasing the frequency exponentially with time will give a pink spectrum. However, there are other properties just as important, especially when comparing with the MLS signals described below.

One advantage that should be mentioned is the reduced sensitivity to time variance e.g. changes in the propagating medium during measurement, for sound waves caused by air movements and temperature changes. The reason is that information is collected for one frequency at the time, whereas for noise signals the information is collected for all frequencies during the whole measurement.

Figure 1.21 shows an example of spectrum and autocorrelation function for a swept sine signal where we have chosen a starting frequency of 100 Hz and made it sweep linearly to 300 Hz. As expected we get a reasonably flat band-like spectrum but as seen from the diagram at b) one certainly needs to be aware of the side lobes in the spectrum.

The second type of deterministic signals extensively used in modern measurement technique is the MLS signals (maximum length sequences). The advantages of using this type of noise-like signals will be clear when treating transfer function measurements in the following chapter. Here we shall give an overview of the most important properties only.

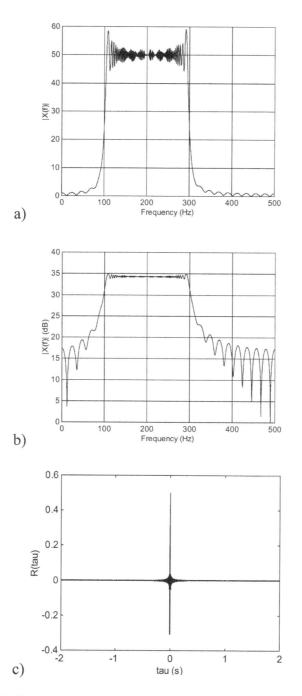

Figure 1.21 Chirp signal, linear sweep 100–300 Hz in 2 seconds. a) Spectrum, linear scale. b) Spectrum, logarithmic scale (arbitrary). c) Autocorrelation function.

The literature on MLS is relatively large but for a discussion of basic properties one may look at the article by Rife and Vanderkooy (1989). Summing up we may state the following: The MLS is a *periodic binary sequence* having an approximately flat spectrum. Such sequences are easily generated using an arrangement of shift registers. The length L of a sequence is given by

$$L = 2^m - 1, \qquad (1.30)$$

where m is the number of steps in the shift register, a number that also denotes the *order* of the sequence. One may generate several sequences of the same order but each of them is unique. An example is shown in Figure 1.22, being the first 65 samples of a sequence of order 12. The whole length is thereby 4095. In practice sequences up to the order of 20–22 are used, i.e. a length of several millions.

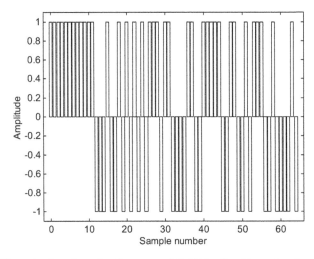

Figure 1.22 Example of maximum length sequence (MLS). The first 65 samples of a sequence of order 12.

When listening to such a sequence and a common white noise signal played through a loudspeaker, it is very difficult to distinguish between them. Technically, they are used in the same way. The MLS is however, not only deterministic but assuming it is replayed periodically, we will get a *line spectrum* with a distance between lines given by

$$\Delta f = \frac{f_s}{L} \qquad (Hz), \qquad (1.31)$$

where f_s is the sampling frequency, i.e. the fixed rate of outputting the binary samples. Using a sampling frequency of 10 000 Hz the time for outputting a sequence of order 12 will be 0.41 seconds. One therefore has the ability to adapt the signal to the given task by choosing sequence length and sampling frequency. The main advantage is, however, the deterministic property; we may repeat the measurement many times to improve the accuracy of the measured variable. This is true only when assuming the system to be constant during the measurement; the system must be *time invariant*. As pointed out above, this may be a serious limitation in using MLS as opposed to the swept sine technique.

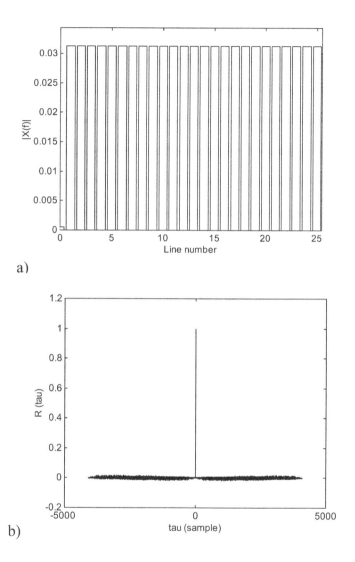

a)

b)

Figure 1.23 Example of maximum length sequence (MLS). a) The first 25 lines in the spectrum of the sequence shown in Figure 1.22. b) Autocorrelation function.

Finally it should be pointed out that, since the MLS has a flat frequency or line spectrum, the autocorrelation function is very near to a Dirac δ-function. Figure 1.23 shows the spectrum and the autocorrelation function for the sequence partly shown in Figure 1.22. For the spectrum, only the first 25 lines are shown. (How many lines are there?)

1.6 REFERENCES

IEC 61260: 1995, Electroacoustics – Octave band filters and fractional-octave band filters.

IEC 61672: 2002, Electroacoustics – Sound level meters. Part 1: Specifications.

ISO 2041: 1990, Vibration and shock – Vocabulary.

ISO 18431–1: 2005, Mechanical vibration and shock – Signal processing. Part 1: General introduction.

ISO 18233: 2006, Acoustics – Application of new measurement technique in building and room acoustics.

Bendat, J. S. and Piersol, A. G. (1980) *Engineering applications of correlation and spectral analysis*. John Wiley & Sons, New York.

Bendat, J. S. and Piersol, A. G. (2000) *Analysis and measurement procedures*, 3nd edn. John Wiley & Sons, New York.

Newland, D. E. (1993*) An introduction to random vibrations, spectral and wavelet analysis*, 3rd edn. Longman, Harlow.

Rife, D. D. and Vanderkooy, J. (1989) Transfer-function measurements with maximum-length sequences. *J. Audio Eng. Soc.*, 37, 419–443.

Tohyama, M. and Koike, T. (1998) *Fundamentals of acoustic signal processing*. Academic Press, London.

CHAPTER 2

Excitation and response of dynamic systems

2.1 INTRODUCTION

The main purpose of Chapter 1 was to give a general description of different types of oscillating motion, how it can be described and how it can be measured when representing the motion of a real physical system. We also presupposed that when working with such a system, either in acoustics or vibration, we could convert the actual physical oscillation variable (force, acceleration, sound pressure etc.) into an electrical signal to be used in the further processing. The concept of signal and oscillation are in this connection synonymous.

In general, oscillations in a physical elastic system will arise when dynamic forces or moments excite the system. A pure acoustic system could be an air-filled enclosure or a collection of such enclosures; a room inside a building, a reactive silencer (exhaust silencer for a car), a driver's cabin and so on. Mechanical systems are often associated with solid structures such as beams, plates or shells. Dealing with building acoustics the actual system is normally a combined or coupled system containing acoustical and mechanical elements. This may easily be illustrated using the transmission of sound energy from your neighbour's TV or hi-fi system. The loudspeakers set up a sound field in your neighbour's room, which excites the separating wall into mechanical oscillation. This motion will cause motion in the air next to the wall, thereby setting up a sound field in the room that in turn will excite our eardrums.

Our interest will therefore be concerned with the coupling between an oscillation variable describing the excitation or input to the system and the corresponding variable describing the response or output. In the following we shall use these words alternatively because it is quite common to talk about the input–output relationship of a system, in particular when dealing with electric circuits.

Assuming that our physical system is linear and that the physical parameters are constant, we may always define a *transfer function*, a frequency function giving the relationship between the input and output variables. Several transfer functions have their own names; impedance and mobility are important examples. Assuming that the parameters are constant means that the system properties are independent of time, i.e. the system is *time invariant.* Linearity means the *principle of superposition* is valid, which implies that if the excitation contains several frequency components, expressed by a Fourier series or transform, the response will be the summed response caused by each component alone.

The assumption that the parameters are constant in time may often be a reasonable one, at least when the time span of the measurements is relatively short. The assumption concerning linearity could be more critical as all physical systems will give a non-linear response when driven too hard. The transition to non-linearity will normally occur gradually, which does not lessen the problem. Nevertheless, we may in most cases assume that our physical system is linear within a certain magnitude range of excitation. This applies to the sound pressure levels we normally encounter within a building. This may not be the case when exciting structures using a mechanical source. Testing the transmission of impact sound, one uses a hammer blow or several hammers (the standardized tapping machine). In such cases one should be aware that the result might depend on the applied force. The only way to control it is to perform measurements varying the force level.

In the following we shall refer to examples using simple mechanical systems. As an introduction, however, it could be useful to address a practical case containing the aforementioned couplings between acoustical and mechanical components.

2.2 A PRACTICAL EXAMPLE

Figure 2.1 shows a machine or generator of some kind mounted on a base plate, placed on the floor in a building. Additionally, due to noise emission it is placed within an enclosure. The machine needs cooling and therefore openings in the enclosure are required, the openings are equipped with silencers. The problem to be addressed could be of two kinds; there could either be an existing problem or there is a question of designing a system, with the sketched components, to reduce the noise. In the latter case prediction tools must be available, which again presuppose knowledge of all sound sources.

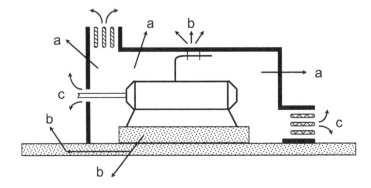

Figure 2.1 A machine unit with enclosure and supplied with silencers. Transmission paths for sound energy are indicated: a) direct transmission through enclosure walls. b) by way of structural connections. c) by way of openings (silencers and leakages).

Some pertinent questions in this connection will be: how large is the force amplitude input to the foundation and floor due to the vibration of the machine? How large is the sound power radiated from the floor into the room below due to these forces? What will the sound pressure level inside the enclosure be and furthermore, what is the relation between this sound pressure level and the sound pressure level outside the enclosure, partly due to transmission through the walls, partly through the silencers? What is the possibility of "short-circuiting" any of the noise measures by unwanted mechanical couplings, air leaks etc.?

All these questions require knowledge of transfer functions of some kind, the relationship between physical variables that will normally be a function of frequency and we must be able to calculate these functions, analytically, numerically or by using empirical methods. However, it should be observed that these variables normally are space dependent. As an example, the sound pressure level, both inside and outside the enclosure, could be expected to vary with position. One has then to deal with a representative set of transfer functions between the variables.

On the other hand, the noise problem could be an existing one. A solution then requires the analogous possibility using measurement techniques to map the noise field. The complexity of such mapping will be highly dependent on the number of sources, i.e. if there is just one dominating source or if there is a complex interaction of many sources.

2.3 TRANSFER FUNCTION. DEFINITION AND PROPERTIES

When using the term transfer function we will, as pointed out above, assume that the actual system is linear and stable. The purpose of such a function is that when known, one may not only determine the response of a harmonic input with a given frequency but also find the response following an arbitrary input time function $x(t)$. This could be a transient or stochastic time function or a combination of such functions. To make it simple we shall as much as possible use a harmonic signal input for the illustrations in this chapter.

2.3.1 Definitions

Strictly speaking, the term transfer function applies to the ratio of the *Laplace transforms* of the input and output signals, the *frequency response function* $H(f)$ or $H(\omega)$ being a special case. When writing

$$Y(f) = H(f) \cdot X(f),   \tag{2.1}$$

$Y(f)$ and $X(f)$ are the Fourier transform of the output signal $y(t)$ and the input signal $x(t)$, respectively, as shown in Figure 2.2. Normally we shall use the term *transfer function* for the function $H(f)$ but we shall also sometimes apply the term *frequency response*. The latter is often used when the input signal amplitude is frequency independent but has a constant arbitrary value.

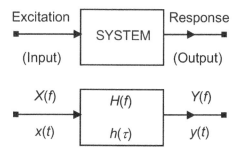

Figure 2.2 A system having one input and one output. Mathematical representations in time and frequency domains.

A description giving the relationship between the input and output in the time domain is also appropriate. The *impulse response* gives us the relation between the time signals $x(t)$ and $y(t)$. The $x(t)$ being an infinite sharp pulse, i.e. a Dirac δ-function, the signal $y(t)$ will give us the impulse response. The transfer function $H(f)$ and the impulse response $h(\tau)$ is also a Fourier transform pair and they therefore contain the same information. Modern measurement technique using signal types such as MLS or swept sine, utilize this property as they may easily determine the impulse response and from there use the Fourier transform to find the corresponding transfer function. Concerning the different types of test signals being used, see section 1.5.2.

A transfer function $H(f)$ is normally a complex function, which in other words tells us that the variables describing the input and output have a phase difference. We may therefore choose between having the $H(f)$ expressed by its real and imaginary parts or by its modulus and phase:

$$H(f) = \text{Re}\{H(f)\} + j \cdot \text{Im}\{H(f)\} = A(f) + j \cdot B(f)$$

$$\text{or} \qquad H(f) = |H(f)| \cdot e^{j\theta(f)} = \sqrt{A^2(f) + B^2(f)} \cdot e^{j\theta(f)}, \qquad (2.2)$$

where $\theta(f)$ is given by

$$\tan(\theta(f)) = \frac{B(f)}{A(f)}. \qquad (2.3)$$

The modulus $|H(f)|$ is often referred to as the gain factor of the system and $\theta(f)$ the corresponding phase factor. It is important to note that $H(f)$ is a function of frequency only, which is a consequence of assuming the system to be linear and stable. If the system does not fulfil these conditions the transfer function will, in the former case, depend on the input amplitude. In the latter case, there will be a time dependency.

2.3.2 Some important relationships

Knowing the transfer function we may calculate the response (output) for any type of excitation (input) if the conditions, as mentioned above, are fulfilled. A Fourier series or integral may express the excitation and the response will be a sum of the responses for each of the components in the excitation. From the response we may then calculate the mean square value or the RMS-value.

Associated with the general treatment of oscillations or signals in Chapter 1 we shall state some important relationships concerning "two-signal" or "two-channel" analysis. We will again assume that the system has *one* input and *one* output as shown in Figure 2.2. For systems having several inputs and/or several outputs the reader should consult the specialized literature on the subject, e.g. Bendat and Piersol (2000).

2.3.2.1 Cross spectrum and coherence function

Broadband stochastic signals are suitable for investigating an actual system. Assuming that the excitation (input signal) is such a signal we may show that the following two relationships are valid, equations linking the transfer function and the spectral density:

$$G_{yy} = |H(f)|^2 \cdot G_{xx} \qquad (2.4)$$

$$\text{and} \qquad G_{xy} = H(f) \cdot G_{xx}. \qquad (2.5)$$

The first equation links together the *power spectrum*, or more correctly the *power spectral density*, on the input and output side of the system. Only the gain factor appears in the equation and all quantities are real. Equation (2.5), however, is complex and here we have got a new spectral function G_{xy}. This is the *cross spectral density function*, shortened to *cross spectrum* and defined by

$$G_{xy}(f) = \lim_{T \Rightarrow \infty} \left[\frac{2}{T} X_T^*(f) \cdot Y_T(f) \right], \qquad (2.6)$$

where the star symbol * signifies that the complex conjugate of the Fourier transform of the input signal shall be used. A very important application of this function is to determine the transfer function $H(f)$ by way of Equation (2.5). This is the preferred instrument technique instead of using the direct definition given in Equation (2.1). The reason is that it may be shown that with the former the expected value will be more correct, which is related to the ideal situation expressed by the Equations (2.4) and (2.5). Apart from the assumptions concerning linearity and time invariance, we have tacitly assumed that there are no external noise signals either in the input or output to disturb our measurements. In practice, there certainly will be such disturbances. A method to control this, i.e. finding out if the signal in the output really is caused by the excitation signal and not by any disturbing signal, is by measuring the so-called *coherence function* γ_{xy}, given by

$$\gamma_{xy}(f) = \frac{|G_{xy}(f)|^2}{G_{xx}(f) \cdot G_{yy}(f)} \qquad 0 \le \gamma_{xy}(f) \le 1. \qquad (2.7)$$

A coherence function identically equal to 1.0 implies that the output signal is caused by nothing other than the applied input signal. If less than 1.0, this means that there are systematic and/or random errors in the measured transfer function. This need not be caused by external noise only but could be linked to the problem of using too few lines in the discrete Fourier transform, i.e. too few lines when the transfer function varies strongly within narrow frequency intervals. An example is when the system exhibits strong resonances (see below). This type of systematic error, which is caused by inadequate frequency resolution, is called *leakage*, a subject treated in section 1.4.4.

2.3.2.2 Cross correlation. Determination of the impulse response

In an analogous way as the power spectral density G (power spectrum) has its time domain equivalent description in the autocorrelation function $R(\tau)$, the cross spectrum G_{xy} has its equivalent, its Fourier transform, in the *cross correlation function* $R_{xy}(\tau)$. This is defined by

$$R_{xy}(\tau) = \lim_{T \Rightarrow \infty} \frac{1}{T} \int_0^T x(t) \cdot y(t+\tau) \, d\tau. \qquad (2.8)$$

One might say that this function experienced its renaissance in measurements of sound and vibration by the introduction of MLS as a test signal. Using such signals one may easily and accurately determine impulse responses for a system and thereby one may determine the corresponding transfer functions. The method is widely used in building acoustics, in room acoustic measurements as well as in measurements of sound insulation. In recent years, however, the swept sine technique has become a rival to the technique using MLS (see below). An important feature in both methods is the larger dynamic range as compared with earlier conventional methods.

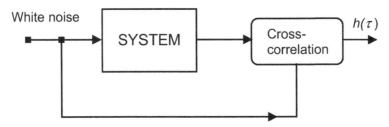

Figure 2.3 Principle determination of the impulse response of a system by cross correlation.

The measurement principle using MLS is based on the fact that when using a stochastic white noise input signal one will obtain the impulse response when cross correlating the output and input signals. The principle set-up is shown in Figure 2.3. The crucial point is that one may get an accurate estimate of the impulse response by substituting the white noise signal with an MLS signal. Since the latter is deterministic one may, assuming as before that the system is time invariant, constantly improve the estimate by increasing the number of sequences used in the averaging process.

Obtaining the impulse response using the swept sine technique may be implemented in different ways. Maybe the simplest one to grasp is the one illustrated in Figure 2.4, where one as a first step performs a Fourier transform of both the output and the input signal. By a spectral division we obtain the transfer function directly, and the impulse response may be obtained by an inverse Fourier transform. As pointed out in Chapter 1 the swept sine technique has some important advantages compared with the technique using MLS, e.g. more robust in terms of time variance and an even greater dynamic range may be obtained.

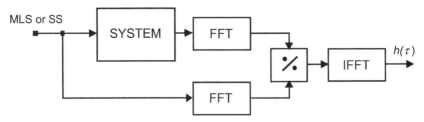

Figure 2.4 Principle determination of the impulse response of a system by spectral division of the Fourier transforms.

2.3.3 Examples of transfer functions. Mechanical systems

A transfer function is certainly linked to the specific variables defined as being the excitation (input) and the response (output). In acoustics and vibration there are a number of transfer functions we shall use and many of these functions have special names. In this chapter, we shall draw on examples from mechanical systems. The analogous acoustic quantities will be introduced in the following chapter.

For mechanical systems we shall define the following quantities; see also ISO 2041:

Mechanical impedance, the complex ratio of force F, applied to a point in the system, and the resulting velocity v:

$$Z_{mech} = \frac{F}{v} \qquad \left(\frac{N \cdot s}{m} \right). \tag{2.9}$$

This definition presupposes simple harmonic motion. If this is not the case, the variables must be interpreted as functions of frequency, i.e. Fourier transforms. The force and velocity may be taken at the same or different points. One normally distinguishes between these two cases by applying different names. In the former case the name *point impedance*, or more precisely *driving point impedance*, is used whereas one uses the name *transfer impedance* in the latter case. Do note that the impedance is *not* the transfer function when interpreting the force as the excitation. In that case the inverse quantity represents the transfer function, which is called

Mechanical mobility and thus

$$M_{mech} = \frac{v}{F} \qquad \left(\frac{m}{N \cdot s} \right). \tag{2.10}$$

The quantity is in some cases referred to as *admittance.* We shall also define the

Transmissibility, which is the non-dimensional ratio of the response amplitude of a system in steady-state forced motion to the excitation amplitude. The ratio may be one of forces, displacements, velocities or accelerations.

Analogous quantities are used for mechanical moments, where one finds *moment impedance* and *moment mobility*. Then the force and velocity are replaced by moment and angular velocity. It should also be mentioned that one may also find data represented as *apparent mass*, i.e. the velocity v in Equation (2.9) is replaced by the acceleration a.

2.3.3.1 Driving point impedance and mobility

The fundamental physical characteristics of mechanical systems are mass, stiffness and damping. We shall in the first place assume that our masses, springs and dampers are ideal concentrated (or lumped) elements. When driving these elements using a harmonic force $F \cdot \exp(j\omega t)$ the resulting velocity v will also be harmonic and we may create the Table 2.1 below.

The fundamental and important difference between the impedances, maybe looking at the impedance as resistance against movement, increase with frequency for a mass whereas decrease with a spring. For a viscous damper the damping is proportional to velocity, thereby making the impedance frequency independent. The damping in elastic media, e.g. metal, rubber and plastics, is however better characterized as hysteretic. This kind of damping is proportional to the displacement and described, as shown later on, by using complex spring stiffness, for materials using a complex modulus of elasticity.

Table 2.1 Impedance and mobility of the simple elements mass, spring and viscous damper. F – force, v – velocity, a – acceleration, k – spring stiffness, c – damping coefficient.

Element	Symbol	Calculation	Impedance	Mobility
Mass	$F \rightarrow m$	$F = m \cdot a = m\dfrac{dv}{dt} = j\omega m v$	$j\omega m$	$1/j\omega m$
Spring	$F \rightarrow k$	$F = k \cdot x = k\displaystyle\int v\,dt = -j\dfrac{k}{\omega}v$	$-j\dfrac{k}{\omega}$	$j\dfrac{\omega}{k}$
Damper (viscous)	$F \rightarrow c$	$F = c \cdot v$	c	$1/c$

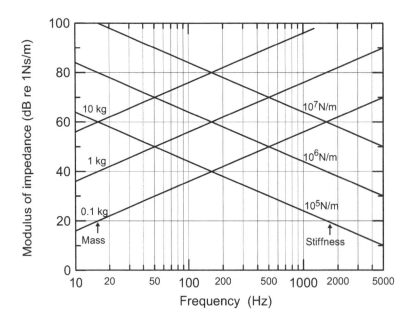

Figure 2.5 Modulus of impedance for the components mass and spring, given as a function of frequency. Mass weight and spring stiffness are indicated on the curves.

When depicting an impedance or mobility in a frequency diagram one usually uses a logarithmic scale and the modulus of the actual quantity is given in decibels (dB) with an agreed reference value. In the impedance diagram, shown in Figure 2.5, a reference value of 1N·s/m is used and the solid lines show the impedance for lumped mass and spring elements. Using such diagrams for plotting the impedance of composite and more complex systems one will immediately observe in which part of the frequency range we may characterize the systems behaviour in terms of mass, a spring or damping. (What would the corresponding mobility diagram look like?)

2.4 TRANSFER FUNCTIONS. SIMPLE MASS-SPRING SYSTEMS

The simple mass-spring system is the classical example for illustrating transfer functions in a physical system, a system also called a linear harmonic oscillator. We shall model the system using three concentrated (lumped) elements, a mass, a spring and a damper. The system is assumed to have one degree of freedom only, a transverse movement as shown in Figure 2.6. The justification for using such a simple system is twofold: it exhibits most of the phenomena found in more complex systems and furthermore, it gives us the opportunity to define and clarify the concepts and quantities to be used later on. We shall either use the displacement or the velocity as the response quantity, the latter when dealing with the impedance of the system.

As seen from the figure we assume that an outside force F drives the mass. The first task will be to calculate the movement of the mass as a function of frequency, second, we shall calculate the transmitted force F' to the base or foundation. The latter task is fundamental to the subject area of *vibration isolation.*

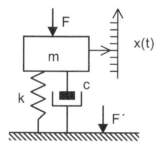

Figure 2.6 Simple mass-spring system (harmonic oscillator).

2.4.1 Free oscillations (vibrations)

We shall start by assuming that the system in one way or other is displaced from its equilibrium position and thereafter left to move freely. Without any outside forces operating, i.e. F equal to zero, the sum of inertial forces F_m, spring forces F_k and viscous damping forces F_c will be equal to zero:

$$F_m + F_c + F_k = 0, \qquad (2.11)$$

giving

$$m \cdot a + c \cdot v + k \cdot x = 0 \quad \text{or}$$

$$m\frac{\mathrm{d}^2 x}{\mathrm{d}t^2} + c\frac{\mathrm{d}x}{\mathrm{d}t} + k \cdot x = 0, \tag{2.12}$$

where x, v and a are displacement, velocity and acceleration, respectively. The damping coefficient c will be the prime factor in the solution of the differential equation describing the transient motion when the system is left to vibrate freely. A suitable variable characterising the damping is the *damping ratio* ζ. This quantity gives the damping coefficient c relative to the *critical damping coefficient* $c_{critical}$ of the system,

$$\zeta = \frac{c}{c_{\text{critical}}} = \frac{c}{2m\omega_0} = \frac{c}{4\pi m f_0}. \tag{2.13}$$

f_0 is the fundamental frequency (*eigenfrequency*), the natural frequency of oscillation for the undamped system, i.e. when the damping coefficient c is equal to zero. This frequency is given by

$$f_0 = \frac{\omega_0}{2\pi} = \frac{1}{2\pi}\sqrt{\frac{k}{m}}. \tag{2.14}$$

We shall in the following use both the frequency f and the angular frequency ω, choosing the one most suitable in each case, but without changing the name of quantities such as natural frequency etc. Inserting the damping ratio into Equation (2.12) we may solve the equation for the following three cases:

$\underline{\zeta < 1}$ will give us a damped oscillatory motion where the displacement x may be expressed as:

$$x(t) = \mathrm{e}^{-\zeta\omega_0 t}\left[A\sin(\omega_d t) + B\cos(\omega_d t)\right] \quad \text{or}$$

$$x(t) = C \cdot \mathrm{e}^{-\zeta\omega_0 t} \cdot \cos(\omega_d t + \theta), \tag{2.15}$$

where

$$\omega_d = \omega_0\sqrt{1-\zeta^2}, \quad C = \sqrt{A^2 + B^2} \quad \text{and} \quad tg\theta = \frac{B}{A}.$$

The coefficients A and B, or alternatively C and θ, must be determined from the initial conditions. In section 1.4.2.2, we used this transient motion for illustrating the calculation of the Fourier transform.

$\underline{\zeta = 1}$ indicates that the system is critically damped. The movement is no longer oscillatory, the system is returning to its stable position in a minimum time. (Make a comparison with the spring system of e.g. a car.) The solution is now

$$x(t) = (A + B \cdot t) \cdot \mathrm{e}^{-\xi\omega_0 t}. \tag{2.16}$$

$\zeta > 1$ shows that the system is more than critically damped giving a solution

$$x(t) = e^{-\xi\omega_0 t}\left[A\, e^{\omega_0 \sqrt{\zeta^2 - 1}\, t} + B\, e^{-\omega_0 \sqrt{\zeta^2 - 1}\, t} \right]. \tag{2.17}$$

2.4.1.1 Free oscillations with hysteric damping

As stated in section 2.3.3.1, using viscous damping is not appropriate in modelling a system with elastic components such as rubber, plastics etc. The damping is better described as hysteretic, normally using the *loss factor* η as a characteristic quantity. In our simple mass-spring system we shall remove the viscous damper and introduce damping through a complex spring stiffness $\underline{k}$

$$\underline{k} = k(1 + j \cdot \eta). \tag{2.18}$$

The loss factor η will always be much less than one. For metals one will find η in the range of $10^{-4} - 10^{-3}$, for rubber of the order 10^{-2}. Equation (2.12) will then be replaced by the following

$$m \cdot \frac{d^2 x}{dt^2} + k(1 + j \cdot \eta)x = 0. \tag{2.19}$$

We now assume that the solution of this equation will have the same form as Equation (2.15) but we shall express it using the complex form, $x(t) = A \cdot \exp(j\gamma\, t)$. By insertion into Equation (2.19) we easily solve for the exponent γ

$$\gamma = \sqrt{\frac{k}{m}(1 + j \cdot \eta)} \underset{\eta \ll 1}{\approx} \sqrt{\frac{k}{m}}\left(1 + j\frac{\eta}{2}\right) = \omega_0\left(1 + j\frac{\eta}{2}\right). \tag{2.20}$$

Hence, we obtain

$$x(t) \approx A\, e^{j\omega_0\left(1 + j\frac{\eta}{2}\right)} \quad \text{having a real solution: } x(t) = A\, e^{-\frac{\eta}{2}\omega_0 t} \cdot \cos(\omega_0 t). \tag{2.21}$$

Compared with the solution (2.15) the damping ratio ζ is replaced by $\eta/2$ in the exponential term. It should also be mentioned that other quantities are in use for expressing the damping, such as the *logarithmic decrement* δ and the *Q factor*. Assuming that the damping is small the relationship between all these quantities is as follows

$$2\zeta = \frac{\delta}{\pi} = \eta = \frac{1}{Q}. \tag{2.22}$$

The *reverberation time T* is used in building acoustics to express the damping of sound in rooms. However, the concept is useful when dealing with vibration as well. By definition the reverberation time T is the time elapsed before the energy in an oscillating system is reduced to 10^{-6} of the initial value. As the energy is proportional to the square of the vibration amplitude we may represent the definition by

$$\frac{x^2(t=0)}{x^2(t=T)} = 10^6,$$
(2.23)

and if further applied to Equation (2.21) we have

$$T = \frac{\ln(10^6)}{\omega_0 \cdot \eta} \approx \frac{2.2}{f_0 \cdot \eta}.$$
(2.24)

This expression is applied in measurement methods for determining the loss factor for materials where the energy losses are relatively small. For materials with high losses the reverberation time will be too short to obtain reasonable accuracy. A better method is then to excite the material specimen into resonance and measure the Q factor.

2.4.2 Forced oscillations (vibrations)

Driving our simple mass-spring system using an external force F we now obtain

$$F_m + F_c + F_k = F \quad \text{or}$$
$$m\frac{d^2x}{dt^2} + c\frac{dx}{dt} + k \cdot x = F(t),$$
(2.25)

where we have, in the last equation, indicated that the external force could be an arbitrary function of time. There are several available procedures for solving the equation. Our aim is to determine the transfer function between applied force and displacement and, later, between force and velocity, which is the driving point impedance. Letting the input force be a simple harmonic force is the easiest way to solve the differential equation. Even then the solution will contain two terms. One of these terms will represent a transient motion as we start when the system is in a stable position. This will be the solution of the homogeneous Equation (2.12) whereas the other term will be the stationary part, which will be of primary interest. We may solve this term by expressing the force as $F(t) = F_0 \cdot \exp(j\omega t)$ and then assuming a solution having the form $x(t) = x_0 \cdot \exp(j\omega t)$. Inserting this into Equation (2.25) we get

$$x_0 = \frac{F_0}{-\omega^2 + j\omega c - k}$$
(2.26)

and therefore we may write

$$x(t) = \frac{F_0 \cdot e^{j\omega t}}{j\omega\left[c + j\left(\omega m - \frac{k}{\omega}\right)\right]} = \frac{F_0 \cdot e^{j\omega t}}{j\omega Z_m}.$$
(2.27)

When differentiating we get the velocity $v(t)$ as

$$v(t) = \frac{dx(t)}{dt} = j\omega x(t) = \frac{F_0 \cdot e^{j\omega t}}{Z_m}. \tag{2.28}$$

The quantity Z_m is the mechanical impedance, the driving point impedance at the point where the force is applied. In this case, this will be the sum of the impedances for the three elements because they all have the same velocity and the force is the sum of the forces on the elements. Alternatively, expressing the impedance by its modulus and phase as shown in Equation (2.2), we write

$$|Z_m(\omega)| = \sqrt{c^2 + \left(m\omega - \frac{k}{\omega}\right)^2} \qquad \text{and}$$

$$\tan\theta(\omega) = \frac{m\omega - \frac{k}{\omega}}{c}. \tag{2.29}$$

It should be noted that the phase changes sign when the frequency ω of the applied force is equal to the fundamental frequency (eigenfrequency) ω_0 of the system, i.e. when we excite the system into *resonance*. One possible method of mapping the resonance frequencies of a system is therefore by detecting the frequencies where one finds the maximum relative phase changes.

We shall use an impedance diagram to depict the modulus given in Equation (2.29). Figure 2.7 shows an example using the following data for the system components: the mass weight m is 1.0 kg, the spring stiffness k is $4 \cdot 10^5$ N/m and the damping coefficient c is equal to 18 kg/s. It should be fairly obvious why one normally divides the response to the force into three main ranges, which are called *stiffness-controlled*, *damping-controlled* and *mass-controlled* ranges.

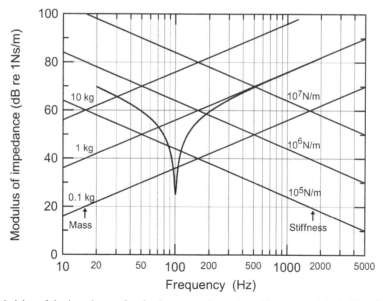

Figure 2.7 Modulus of the impedance of a simple mass-spring system. The mass weight is 1 kg, the spring stiffness is $4 \cdot 10^5$ N/m and the damping coefficient is 18kg/s.

Stiffness-controlled range, $\omega << \omega_0$
In this case, we have $\omega^2 << k/m = \omega_0^2$ and $|Z_m| \approx k/\omega$. The spring stiffness is the determining factor for the impedance. At very low frequencies the conditions are the same as when the force is a simple static load. The displacement is proportional to the force and in phase with it. Assuming a frequency independent force the velocity amplitude will increase with frequency and arrive at its maximum amplitude in the damping-controlled range.

Damping-controlled range, $\omega \approx \omega_0$
Here we find $|Z_m| \approx c$, telling us that it is damping only that controls the amplitude at resonance. The velocity and force are in phase whereas the displacement and force are 90° out of phase.

Mass-controlled range, $\omega >> \omega_0$
Above the resonance frequency the mass will start to be the controlling factor, $|Z_m| \approx \omega m$. Again assuming a frequency independent force the velocity will be inversely proportional to frequency and there will be a 90° phase difference between force and velocity. The acceleration will be constant and in phase with the force. (How will the displacement behave?)

Most people will be more familiar with a description showing a resonance to be a maximum value and not a minimum as shown here. We shall therefore use the mobility as the descriptive quantity when giving further examples.

2.4.3 Transmitted force to the foundation (base)

Using the equations above we are now in a position to calculate the force transmitted to the foundation or base, i.e. the ratio of the transmitted force and the applied force. This is of great interest in the field of vibration isolation of rotating machinery exhibiting unbalanced forces. The task here is to design an elastic supporting system reducing the transmission of forces to the foundation and thereby prevent harmful vibrations to be transmitted to the environment. We shall again use the simple model depicted in Figure 2.6 where the machine is modelled as a lumped mass and is placed on an elastic element, a simple spring. This model is, for several reasons, not a very realistic one especially due to the assumption of a foundation of infinite stiffness. If that is the case in practice, we should not need the isolation! We shall however, treat the more general case later.

From the figure we obtain for the force F' transmitted to the base

$$F' = k \cdot \left(-j\frac{v}{\omega} \right) + cv = v \left(c - j\frac{k}{\omega} \right). \tag{2.30}$$

Using the expression for the velocity v from Equation (2.28) we find

$$F' = \frac{F_0 \cdot e^{j\omega t}}{Z_m} \left(c - j\frac{k}{\omega} \right) = \frac{c - j\dfrac{k}{\omega}}{c + j \cdot \left(\omega m - \dfrac{k}{\omega} \right)} \cdot F. \tag{2.31}$$

We shall introduce the fundamental frequency ω_0 and the damping ratio ζ arriving at

$$F' = \frac{1 + j \cdot 2\zeta \dfrac{\omega}{\omega_0}}{1 - \left(\dfrac{\omega}{\omega}\right)^2 + j \cdot 2\zeta \dfrac{\omega}{\omega_0}} \cdot F. \tag{2.32}$$

If we only are interested in the ratio of the force amplitude transmitted and the exciting force amplitude and not their phase relationship, we calculate the modulus of the ratio of the complex quantities, i.e. the transmissibility T as

$$T = \left|\frac{F'}{F}\right| = \frac{|F'|}{|F|} = \left[\frac{1 + \left(2\zeta \dfrac{\omega}{\omega_0}\right)^2}{\left(1 - \left(\dfrac{\omega}{\omega_0}\right)^2\right)^2 + \left(2\zeta \dfrac{\omega}{\omega_0}\right)^2}\right]^{\frac{1}{2}}. \tag{2.33}$$

Figure 2.8 shows the transmissibility for four values of the damping ratio ζ. At low frequencies we find the force transmitted is the same as the exciting force; there will be no vibration isolation. The force amplification at resonance will depend on the damping and to obtain any isolation at all, the frequency must be higher than $2^{1/2} \cdot f_0$. When designing for vibration isolation one therefore must ensure that the frequencies of the unbalanced forces are relatively high compared to the natural frequencies of the system.

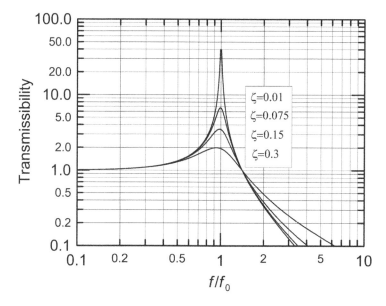

Figure 2.8 Transmissibility, the ratio of transmitted force (to the foundation) and the applied force, of a simple mass-spring system with viscous damping. The damping ratio ζ for the curves is given in the legend.

In the above model viscous damping was assumed, which will give a transmissibility dependent on damping also at the higher frequencies. As an alternative model we shall assume that damping is hysteretic (see section 2.4.1.1). Instead of Equation (2.30), we now have

$$F' = k(1 + \mathrm{j} \cdot \eta) x = -\mathrm{j} \frac{k}{\omega} (1 + \mathrm{j} \cdot \eta) v. \qquad (2.34)$$

Hence

$$F' = \frac{\dfrac{k}{\omega}(1 + \mathrm{j} \cdot \eta)}{\dfrac{k}{\omega}(1 + \mathrm{j} \cdot \eta) - \omega m} \cdot F = \frac{1 + \mathrm{j} \cdot \eta}{1 - \left(\dfrac{\omega}{\omega_0}\right)^2 + \mathrm{j} \cdot \eta} \cdot F. \qquad (2.35)$$

The transmissibility T_h will then be given by

$$T_\mathrm{h} = \left[\frac{1 + \eta^2}{\left(1 - \left(\dfrac{\omega}{\omega_0}\right)^2\right)^2 + \eta^2} \right]^{\frac{1}{2}}. \qquad (2.36)$$

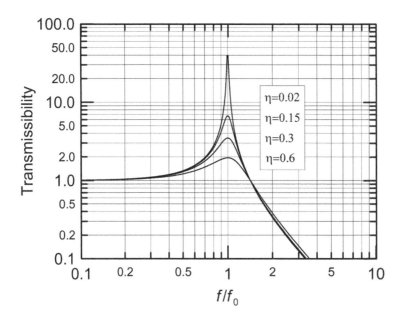

Figure 2.9 Transmissibility, the ratio of transmitted force (to the foundation) and the applied force, of a simple mass-spring system with hysteretic damping. The loss factor η is indicated on the curves.

This transmissibility is shown in Figure 2.9 for four values of the loss factor η. These values are, according to Equation (2.22), chosen equal to twice the damping ratios ζ used in Figure 2.8. It should be noted that the transmissibility is independent of the loss factor at the higher frequencies under the condition where η is not unrealistically high, thereby violating the assumption $\eta \ll 1$ as indicated in Equation (2.20).

In addition to the transmissibility T other quantities are in use to characterize the usefulness of elastic supporting systems. An appropriate question one may ask is: What do I gain by using an elastic system compared to mounting my machine directly on to the foundation? In that case one must of course model the foundation in a more realistic way than one of infinite stiffness. We shall define the *efficiency E* of the isolating system by the ratio[1]

$$E = \frac{\left| v_{\text{foundation}} \right|_{\text{without isolator}}}{\left| v_{\text{foundation}} \right|_{\text{with isolator}}}. \tag{2.37}$$

Later on we shall give examples using this quantity but this requires us to extend our knowledge to systems having several degrees of freedom.

2.4.4 Response to a complex excitation

Up until now we have assumed that the exciting force could be described by a simple harmonic time function, i.e. there is only one frequency operating at a time. We have, however, pointed out that knowing the transfer function we may calculate the response for any type of excitation. For this to be possible the condition of linearity must be fulfilled, the principle of *superposition* must apply. This means that, when expressing the excitation in a Fourier series or transform, the response will be the sum of the responses for each component in the excitation. A periodic excitation will give a periodic response and the system will resonate when a component in the excitation coincides with one of the natural frequencies (eigenfrequencies) of the system.

We shall give an example again using the simple mass-spring system in Figure 2.6 but we will exchange the viscous damping with the structural or hysteric one. We furthermore exchange the harmonic time function of the force with a stochastic one. This implies describing the force excitation using a spectrum, a spectral density function $G_F(f)$, and our task is to calculate the resulting velocity of the mass. We certainly know that the velocity $v(t)$ will be a stochastic function but what is the resulting RMS-value?

To answer the question we shall use the general Equation (2.4) linking together the spectral densities (power spectrum) of the input and output quantities. In our system, the input quantity is the force F with the power spectrum $G_F(f)$, the output being the velocity v with power spectrum $G_v(f)$. The transfer function $H(f)$ is therefore $1/Z_{\text{m}}$, i.e. the mobility M of the system. The task is therefore to calculate the integral

$$\tilde{v}^2 = \int_0^\infty G_v(f)\,\mathrm{d}f = \int_0^\infty \left| M(f) \right|^2 \cdot G_F(f)\,\mathrm{d}f. \tag{2.38}$$

In this case the mobility will be

---

[1] The term *insertion loss* is also commonly used, indicating the difference in some quantity when inserting a new member or device into an existing system.

$$M = \frac{1}{Z} = \frac{1}{\mathrm{j}\left(\omega m - \dfrac{k\left(1 + \mathrm{j}\cdot\eta\right)}{\omega}\right)} \qquad (2.39)$$

and we must solve an integral, which after some rearranging may be written

$$\tilde{v}^2 = \frac{4\pi^2}{k^2} \int\limits_0^\infty \frac{f^2 \cdot G_F(f)}{\eta^2 + \left(\left(\dfrac{f}{f_0}\right)^2 - 1\right)^2} \cdot df. \qquad (2.40)$$

The result is certainly wholly dependent on the force power spectrum $G_F(f)$. For simplicity we shall assume that the spectrum is white, in any case constant in a frequency range where the response is high. Setting $G_F(f)$ equal to G_0 we may show that

$$\tilde{v}^2 = \frac{G_0}{4\eta} \cdot \frac{1}{\sqrt{k \cdot m^3}}. \qquad (2.41)$$

The importance of having a high loss factor η is certainly expected as we then get a low response at resonance. The way the mass and stiffness influence the result are, however, not easy to guess.

2.5 SYSTEMS WITH SEVERAL DEGREES OF FREEDOM

We have in the derivations above used a very simple mechanical system having *one degree of freedom* to illustrate the general term transfer function and its special forms impedance and mobility. The system has just *one* natural frequency and *one* natural way of motion, the latter usually called a *mode*. An extensive coverage of mechanical systems having several degrees of freedom is outside the scope of this book. For the interested reader there are a number of textbooks on the subject to consult, e.g. Meirovitch (1997). Here we shall just give a short general description of systems described either as *discrete* or *continuous*. In the former case, we are able to model the system composed of concentrated or lumped elements, such as masses, springs and dampers. This is opposed to continuous systems where wave motion must be taken into consideration; where the wavelength is becoming comparable with or less than the dimensions of the system.

We shall, in this main section, give some examples of models using lumped elements, to present a more realistic treatment of the theme vibration isolation than that given in section 2.4.3. Examples on continuous systems will be presented in the following chapters.

2.5.1 Modelling systems using lumped elements

The modelling of a system as a collection of lumped elements (masses, springs and dampers) may be applied in practice both to mechanical and acoustical systems. If the task is to calculate the motion of each element in the system we have to find all the natural frequencies of the system as well as the *modal vectors* (*eigenvectors*). The latter defines the natural forms of vibration, the *natural modes*. The important point is that any possible type of motion, resulting from a given force excitation, may be described by a linear combination of these modal vectors. This is the background for calculating the response using so-called *modal analysis*; see section 2.5.3.1 below.

In many cases, however, we are not interested in performing a detailed analysis of all parts of the system. The task may only be to calculate the impedance (or mobility) at a place where the force (or moment) is attacking, making us able to control the mechanical power transmitted to the system. Such an analysis may be performed in a simple way dealing with systems having a small number of elements. There exists, however, several commercial computer programs that will be helpful if one is able to model the actual system using an analogue electrical system.

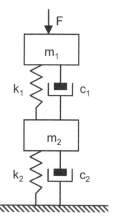

Figure 2.10 System with two masses, springs and dampers. On the calculation of mobility, see text.

We shall present an example using this kind of modelling applying the system shown in Figure 2.10, which is a combination of two simple mass-spring systems. The system with elements using suffix 1 has identical data as used for calculating the impedance shown in Figure 2.7. In the other system, suffix 2, the mass is increased by a factor of 10 and the damping is also increased. The total or combined system has now two natural frequencies. For the frequencies in this case, and for several other simple systems, one may find analytical expressions in the literature, e.g. Blevins (1979). These are f_i ($i = 1, 2$) given by

$$f_i = \frac{1}{2^{\frac{3}{2}}\pi}\left\{\frac{k_1}{m_1} + \frac{k_1}{m_2} + \frac{k_2}{m_2} \mp \left[\left(\frac{k_1}{m_1} + \frac{k_1}{m_2} + \frac{k_2}{m_2}\right)^2 - 4\frac{k_1 k_2}{m_1 m_2}\right]^{\frac{1}{2}}\right\}^{\frac{1}{2}}. \quad (2.42)$$

For this case we have chosen to calculate the input point mobility and Figure 2.11 shows the result compared with the result setting the mass m_2 infinitely large. Comparing with the latter result, which we calculated earlier on and where the natural frequency is 100 Hz, we have now two resonances. The corresponding natural frequencies, calculated using Equation (2.42) are 30.2 and 106.0 Hz, respectively. (It should be observed that the natural frequencies are calculated for systems without damping).

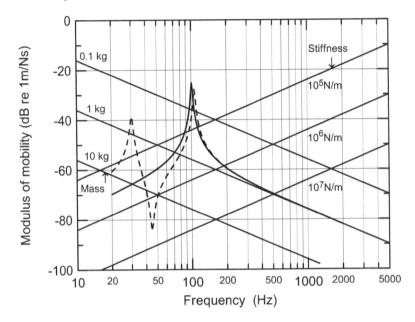

Figure 2.11 Input mobility of system shown in Figure 2.10. Dashed curve: m_1=1kg. k_1=4·10^5 N/m. c_1=18 N·s/m. m_2=10kg. k_1=4·10^5 N/m. c_1=100 N·s/m. Solid curve: m_2= ∞.

2.5.2 Vibration isolation. The efficiency of isolating systems

We shall again use the system depicted in Figure 2.10 to give a more realistic illustration of vibration isolation, and in particular using the efficiency E, defined in Equation (2.37), to characterize the effect of the elastic support. Elastic supports are normally introduced to reduce vibrations from a machine transmitting vibrations to the foundation, alternatively to reduce vibrations in the foundation being transmitted to e.g. sensitive equipment. The former case is sketched in Figure 2.12, where we wish to isolate any unbalanced forces in the upper structure (machine) from the lower structure (foundation or floor). As a measure of success we shall compare the two situations marked a) and b). In general, this may be a complicated task. We normally have to apply more than one isolator, which implies several degrees of freedom. Furthermore, we may as a worst case have to model the machine and foundation as continuous systems.

    We shall, however, assume that each of the three structures may be characterized by the mobility applicable for a movement in only one direction, here specifically in the vertical direction. Using indices m, i and f indicating the mobility of structure 1 (machine), isolator and structure 2 (foundation) respectively, we may express the efficiency E as

$$E = \left|\frac{v_2'}{v_2}\right| = \left|\frac{M_m + M_f + M_i}{M_m + M_f}\right|. \tag{2.43}$$

To make the isolator system efficient it is not sufficient to make the isolator "soft", i.e. choose one with a high mobility. It must also be high compared with the sum of the motilities of the attached structures.

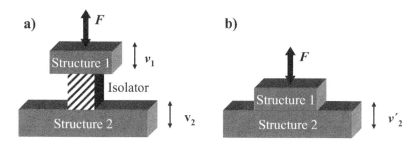

Figure 2.12 Sketches for the calculation of the efficiency of a vibration isolating system. a) Structures coupled through isolator. b) Structures in direct contact.

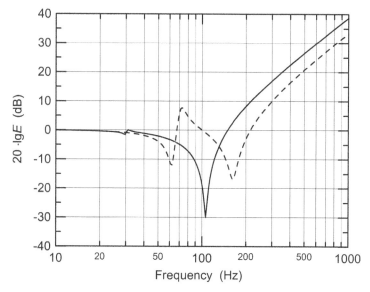

Figure 2.13 The vibration isolation given by the efficiency on a logarithmic scale. The system is shown in Figure 2.10, where the mass m_1 represents the "machine" and the components with suffix 2 represents the "foundation". Solid curve – component values as in Figure 2.11. Dashed curve – $m_2 = m_1 = 1.0$ kg.

We shall use the model shown in Figure 2.10 to illustrate the behaviour of the efficiency E given by Equation (2.43). We shall let the machine be represented by the mass m_1 and the foundation by the mass m_2, spring stiffness k_2 and damping coefficient c_2. The isolating element has the corresponding spring stiffness k_1 and damping coefficient c_1. The efficiency, given on a logarithmic scale, is shown in Figure 2.13 using two different values for the mass m_2. For the solid line curve all component data are the same as used for calculating the input mobility shown in Figure 2.11. As seen from Figure 2.13 we do not obtain any isolation for frequencies less than approximately $\sqrt{2}$ times the highest natural frequency, which is around 150 Hz. In fact, below this frequency the conditions are worse than without the isolator. On the other hand, the influence of the foundation is negligible.

However, by reducing the mass of the foundation by a factor of 10, as shown by the dashed curve, the situation is quite different. In fact, we do get $E > 1.0$ ($20 \cdot \lg E > 0$) in the interval between the two resonance frequencies but at the same time the frequency must be well over 200 Hz to obtain good isolation. (E will be > 1.0 for frequencies above 220 Hz).

2.5.3 Continuous systems

A model using lumped elements will, however, be useless when wave phenomena start appearing, which applies when the wavelength of an actual wave type becomes comparable with the physical dimensions of the elements. We then have to deal with systems having distributed mass and stiffness, the number of freedoms will in principle be infinite.

The above does not imply that discrete and continuous systems in principle represent different types of dynamical system exhibiting dissimilar dynamical characteristics. It should, as indicated above, merely be regarded as two different mathematical models for the same physical system. The behaviour is analogous even though the one is described using ordinary differential equations, the other by partial differential equations.

Comparing with the general description of discrete systems we now mathematically express the response of a continuous system to a given excitation by its *eigenfunctions* and its associated *eigenfrequencies* (or natural frequencies). The eigenfunctions are functions of the space coordinates. Analogous to the eigenvectors of discrete systems, which describe the *natural modes* of vibration, the eigenfunctions describe the natural *modal shapes* of the continuous system. Furthermore, expressing a complex vibration pattern with these functions is wholly analogous to the use of Fourier series or transforms on oscillations in the time domain.

We shall treat these subjects in more detail when dealing with wave and wave phenomena in Chapter 3 and also further on when dealing with sound transmission in Chapter 6. At this point we shall give a short overview on calculation and measurement methods relevant to continuous systems.

2.5.3.1 Measurement and calculation methods

For a number of simple structures having idealized boundary conditions we may find explicit analytical expressions for the eigenfunctions. As an example, we may for a panel (wall) assume that it is simply supported (or, alternatively, clamped) along the edges. In approximation, this could be true but such boundary conditions are always an idealization. In practice, performing calculations of either transfer functions or vibration modes for a certain wave type we therefore have to rely on various approximate methods. *Finite element methods* (FEM, BEM) implemented in advanced computer programs, such as e.g. ANSYS™, ABAQUS™ and FEMLAB™, have made such methods powerful tools. At the same time one has the ability to use advanced experimental methods, *modal testing* or *modal analysis*, to compare and to give feedback on the calculated results. In an interactive way, one may then improve on the mathematical model of the structure under investigation.

Another important analytical tool is *statistical energy analysis (SEA)*. We shall therefore give an overview of this method later in Chapter 7. As distinct from the finite element methods, SEA is what we might call an *energy flow method*. It will not give any detailed description of the oscillating motion at a given frequency but gives us a picture of the energy flow in the system averaged over wide frequency bands. The fundamental basis of the method is that each element or subsystem making up the structure under investigation exhibits a number of natural frequencies inside these frequency bands. The calculated response to an excitation is therefore always some mean value for one or more frequency bands. In a vast number of cases, not only in building acoustics but also in general noise problems, this is sufficient information. Several commercial computer program packages are available, e.g. AutoSEA™, SEADS™ and SEAM™.

In general, we know that the response to an input to a mechanical system such as a plate, a beam or a shell will be dependent on position. Detecting a resonance when the driving frequency coincide with one of the natural frequencies then presupposes that the amplitude of the vibration mode, associated with this natural frequency, is different from zero in the driving point. In a measurement situation for a proper mapping of natural frequencies several driving input points have to be used.

An experimental *modal analysis* does not map the natural frequencies of a structure only but determines the natural vibration patterns, the *modal shape* of the structure. Putting it simply, determining the modal shape is based on the measurement of a number of transfer functions for the structure. A force is applied in one or more points and for each driving point the response is measured at a number of positions distributed over the whole structure. From the measured resonance frequencies, one may estimate the natural frequencies. Simultaneously, one has at each of these frequencies and at each measuring point the information on how the structure vibrate both in amplitude and phase, i.e. one has an estimate of the associated *modal shape*. From this information it is possible to construct a model of the structure for solving the inverse problem: calculating the response to an arbitrary excitation. This is possible due to the response being a combination of the responses of the separate modes. An introduction to this technique can be found in the book by Ewins (1988). Modal analysis is, as mentioned above, an important measurement method giving feedback to finite element methods.

2.6 REFERENCES

ISO 2041: 1990, Vibration and shock – Vocabulary.

Bendat, J. S. and Piersol, A. G. (2000) *Random data: Analysis and measurement procedures*, 3rd edn. John Wiley & Sons, New York.
Blevins, R. D. (1979) *Formulas for natural frequency and mode shape*. Van Nostrand Reinhold Company, New York.
Ewins, D. J. (1988) *Modal testing: Theory and practice*. Research Studies Press Ltd, Taunton; John Wiley & Sons, New York.

Meirovitch, L. (1997) *Principles and techniques of vibration*. Prentice-Hall International Inc., Thousand Oaks, NJ.

CHAPTER 3

Waves in fluid and solid media

3.1 INTRODUCTION

This chapter is devoted to the fundamental properties of waves in fluids as well as in solid media, the latter being metal, concrete, plastics etc. Concerning fluid media we shall be considering gases only, which in normal cases in building acoustics will be air.

In addition to the treatment of the various types of wave motion we shall deal with the way waves are generated, i.e. examine sound sources and the way we can calculate the sound field generated by these sources. Furthermore, we shall use some simple cases of sound reflection from surfaces as an introduction to the later treatment of sound absorption and sound transmission.

As pointed out earlier, a wave is characterized by an oscillating motion propagated through the actual medium by virtue of its physical characteristics. Energy is transported by the wave but there is no net transfer of the medium. This does not imply that the medium, in a global sense, cannot be transported along with the wave, e.g. by wind, air movement in a ventilation duct etc. We shall, however, limit the treatment to cases where the medium is at rest.

Sound waves in solids can, as opposed to sound waves in fluids, store energy in shear motion. Whereas only compressional waves can exist in fluids, several other types of wave and combinations thereof are possible in solids. Of special importance in sound transmission in buildings is *bending waves*, also called *flexural waves*. Bending waves in plate-like structures will therefore be an important subject.

We shall presuppose that the acoustic phenomena are linear. Simply stated, this implies that the excursions in value of the physical quantities during wave motion are small compared with the value in a state of equilibrium. Non-linear phenomena occurring due to large deformations or at very high pressures are outside the scope of this book.

3.2 SOUND WAVES IN GASES

A sound wave propagating through a gas gives space and time variations in pressure, density and temperature as well as relative displacement from equilibrium of the particles of the gaseous medium. Observing the instantaneous values of pressure, density and particle velocity we may split these into an equilibrium part (or "direct current" part) and a fluctuating part due to the wave. We may write

$$P_{total} = P_0 + p, \qquad \rho_{total} = \rho_0 + \rho, \quad \text{and} \quad v_{total} = V_0 + v, \qquad (3.1)$$

where P_0 and ρ_0 are the equilibrium value for the pressure (the atmospheric pressure) and density, respectively. The acoustic "disturbances" are the *sound pressure p* and the *density variation ρ*. Given the loudness of sound we are normally exposed to means that the acoustic fluctuating parts are relatively small; we may assume the phenomena to be linear. Also note, as pointed out in the introduction, that we will assume V_0 to be zero. It should also be observed that this quantity as well as the particle velocity v is a vector quantity.

Given these approximations we can write down the linear or the so-called acoustic approximations for the governing fluid equations. These equations concern the conservation of mass, the fluid forces and the relationship between changes in pressure and density. We get

$$\frac{\partial \rho}{\partial t} = -\rho_0 \nabla \cdot v, \tag{3.2}$$

$$\nabla p = -\rho_0 \frac{\partial v}{\partial t} \tag{3.3}$$

and

$$p = c^2 \rho, \tag{3.4}$$

where c is the sound speed (phase speed) in the actual medium. Why it is termed phase speed is due to the fact that travelling along with the wave at the same speed c one always sees the same pattern; there is no change of phase. Eliminating the variables v and ρ from these equations we get the *wave equation*

$$\nabla^2 p - \frac{1}{c^2} \frac{\partial^2 p}{\partial t^2} = 0. \tag{3.5}$$

We now assume that the sound wave is split into partial waves having a harmonic time dependency. This means that it is sufficient to observe just one frequency component at the time. The sound pressure p and the particle velocity v at an arbitrary point in the sound field may then be expressed as

$$p(r,t) = \text{Re}\{\hat{p}(r) \cdot e^{j\omega t}\}$$

and

$$v(r,t) = \text{Re}\{\hat{v}(r) \cdot e^{j\omega t}\}, \tag{3.6}$$

where $\text{Re}\{...\}$ signify that we shall use the real part of the expression, further that the amplitudes of p and v, indicated by the "hats", are in general complex values. As a rule one implicitly uses the real value to obtain the real physical quantity. It is therefore common practice to leave out the $\text{Re}\{...\}$ in the expressions. When introducing the harmonic time dependency Equation (3.5) will transform into the *Helmholtz equation*

$$\nabla^2 p + \frac{\omega^2}{c^2} p = \nabla^2 p + k^2 p = 0, \tag{3.7}$$

where k (m^{-1}) is denoted *wave number*.

The pressure and the particle velocity in a sound field may vary in a very complex manner but still obeying the wave equation. Plane wave and spherical wave fields are examples of idealized types of wave field that are not only important theoretically but also in practical measurement situations.

3.2.1 Plane waves

Using the notion *plane wave* implies a wave where the sound pressure varies in *one* direction only. We have not yet introduced any energy losses in the Equations (3.2) through (3.4), which means that the pressure amplitude will be constant, and we may write

$$p(\boldsymbol{r},t) = \hat{p} \cdot e^{j(\omega t - \boldsymbol{k} \cdot \boldsymbol{r})}, \tag{3.8}$$

where $\boldsymbol{k} = \boldsymbol{n} \cdot 2\pi/\lambda$ is the *wave number vector* and λ is the *wavelength*. It should be observed that the last term in the exponent is the vector product of the wave number vector and the coordinate vector $\boldsymbol{r}$. For a plane wave in the positive x-direction we obtain

$$p(x,t) = \hat{p} \cdot e^{j(\omega t - k_x x)}, \tag{3.9}$$

where k_x is the component of the wave number in the x-direction. The equation tells us that either we observe the pressure as a function of time or as a function of location at a given time, it represents a simple oscillatory motion. As for other types of oscillation we shall use the RMS-value as a characteristic quantity,

$$\tilde{p}^2 = \frac{1}{T} \int_0^T p^2(x,t) \, dt, \tag{3.10}$$

which further may be expressed in decibels (dB) by the *sound pressure level L_p*

$$L_p = 10 \cdot \lg \left(\frac{\tilde{p}^2}{p_0^2} \right) \ (dB). \tag{3.11}$$

The reference value p_0 is equal to $2 \cdot 10^{-5}$ Pa, an international standard value for sound in air.

In practice, one may generate a plane wave in a duct or tube under the condition that the diameter is much less than the wavelength. In addition, the wave must not be reflected back from the end of the duct or tube; it must be equipped with a so-called anechoic termination. This technique is especially used for determination of sound power emitted by sources in duct systems, e.g. air conditioning fans as specified in ISO 5136.

3.2.1.1 Phase speed and particle velocity

The most common form expressing the pressure in a plane wave is given by Equation (3.8). For a plane wave, however, the wave Equation (3.5) will be satisfied by any function having the argument $(t - x/c)$ or $(t + x/c)$. The sound pressure will be constant as long as this argument has a constant value, which makes us realize that the quantity c really represents the propagation speed. As mentioned above, when travelling along with the wave at a speed $dx/dt = \pm c$ one will always "see" the same phase of the wave and the pressure will be constant. An analogous example is when surfing in the sea. This is the reason for calling c the phase speed.

According to the relationship we have used concerning changes in pressure and density, Equation (3.4), we have implicitly assumed that the changes take place adiabatically. This implies that the changes happen so fast that the temperature exchange with the surrounding medium is negligible. This is opposed to wave propagation in capillary tubes or generally in porous media, a theme we will treat in Chapter 5.

Starting from the general adiabatic gas equation

$$P \cdot V^{\gamma} = \text{constant},\tag{3.12}$$

where P and V are the pressure and volume of the gas, respectively and where γ is the adiabatic constant (≈ 1.4 for air), we may show that Equation (3.4) gives

$$p = c^2 \cdot \rho = \frac{\gamma P_0}{\rho_0} \cdot \rho.\tag{3.13}$$

This also implies that the phase speed is proportional to the absolute temperature T (°K) because we have

$$c_0 = \sqrt{\frac{\gamma P_0}{\rho_0}} \propto \sqrt{T}.\tag{3.14}$$

Due to our application here we have here added an index zero to the phase speed. In the literature several approximate expressions may be found. A more accurate one is:

$$c_0(\text{air}) = 20.05 \cdot \sqrt{273.2 + t},\tag{3.15}$$

where the temperature t is given in degree Celsius (°C). When it comes to the particle velocity v, dealing with linear acoustics, it is implicitly assumed that its absolute value is much less than the phase speed. That the assumption is fulfilled for the sound pressures normally experienced in our daily life is illustrated in the example below. (We disregard sound pressure levels that may even briefly damage our hearing). For the one-dimensional case we may write Equation (3.3) as

$$\frac{\partial p}{\partial x} = -\rho_0 \frac{\partial v_x}{\partial t},\tag{3.16}$$

which for a harmonic time dependency gives

$$-jk \cdot p = -j\omega\rho_0 \cdot v_x \qquad \text{and thereby}$$
$$p = \rho_0 c_0 \cdot v_x.\tag{3.17}$$

The quantity $\rho_0 c_0$ is the *characteristic impedance* of the medium, and it is a special case of the *specific acoustic impedance* defined by

$$Z_s = \frac{p}{v} \qquad \left(\frac{\text{Pa} \cdot \text{s}}{\text{m}}\right).\tag{3.18}$$

Example What is the magnitude of the particle velocity at a sound pressure of 1.0 Pa, being equivalent to a sound pressure level of ≈ 94 dB?

At a temperature of 20 °C the characteristic impedance of air will be 415 Pa·s/m, which inserted into Equation (3.17) will give $\approx 2.5 \cdot 10^{-3}$ m/s = 2.5 mm/s.

3.2.2 Spherical waves

Assuming spherical symmetry we arrive at the second idealized type of wave, the spherical wave. The wave equation may then be expressed as

$$\frac{\partial^2 (r \cdot p)}{\partial r^2} - \frac{1}{c_0^2} \frac{\partial^2 (r \cdot p)}{\partial t^2} = 0. \tag{3.19}$$

Analogous to the plane wave we may then express a partial spherical wave propagating from a centre (coordinate $r = 0$) as

$$p(r,t) = \frac{\hat{p}}{r} \cdot e^{j(\omega t - kr)}. \tag{3.20}$$

In this case, the coordinate vector has the same direction as the wave number vector and we may omit the vector notion. In contrast to the case of plane waves, the specific impedance will not be constant but will depend on the ratio of wavelength and distance from the source point. Using Equation (3.3), with the gradient expressed in spherical coordinates, in Equation (3.20) we get

$$Z_s = \rho_0 c_0 \frac{jkr}{1 + jkr}. \tag{3.21}$$

As seen from the equation, there will be a phase difference between sound pressure and velocity. Only in the case where the distance r is much larger than the wavelength, i.e. when $kr \gg 1$, we may set $Z_s \approx \rho_0 c_0$.

3.2.3 Energy loss during propagation

In the expressions for the sound pressure, given in Equation (3.8) for a plane wave and in Equation (3.20) for a spherical wave, we presupposed that the wave number k was a real quantity. In the physical sense this implies that the wave suffers no energy loss; the wave is not *attenuated* during propagation through the medium. However, in real media there will always be some energy losses caused by various mechanisms. Furthermore, in many cases one does try to optimize such losses; e.g. by the design of sound absorbers to be applied in rooms or to be used in various types of silencer. In other cases, e.g. during outdoor sound propagation over large distances natural losses will occur due to so-called relaxation phenomena. These losses, which are strongly frequency dependent, will be treated in Chapter 4.

It must be stressed that the attenuation we are concerned with here represents a real energy loss as opposed to a purely spherical spreading of sound energy over an increasing volume. We shall, whenever necessary, use the term *excess attenuation* to distinguish such losses from the latter type. Formally, we shall introduce such losses either by using a complex wave number k' or a complex *propagation coefficient* Γ.[1] We shall introduce these quantities by writing $\Gamma = j \cdot k' = \alpha + j \cdot \beta$. For the sound pressure in a plane wave, propagating in the positive x-direction, we may then write

$$p(x,t) = \hat{p}_0 \cdot e^{-\Gamma x} \cdot e^{j\omega t} = \hat{p}_0 \cdot e^{-(\alpha + j\beta)x} \cdot e^{j\omega t}, \qquad (3.22)$$

where the components α and β are the *attenuation coefficient* and the *phase coefficient*, respectively. Comparing with Equation (3.9) we immediately see that the phase coefficient β is equal to our real wave number k_x, whereas the attenuation coefficient represents the energy losses. The latter is often specified by the number of decibels per metre, which by using Equation (3.22) is given by

$$\text{Attenuation (dB/m)} \approx 8.69 \cdot \alpha. \qquad (3.23)$$

In this book we shall reserve the symbol α for the *absorption factor*. Therefore later on we will replace the attenuation coefficient α with the quantity $m/2$, where m is called *power attenuation coefficient*.

Figure 3.1 may be used as an illustration of the sound pressure amplitude of an ideal plane wave and a wave being attenuated during propagation, respectively. These may be regarded as sections of the wave fronts. One must, however, be aware that the actual physical waves are compressional and not transverse types of waves, the former exhibiting alternating condensation and rarefaction.

a)   b)

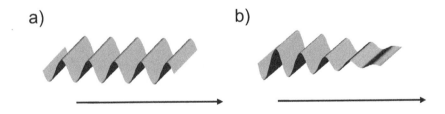

Figure 3.1 Sketch of the sound pressure in a wave front. a) Ideal plane wave. b) Attenuated plane wave.

3.2.3.1 Wave propagation with viscous losses

In our illustrations using the simple mass-spring system, we introduced viscous losses in the equation of force. In an analogous way we will do the same with Equation (3.3), the so-called *Euler equation*. For simplicity, we shall assume that the wave is a plane one, and we shall add a loss term $r \cdot v_x$ where v_x is the particle velocity in the x-direction. The quantity r is the *airflow resistivity* of the medium of propagation having a dimension of $Pa \cdot s/m^2$. As we shall see later, this is an important parameter when characterizing porous materials. Equation (3.16) will then be modified to

---

[1] In ISO 80000 Part 8, the small Greek letter γ is used.

$$\frac{\partial p}{\partial x} = -\rho_0 \frac{\partial v_x}{\partial t} + r \cdot v_x.$$ (3.24)

Combining this equation with the corresponding one-dimensional versions of Equations (3.2) and (3.4), we may show that the propagation coefficient Γ and the specific impedance Z_s can be written as

$$\Gamma = j\frac{\omega}{c_0}\sqrt{1-j\frac{r}{\rho_0\omega}} \qquad \text{and}$$

$$Z_s = \frac{p}{v_x} = \rho_0 c_0 \sqrt{1-j\frac{r}{\rho_0\omega}}.$$ (3.25)

If we compare with the corresponding expressions for a lossless plane wave (for Z_s see Equation (3.17)), we have now got an additional complex root expression. A medium having this property is moreover the simplest model for a porous material, a *Rayleigh model* (named after physicist Lord Rayleigh). This model will, together with other models for porous material, be treated in Chapter 5. At this point, however, we shall only give an example of the attenuation brought about by such a resistive component. We shall assume a high frequency and/or a low flow resistivity such that the imaginary part in the root expression is $\ll 1$. The propagation coefficient will then be

$$\Gamma = j\frac{\omega}{c_0}\sqrt{1-j\frac{r}{\rho_0\omega}} \approx j\frac{\omega}{c_0}\left(1-j\frac{r}{2\rho_0\omega}\right) = \frac{r}{2\rho_0 c_0} + j\frac{\omega}{c_0}.$$ (3.26)

The attenuation coefficient α is the real part of the expression, and the attenuation ΔL in decibels per metre will be given by

$$\Delta L = 8.69\frac{r}{2\rho_0 c_0} \quad (\text{dB/m}).$$ (3.27)

Having a flow resistivity of 1000 Pa·s/m^2 will then give us an attenuation ≈ 10 dB/m. We may add that the normal quality of mineral wool used in buildings has a flow resistivity 10 times higher but we shall be reminded of the assumption introduced above.

(You may try the analogous calculation assuming a high flow resistivity and/or low frequencies. Hint: $\sqrt{1-j\cdot x} \approx (1-j)\sqrt{x/2}$ for $x \gg 1$.)

3.3 SOUND INTENSITY AND SOUND POWER

A sound wave involves transport of energy and the energy flow per unit time through a given surface is called *sound power*. If this surface encloses a given sound source completely we will determine the total power, which is a characteristic quantity of the source.

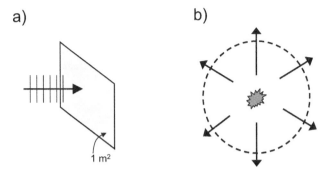

Figure 3.2 a) Intensity: power through a surface of area 1 m² normal to the direction of wave propagation. b) The intensity integrated over a closed surface gives the total emitted power of a source.

With the term *sound intensity* is meant the sound power transmitted through a surface area of 1 m² perpendicular to the direction of propagation (Figure 3.2). When using the terms sound intensity and sound power it is normally understood that these are time-averaged quantities. However, for completeness we shall also introduce the instantaneous quantities as well. The intensity is, analogous to the particle velocity, a vector quantity and is given by the product of the sound pressure at a point and the associated particle velocity, i.e. expressed as

$$I(t) = p(t) \cdot v(t) \quad (\text{watt/m}^2).$$ (3.28)

The time-averaged sound intensity is ideally defined by the expression

$$I_T = \lim_{T \Rightarrow \infty} \frac{1}{T} \cdot \int_0^T p(t) \cdot v(t) \, dt,$$ (3.29)

where T is the measuring time, which in practice must certainly be finite. The total sound power emitted from a given source is found by integrating the time-averaged intensity over a surface completely enclosing the source

$$W = \oint I_T \cdot n \, dS = \oint I_{Tn} \cdot dS \quad (\text{watt}),$$ (3.30)

where n denotes the unit vector normal to an element dS of the surface. The quantity I_{Tn} is then the component of the intensity in the normal direction, the normal time-averaged sound intensity often being abbreviated to *normal sound intensity*. It should be noted that this quantity is a signed one. In the same way as for the sound pressure level we define a *normal (time-averaged) sound intensity level* as

$$L_{ITn} = 10 \cdot \lg \left[\frac{|I_{Tn}|}{I_0} \right] \quad (\text{dB}).$$ (3.31)

In this equation $|I_{Tn}|$ is the absolute value of the normal sound intensity and I_0 is the reference value for the intensity, equal to 10^{-12} watt/m^2. When the normal intensity in a measurement situation is negative, the level is expressed as $(-)$ *XX* dB. In the same way we define the *sound power level*

$$L_W = 10 \cdot \lg \left[\frac{W}{W_0} \right] \text{ (dB)}, \qquad (3.32)$$

where the reference value W_0 is 10^{-12} watt. In a plane wave field we may use the simple relationship between sound pressure and particle velocity to write

$$W = \oint \frac{\tilde{p}^2}{\rho_0 c_0} \cdot dS, \qquad (3.33)$$

where the wavy symbol above *p* indicates an RMS-value. The expression is also valid for an ideal spherical wave field and is the basic equation used in several ISO standards for determination of sound power in a *free field*. The term signifies a sound field without reflections, e.g. in anechoic rooms, in ducts having an anechoic termination etc. (References on the subject are given at the end of the chapter.)

Using intensity, one is not dependent on having a free field to determine the sound power. The common procedure is first to define a closed measuring surface around the source. One then divides this surface into smaller subareas S_i, thereby measuring the normal sound intensity by placing an intensity probe normal to each of these smaller surfaces. In this way one determines an average intensity value, both in space and time, for each surface and finally sums up the result using the expression

$$W = \sum_i^N \langle I_{Tni} \rangle \cdot S_i. \qquad (3.34)$$

N is the number of subareas used and it should again be noted that the space average $\langle I_{Tni} \rangle$ is a signed scalar value.

3.4 THE GENERATION OF SOUND AND SOURCES OF SOUND

Up to now we have described acoustic waves and wave motion without touching on how waves are being generated. In one way or another we shall have to feed energy into the system to start a wave motion, mathematically expressed; the right hand side of the wave Equation (3.5) cannot be equal to zero everywhere.

Sound generation is normally linked to processes involving mechanical energy, thereby resulting in a transformation of a part of this energy into acoustical energy. Concerning building acoustics the most common processes occurring are: 1) Buildings elements or whole constructions are excited into vibrations due to impacts, by friction, by sound pressure etc. Due to the fact that they are in contact with the surrounding medium (air) they will transfer this motion to the medium, and a sound field is generated due to this volume displacement. 2) Liquid flow or gas flow results in pressure and velocity variations in the medium and/or one has turbulent flows interacting with solid surfaces.

These two main types of process are denoted mechanical and aerodynamic/ hydrodynamic generated sound, respectively. The latter type is normally connected with the building service equipment such as the air-conditioning system, pumps, compressors etc.

There are also other types of sound generating mechanism such as explosions, just to mention another common type, and there are instances where thermal energy may transform directly into acoustic energy. The so-called Rijke tube is an example of the latter form of sound generation. Conversely, acoustic energy may transform into other types of energy, e.g. by the phenomenon of *sonoluminence*. We shall, however, not delve further into this phenomenon but concentrate on the mechanisms coupled to sound transmission in buildings.

The most important aspects concerning sound transmission are the mechanisms by which building elements as beams, plates and shells generate sound when set in motion. Questions to be asked could be: Why does a thin panel radiate less sound than a thicker one even when the vibration amplitude is the same? Why does an additional thin panel mounted on to a thicker wall being called an *acoustical lining*? Why is the amount of sound energy produced by a building element dependent not only on dimensions and material parameters but also on the way it is excited, by point forces, moments or sound pressure?

Before moving on to the themes concerning the dynamics of buildings elements, i.e. excitation, response and sound radiation, we shall use some elementary or idealized types of sound sources to illustrate the basic properties of sources in general. One will find the terms monopole, dipole, quadrupole, octopole and so on. Normally, however, one uses the term multipole when the number of elementary sources exceeds four. Multipoles of different order are useful in modelling sound radiation from plate-like structures. However, through the so-called Rayleigh integral we have another efficient basis for calculating the sound radiation from plane surfaces. A classic illustration of the use of this integral is in calculating the radiated sound from a plane surface of circular shape set into an infinite large wall. This type of source is called a baffled piston, which is used as a first approximation for a loudspeaker in a closed box. Actually, this type of source has much wider application.

3.4.1 Elementary sound sources

Vibrating surfaces in contact with the surrounding medium, which is air for most practical applications in building acoustics, give a volume displacement and thereby a wave motion is generated. A sensible way of calculating the sound radiated from a large vibrating plate or panel could then be to divide it into small elements, calculate the sound field from each element and thereafter sum these contributions, i.e. we make use of the principle of superposition. Normally, this process is not so simple due to the fact that the sound field from each element not only depends on the geometry of the surface, of which the element is a part, but also on other neighbouring surfaces. This does not imply, however, that combinations of elementary sources are not useful for modelling. In addition, some simple expressions arrived at in this way are useful in practice.

3.4.1.1 Simple volume source. Monopole source

The simplest type of source may be envisaged as a pulsating sphere, an elastic ball having radius a and where the volume fluctuates harmonically with a given angular frequency ω. The radial surface velocity of the sphere may then be written as $u = \hat{u} \cdot \exp(\mathrm{j}\omega t)$, which is also indicated in Figure 3.3.

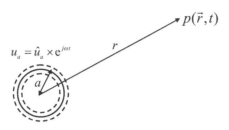

Figure 3.3 Sound radiation from a monopole source.

An outgoing wave is generated and the sound field must, due to symmetry, be equal in all directions. Outside the sphere the sound pressure must satisfy the wave equation using spherical coordinates. The solution must be of the same type as in Equation (3.20) so we may write

$$p(r,t) = \frac{A}{r} \cdot \mathrm{e}^{\mathrm{j}(\omega t - kr)}, \tag{3.35}$$

where r is the distance to the centre of the sphere and A is unknown for the time being. To determine the latter we shall again make use of the Euler equation (3.3), which connects the gradient of the pressure to the particle velocity $v(r,t)$,

$$\frac{\partial p(r,t)}{\partial r} = -\mathrm{j}\rho_0 \omega \cdot v(r,t) = -\mathrm{j}\rho_0 c_0 k \cdot v(r,t). \tag{3.36}$$

We may then calculate the particle velocity, which at the surface of the sphere, i.e. when the distance r is equal to the radius a, must be equal to the velocity u_a of the sphere. The unknown quantity A is thereby determined, giving the pressure

$$p(r,t) = \frac{\mathrm{j}\rho_0 c_0 k a^2 \hat{u}_a}{(1 + \mathrm{j}ka)r} \mathrm{e}^{\mathrm{j}(\omega t - k(r-a))}. \tag{3.37}$$

The question is now how large the radiated sound power will be and, furthermore, what are the controlling parameters? As we now have expressions both for the pressure and the particle velocity we may calculate the intensity and by integrating the intensity over a closed surface around the source we arrive at the total sound power. We shall perform this exercise at the surface of the sphere where the pressure after some algebra may be written as:

$$p(a,t) = \rho_0 c_0 \hat{u}_a \left(\frac{k^2 a^2}{1 + k^2 a^2} + \mathrm{j}\frac{ka}{1 + k^2 a^2} \right) \cdot \mathrm{e}^{\mathrm{j}\omega t}. \tag{3.38}$$

The sound pressure is then represented by two terms, the first term being in phase with the velocity of the sphere and the other 90° out of phase. The latter term will be dominant if the wavelength is large in relation to the dimensions of the source ($ka \ll 1$) but for large values of ka, it will go to zero. The first term will, when multiplied by the velocity of the sphere, give us the "active" intensity, as opposed to second term which only gives a "reactive" intensity resulting in an exponentially decreasing *near field*.

This situation is not unique for this idealized type of source but generally applies to all acoustic sources. This implies that for broadband vibrating sources, at distances from the source less than 1–2 wavelengths, one will experience variations in the spectral content of the sound pressure. In practical measurements, using sound pressure measurements to determine the sound power of sources, one is therefore advised to perform the measurements at distances from the source greater than its largest dimension, at the same keeping the distance from the surface larger than 1–2 wavelengths. These specific requirements do not apply when it comes to direct measurements of intensity but certain recommendations, as to the measurement distances, do apply in this case.

The radiated real power from the monopole source will then be given by

$$W = \frac{1}{2}\rho_0 c_0 \hat{u}_a^2 \frac{k^2 a^2}{1 + k^2 a^2} 4\pi a^2 = \rho_0 c_0 \tilde{u}_a^2 \frac{k^2 a^2}{1 + k^2 a^2} S, \qquad (3.39)$$

where S is the area of the spherical surface and the symbol $\sim$ indicate RMS-value. Later we shall show that the frequency-dependent factor $k^2 a^2 / (1 + k^2 a^2)$ represents the *radiation factor* of the monopole. This quantity is very important in building acoustics and we shall later give a general definition. In the literature one also comes across the notion of *source strength*, which is the effective volume velocity $\tilde{Q} = S \cdot \langle \tilde{u} \rangle$ of the source. The parentheses indicate, as in Equation (3.34), a space averaged value. For the monopole we then get

$$W = \rho_0 c_0 \frac{k^2 \tilde{Q}^2}{4\pi(1 + k^2 a^2)} \xrightarrow{\quad ka \ll 1 \quad} \rho_0 c_0 \frac{k^2 \tilde{Q}^2}{4\pi} . \qquad (3.40)$$

From these expressions we can see that the monopole is not an efficient radiator at low frequencies. Maintaining the sound power when lowering the frequency implies that the surface velocity must be increased in inverse ratio to the frequency. This again means that the displacement amplitude must increase in inverse ratio to the frequency squared. It goes without saying that this will, in the end, be impossible. Sound sources radiating bass sounds efficiently will therefore never be of small dimensions.

3.4.1.2 Multipole sources

When combining several simple monopole sources, assuming that the surface velocity is fixed and equal on all of them, we may show that the combination may radiate more or less power than each of them alone. The simplest case will be to combine two monopoles, vibrating either in phase or in anti-phase. The sound pressure on the surface of each monopole will be equal to the pressure produced by it plus the pressure caused by the vibration of the other. If the distance between them is small (compared to the wavelength) and they are working in phase the pressure may be nearly doubled and the sound power radiated will correspondingly be increased. However, when working in anti-phase the pressure may be small and the sound power may be drastically reduced. This is easily demonstrated by putting two loudspeakers in a stereo system close together and listening to the amount of bass being produced when playing music, coupling the loudspeakers either in phase or anti-phase.

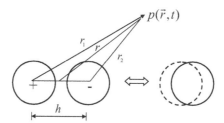

Figure 3.4 Dipole, two monopoles in opposite phase. An oscillating sphere is an equivalent source.

Two simple monopole sources, coupled in anti-phase, are called a *dipole* and are illustrated in Figure 3.4. Assuming that the distance r to the observation point is large compared with the source dimensions we may write the pressure as

$$p(r,t) = \frac{\rho_0 c_0 k^2 \tilde{D}}{4\pi r}\left(1 + \frac{1}{jkr}\right) \cdot \cos\varphi \cdot e^{j(\omega t - kr)}, \quad \text{where} \quad \tilde{D} = \tilde{Q} \cdot h, \qquad (3.41)$$

where h is the centre distance between the monopoles. $\tilde{D}$ is the so-called *dipole moment*. As seen from the equation the dipole will not radiate evenly in all directions; we get a *directivity factor* expressed by the cosine of the angle φ to the point of observation and thereby a *directivity pattern* shaped as a figure-of-eight. When the wavelength is large compared with the source dimensions the sound power will be given by

$$W = \frac{\rho_0 c_0 k^4 \tilde{D}^2}{12\pi}, \qquad (3.42)$$

and here we may see the dramatic reduction in sound power at low frequencies as compared with the monopole source. Comparing with Equation (3.40) we get

$$\frac{W_{\text{dipole}}}{W_{\text{monopole}}} = \frac{k^2 \cdot h^2}{3} \propto f^2 \cdot h^2. \qquad (3.43)$$

An oscillating sphere or ball, i.e. vibrating back and forth, will also act as a dipole and is equivalent to our two monopoles vibrating in anti-phase. This is in fact a very useful example to use later when we will treat the concept of radiation factor.

This kind of dipole action does not apply only to bodies of spherical shape; a vibrating string, a vibrating pipe or beam will also act like a dipole when it comes to sound radiation. Another example is a loudspeaker with an open back. At low frequencies both radiating surfaces will act like a monopole and these will be 180° out of phase with each other. There will be no efficient sound radiation where one does not mount the loudspeaker in a closed box, alternatively provide for a distance between the front and back large in comparison with the wavelength. The latter means that in practice one mounts the loudspeaker in a large baffle. We have tried to illustrate these effects in Figure 3.5.

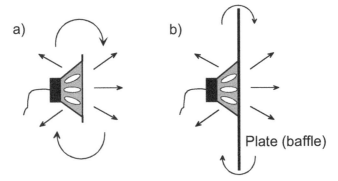

Figure 3.5 A loudspeaker open at the back is a dipole source. a) No efficient radiation due to short distance between the front and back. b) A baffle increases the distance front–back.

Evidently, these results may be extended to poles of higher orders. A quadrupole is made up of four monopoles alternating in anti-phase or it could be two oscillating spheres in anti-phase. This type of sound source will be even less effective than a dipole at low frequencies as the radiated power will be proportional to the wave number with an exponent of six.

A group of sound sources characterised as multipole are vibrating surfaces such as plates, shells etc. that will be treated in detail later. Another important group is connected to fluid flow. A turbulent jet flow is a typical quadrupole source. Turbulent flow interacting with solid surfaces however constitutes sources of dipole character and as such will be found for example in air-conditioning terminal units as grilles and diffusers.

3.4.2 Rayleigh integral formulation

The idealized models treated above are, however, only useful in a qualitative way when it comes to calculating radiation from solid bodies such as plates or shells vibrating in complex patterns, in particular when the wavelength (in air) becomes comparable or less than the dimensions of the source. However, there are tools available to calculate the radiated power in the case when the surface velocity is known, either known in detail or as a space averaged value.

The German physicist von Helmholtz showed well over 100 years ago that the sound pressure outside a vibrating surface (see Figure 3.6) could be expressed as the sum of two integrals:

$$p(R,t) = \oint_S \frac{e^{j(\omega t - kr)}}{4\pi r}\left(\frac{\partial p}{\partial n}\right)_S \mathrm{d}S - \oint_S p(S)\frac{\partial}{\partial n}\left(\frac{e^{j(\omega t - kr)}}{4\pi r}\right)\mathrm{d}S, \qquad (3.44)$$

where n indicate the normal to the surface. The term $\partial p / \partial n$ will then, as in Equation (3.36), be proportional to the normal surface velocity u_n on the vibrating surface S, and can be written

$$\left(\frac{\partial p}{\partial n}\right)_S = j\rho_0 c_0 k u_n(S). \qquad (3.45)$$

When this velocity is known, we may solve the first of these integrals. The second integral requires knowledge of the pressure $p(S)$ on the surface. This pressure distribution cannot be known in advance and involves a generally complicated calculation. Fortunately enough, when dealing with plane surfaces surrounded by another (infinitely) stiff surface, identified as an acoustic baffle, the second integral will be zero. We shall then be left with solving the integral

$$p(R,t) = \mathrm{j}\frac{\rho_0 c_0\, k}{2\pi} \oint_S \frac{u_\mathrm{n}(S)\cdot \mathrm{e}^{\mathrm{j}(\omega t - kr)}}{r}\, \mathrm{d}S. \tag{3.46}$$

It should be seen that we shall be concerned with the radiation to one side of the surface, which explains the number two instead of four in the denominator. In many cases we shall only be interested in the pressure far for the surface, i.e. when the distance r is much larger than the dimensions of the source. In these cases, we may substitute r by the distance R (see Figure 3.6) and, furthermore, place this distance outside the integral. We then get the famous and very useful Rayleigh integral

$$p(R,t) = \mathrm{j}\frac{\rho_0 c_0\, k}{2\pi R} \oint_S u_\mathrm{n}(S)\cdot \mathrm{e}^{\mathrm{j}(\omega t - kr)}\, \mathrm{d}S. \tag{3.47}$$

It should be noted that we still have to keep the distance r in the exponential function. (Why is this?). In Chapter 6, where we treat the problem of sound transmission through walls and floors, we shall use this integral to compute the sound radiation from panels and walls vibrating in certain patterns, in particular from plates vibrating in their natural modes. As an introduction to such application we shall calculate the radiation from a vibrating circular disk or piston set in an infinitely large baffle.

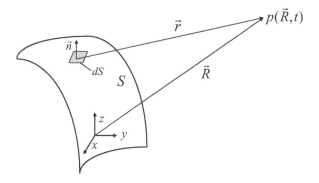

Figure 3.6 Calculating sound radiation from a vibrating surface of area S.

3.4.3 Radiation from a piston having a circular cross section

The circular surface depicted in Figure 3.7 is assumed to be a flat disk set into an infinitely large baffle. The disk, normally denoted a piston source, has a velocity of vibration $u = \hat{u}\cdot\exp(\mathrm{j}\omega t)$, which directly may replace the velocity u_n in Equation (3.46) or (3.47) to find the sound pressure at a point with coordinates (R,φ). At points near to the surface, where Equation (3.46) has to be used, the distance r to surface element dS will be given by

$$r = \left(R^2 + q^2 - 2Rq \sin\varphi \cos\theta \right)^{\frac{1}{2}}, \tag{3.48}$$

where q is the distance between the surface element and the centre of the piston. The solution of the integral is not trivial and it must generally be solved numerically except for points on the axis of the piston. The sound pressure in this *near field* will also fluctuate in a complicated manner due to the changing phase differences of the contributions from the different parts of the surface. The main purpose here is, however, to show the behaviour of the pressure in the *far field* and through this give an example of a source exhibiting a directional pattern quite different from our simple poles, e.g. a dipole.

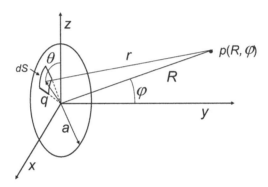

Figure 3.7 Coordinate system for calculation of sound pressure from a piston in a baffle.

At large distances from the surface, where the use of the Rayleigh integral is applicable, we will use the following approximation for r, setting

$$r \approx R - q\sin\varphi\cos\theta \qquad \text{and} \qquad dS = qdqd\theta. \tag{3.49}$$

Inserting these expressions into Equation (3.47) we get

$$p(R,\varphi,t) = j\frac{\rho_0 c_0}{2\pi R}\hat{u}\cdot e^{j(\omega t - kR)}\int_0^a q\,dq\int_0^{2\pi} e^{kq\sin\varphi\cos\theta}d\theta. \tag{3.50}$$

The solution may be expressed as

$$p(R,\varphi,t) = j\frac{\rho_0 c_0 ka^2}{2R}\hat{u}\cdot e^{j(\omega t - kR)}\left[\frac{2J_1(ka\sin\varphi)}{ka\sin\varphi}\right], \tag{3.51}$$

where J_1 is a Bessel function of the first order. It is the term enclosed in parenthesis that determines the directivity distribution of the sound pressure, an example shown in Figure

3.8. The diameter $2 \cdot a$ is chosen equal to 250 mm and the sound pressure level distribution is shown for three frequencies: 1000, 2000 and 5000 Hz. To facilitate the comparison the maximum sound pressure is arbitrarily set to 40 dB for all frequencies. At the higher frequencies, i.e. when the product ka becomes much larger than one, the directivity pattern will be very complicated (ka will approximately be equal to 11.5 at 5000 Hz), whereas the pattern at low frequencies will be little different from the ball-shaped pattern of a monopole.

Thus, assuming that the wavelength is much larger than the dimensions of the piston by setting $ka \ll 1$, we may show that $[2J_1(ka \sin\varphi)/ ka \sin\varphi)] \approx \frac{1}{2}$. Then we may write

$$p(R,t) = j\frac{\rho_0 c_0 ka^2}{4R}\hat{u} \cdot e^{j(\omega t - kR)}$$

(3.52)

or $\qquad p(R,t) = j\frac{\rho_0 c_0 k\tilde{Q}}{2\sqrt{2}\pi R} \cdot e^{j(\omega t - kR)}$.

The latter expression is as expected identical to the one giving the sound pressure from a monopole having source strength Q. (Show that Equation (3.37) gives the same expression when setting $ka \ll 1$ and $r \gg a$.)

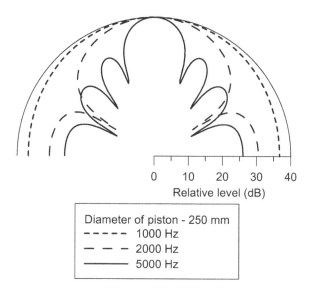

Diameter of piston - 250 mm
- - - - - 1000 Hz
— — — 2000 Hz
——— 5000 Hz

Figure 3.8 Directivity pattern of a piston in a baffle. The maximum sound pressure level is arbitrarily set to 40 dB for all frequencies.

3.4.4 Radiation impedance

In the previous chapter, the concepts of mechanical impedance and mobility were introduced to facilitate the calculation of the response of a mechanical system to a given force input. In connection with sound sources it will be useful to introduce yet another (mechanical) impedance concept, the *radiation impedance*. This represents the ratio of the reaction force of the fluid medium on the source, i.e. caused by the motion of the source, and the source velocity. Denoting this reaction force F_r and the source velocity u we write

$$Z_r = \frac{F_r}{u} = R_r + j \cdot X_r. \tag{3.53}$$

Using the piston source in the last section, assuming that it has mechanical impedance Z_m (in vacuum) and driven by a force F, the radiation impedance will be coupled in series with the mechanical impedance. The velocity of the piston will therefore be

$$u = \frac{F}{Z_m + Z_r}. \tag{3.54}$$

The real part R_r of the radiation impedance will give us the power radiated by the source so we may in general write

$$W_{radiated} = \frac{1}{2} \mathrm{Re}\{F_r \cdot u*\} = \frac{1}{2} \mathrm{Re}\{Z_r \cdot u \cdot u*\} = \tilde{u}^2 \cdot \mathrm{Re}\{Z_r\} = \tilde{u}^2 \cdot R_r. \tag{3.55}$$

Using the monopole as an example we immediately get by using Equation (3.40) that

$$R_r = \rho_0 c_0 \frac{k^2 S^2}{4\pi(1+k^2 a^2)} \xrightarrow{ka \ll 1} \rho_0 c_0 \frac{k^2 S^2}{4\pi} = \frac{\rho_0 \omega^2 S^2}{4\pi c_0}. \tag{3.56}$$

The imaginary part of the radiation impedance will on the other hand represent a load on the source, which, in many cases, may act as a contribution to the mechanical mass of the source. The radiation impedance is therefore an important factor in a number of different cases, not only when considering vibration of solid surfaces, but also generally when a vibrating column of air brings about sound radiation. There is a diverse range of examples one may mention here, ranging from sound radiation from musical instruments to resonance sound absorbers; see Chapter 5 and further on to sound transmission through holes and slits in wall or floors, see Chapter 8.

Finally, we shall consider the radiation impedance of a piston source where we expect that the result for low frequencies will be of the same form as for a monopole. However, in this case the impedance will attain quite another complexity. When calculating Z_r we shall have to use the general Equation (3.46) but in this case we must calculate the pressure on the surface of the piston. Specifically, the pressure p on a surface element dS' is induced by the sum of the movements by all the other elements dS. To arrive at the total pressure on the piston we therefore have to perform yet another integration, namely over the elements dS'. We shall not present this derivation, which may be simplified by using the *principle of reciprocity* outlined in section 3.6, but the end results are important and shall be commented upon. The radiation impedance for a piston placed in a baffle may be written (for a derivation see e.g. Kinsler et al. (2000)).

$$\left(Z_{\mathrm{r}}\right)_{\mathrm{piston}} = \rho_0 c_0 S\left[R_1(2ka) + \mathrm{j} \cdot X_1(2ka)\right],$$

$$\text{where} \quad R_1(x) = 1 - \frac{2\,J_1(x)}{x} \qquad \text{and} \qquad X_1(x) = \frac{2\,H_1(x)}{x}. \tag{3.57}$$

As stated above J_1 is a Bessel function of order one, whereas H_1 is a Struve function of order one. Concerning the definition and properties of these functions we may refer to Abramowitz and Stegun (1970).

The functions R_1 and X_1 are shown in Figure 3.9 as a function of ka going from 0 to a value of 20. For the piston used as an example in Figure 3.8 this implies going up to a frequency of approximately 8700 Hz. As shown the function R_1 will approach the value of 1.0 at the higher frequencies, which means that the radiated power will be given by the expression

$$\underset{ka \gg 1}{W} \approx \rho_0 c_0 S \cdot \tilde{u}^2. \tag{3.58}$$

We shall later on use this expression as a reference when defining the so-called *radiation factor* (or *radiation efficiency*) applying it to all types of sound radiating surface. This will be treated in section 6.3.1.

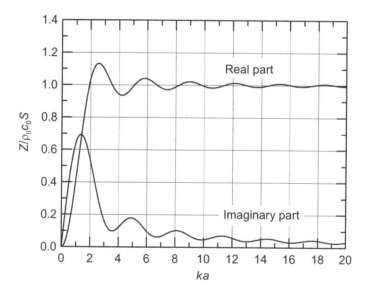

Figure 3.9 Relative radiation impedance of a piston in a baffle. Real part, R_1, and imaginary part, X_1, of the impedance function.

At the other extreme we get $R_1 \xrightarrow{\;x \ll 1\;} \dfrac{x^2}{8}$ and $X_1 \xrightarrow{\;x \ll 1\;} \dfrac{4x}{3\pi}$, which implies that

$$\left(Z_\mathrm{r}\right)_{\substack{\mathrm{piston}\\ka<<1}} \approx \rho_0 c_0 S\left[\frac{\left(2ka\right)^2}{8} + \mathrm{j}\cdot\frac{8ka}{3\pi}\right]. \qquad (3.59)$$

This expression may be written as

$$\left(Z_\mathrm{r}\right)_{\substack{\mathrm{piston}\\ka<<1}} \approx \frac{\rho_0\omega^2 S^2}{2\pi c_0} + \mathrm{j}\cdot\omega\left(\rho_0 S\frac{8a}{3\pi}\right). \qquad (3.60)$$

As expected the real part will, apart from a factor of two, be equal to the impedance of a monopole given in Equation (3.56). The imaginary part will represent the mass type impedance equal to the mass of air contained in a cylinder having the same area as the piston and a height h equal to $8a/3\pi$. This mass impedance will decrease with increasing frequency in accordance with the decrease of X_1 when increasing the frequency.

3.5 SOUND FIELDS AT BOUNDARY SURFACES

The assumption we have used up to now is that the wave propagation is taking place in an infinite space which is homogeneous and isotropic. When dealing with the acoustics inside buildings, however, we shall definitely be more concerned with what is happening at the boundaries between different media, e.g. such as the interface between air and a flexible surface of some kind. When waves are impinging at such a boundary it normally will be diffracted in some way, a part of the energy in the wave will go in another direction. The phenomenon is normally referred to as *reflection* when the boundary surface is much larger than the wavelength. In the opposite case, where the wavelength is much larger than the dimensions of the surface, we shall use the word *scattering*. We shall in this book mainly be concerned with boundary surfaces fulfilling the first condition but certainly when dealing with room acoustics scattering phenomena will be an important aspect.

When dealing with boundary surfaces we shall be interested in the reflected energy as well as the energy transmitted through the surface. The task of designing sound absorbers is to minimize the reflected energy, whereas designing for high sound insulation the aim is to reduce the transmitted energy.

Boundary surfaces will, irrespective of being fixed or set in motion by the sound waves, produce changes in the sound field, which means that some boundary conditions are given. We shall introduce the relevant boundary conditions when they are needed, an example is where the boundary surface is a solid, non-porous wall vibrating due to an outer sound field. The particle velocity of sound normal to the wall surface must then everywhere be equal to the velocity of the wall. If this were not the case, the local density of the fluid would become abnormally high or low, which is highly unlikely.

We shall introduce a complex *pressure reflection factor* R_p giving the ratio, both in amplitude and phase, between the sound pressure in the reflected and the incident wave. We shall write it as

$$R_p(\omega,\varphi) = \frac{\hat{p}_\mathrm{r}}{\hat{p}_\mathrm{i}} = \left|R_p\right|\cdot\mathrm{e}^{\mathrm{j}\delta}. \qquad (3.61)$$

As indicated, the reflection factor will in general be a function of the frequency and the angle of incidence φ of the wave. One will also find a reflection factor defined on the basis of the intensity in the two waves. As the intensity both in a plane and a spherical wave is proportional to the sound pressure squared this reflection factor will be equal to $|R_p|^2$. The part of the incident energy being lost in the reflection process, i.e. $1 - |R_p|^2$, is called the *absorption factor* having the symbol α

$$\alpha = 1 - |R_p|^2. \tag{3.62}$$

Another characteristic quantity to characterise a boundary surface is what we shall denote *surface impedance* Z_g defined as

$$Z_g = \left(\frac{\hat{p}}{\hat{v}_n} \right)_{\text{boundary}}. \tag{3.63}$$

The quantity v_n is the component of the particle velocity normal to the boundary surface. In the example mentioned above, in connection with boundary surfaces, the velocity v_n would be equal to the velocity of the boundary surface. The surface impedance may be considered as a variant of the general quantity specific impedance defined in Equation (3.18). A similar quantity is denoted *transmission impedance* or more commonly *wall impedance*, as one normally will use it as a characteristic for the wall surfaces in a room. However, in this case the quantity $\hat{p}$ is the pressure difference between the two sides, not only the total pressure on one side.

In the following sections, we shall derive expressions for the reflection and absorption factors assuming that the boundary surface is characterized by the surface impedance Z_g. We shall restrict our derivation to plane waves and, in the first place, assume that the wave is incident normally on the surface. At oblique incident there will be an important distinction whether the surface impedance will be a function of the angle of incidence or not. In the latter case, the surface is called *locally reacting*, which means that we need not consider in-plane wave propagation. This implies that the normal component of the particle velocity at a given point on the surface depends on the sound pressure at this point only. In other words, pressure on the surface at a certain point causes no movement elsewhere on the surface. In practice, this is a reasonable assumption for many types of porous absorber, at least in the lower frequency range but in general it may be difficult to decide whether an absorber may be treated as locally reacting or not. One may of course prevent sound propagation along the surface by subdividing the absorber using a lattice of some kind, e.g. a honeycomb core structure but such solutions may not be desirable due to other requirements.

3.5.1 Sound incidence normal to a boundary surface

We shall assume that a plane wave is incident normally on a boundary surface, which is coincident with the plane having the coordinate $x = 0$ (see Figure 3.10).

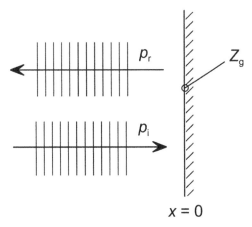

$$x = 0$$

Figure 3.10 Incident and reflected plane wave at a surface.

By using Equations (3.9) and (3.17) we may express the sound pressure and the particle velocity in the incident wave as

$$p_i(x,t) = \hat{p}_i \cdot e^{j(\omega t - kx)}$$

(3.64)

$$\text{and} \qquad v_i(x,t) = \frac{\hat{p}_i}{\rho_0 c_0} \cdot e^{j(\omega t - kx)},$$

where $\rho_0 c_0$ is the characteristic impedance of the medium. For the reflected wave we get

$$p_r(x,t) = \hat{p}_r \cdot e^{j(\omega t + kx)} = R_p \hat{p}_i \cdot e^{j(\omega t + kx)}$$

(3.65)

$$\text{and} \qquad v_r(x,t) = -R_p \frac{\hat{p}_i}{\rho_0 c_0} \cdot e^{j(\omega t + kx)} .$$

There will be a change in sign for the wave number due to the change of direction of the wave. At the same time the particle velocity is changing sign as the gradient of the pressure is changing sign along with the wave number. The total pressure at the boundary surface ($x = 0$) will be

$$p(0,t) = p_i(0,t) + p_r(0,t) = \hat{p}_i \left(1 + R_p \right) \cdot e^{j\omega t} \qquad (3.66)$$

and the particle velocity:

$$v(0,t) = v_i(0,t) + v_r(0,t) = \frac{p_i}{\rho_0 c_0} \left(1 - R_p \right) \cdot e^{j\omega t} . \qquad (3.67)$$

Inserting these expressions into Equation (3.63) we get

$$Z_g = \rho_0 c_0 \frac{1 + R_p}{1 - R_p} \tag{3.68}$$

and furthermore

$$R_p = \frac{\dfrac{Z_g}{\rho_0 c_0} - 1}{\dfrac{Z_g}{\rho_0 c_0} + 1} = \frac{Z_g - Z_0}{Z_g + Z_0} \quad \text{where} \quad Z_0 = \rho_0 c_0. \tag{3.69}$$

Inserting this expression into Equation (3.62) we arrive at the absorption factor for normal incidence expressed as

$$\alpha = \frac{4 \operatorname{Re}\left\{\dfrac{Z_g}{Z_0}\right\}}{\left|\dfrac{Z_g}{Z_0}\right|^2 + 2 \operatorname{Re}\left\{\dfrac{Z_g}{Z_0}\right\} + 1}. \tag{3.70}$$

For most simple illustration using these equations we may assume that the boundary "surface" is dividing two different gases. As a thought experiment we shall have an infinite long tube containing the gases, which are separated by a massless membrane. Letting the gases be air and helium, having at 20°C a characteristic impedance of 415 and 170 Pa·s/m, respectively, this "surface" will give a reflection factor $|R_p|$ equal to 0.42 using Equation (3.69) and an absorption factor α equal to 0.82 by using Equation (3.70).

Another example, which may be more interesting, is the boundary between air and water. Setting the density and sound speed for water as 1000 kg/m³ and 1500 m/s, respectively; we arrive at a characteristic impedance of $1.5 \cdot 10^6$ Pa·s/m. This implies that we obtain a pressure reflection factor approximately equal to 0.9995 and an absorption factor of about $1.1 \cdot 10^{-3}$. A water surface is therefore, practically speaking, a totally reflecting surface or a nearly "infinitely hard" surface.

Some special cases of the equations above may be listed:
- An "infinitely hard" surface, i.e. $Z_g \Rightarrow \infty$ gives $|R_p| = 1$, $\delta = 0$ and $\alpha = 0$.
- A "soft" surface, denoted a *pressure release surface*, i.e. $Z_g \Rightarrow 0$ gives $|R_p| = 1$, $\delta = \pi$ and $\alpha = 0$.
- A totally absorbing surface, i.e. $Z_g = Z_0$ gives $|R_p| = 0$, $\delta = 0$ and $\alpha = 1$.

As is apparent from Equation (3.70), the impedance of the boundary surface uniquely determines the absorption factor but the opposite is not true. Representing the absorption factor by parametric curves in a Cartesian coordinate system, using the real part of the impedance as abscissa and the imaginary part as ordinate, it is relatively easy to show that the curves are circles. The circles have their centres in $(x_0, 0)$ and the their radii will be $(x_0^2 - 1)^{1/2}$, where $x_0 = (2/\alpha) - 1$. This is shown in Figure 3.11 having an elliptical form due to a difference in scale on the two axes. Finally, Figure 3.12 maybe illustrate in a better way how the impedance components should be adjusted to achieve a high absorption factor.

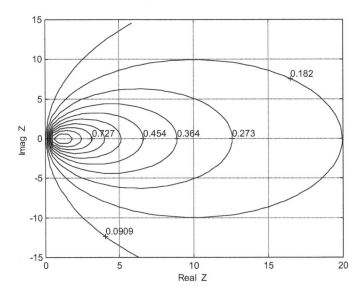

Figure 3.11 Absorption factor as a function of the normalised impedance components ($Z = Z_g/Z_0$) of the boundary surface. Normal incidence.

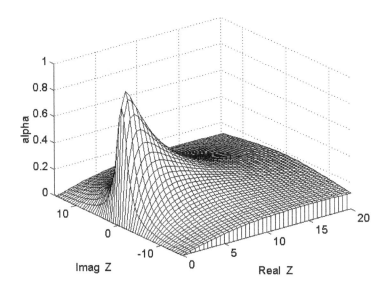

Figure 3.12 Three-dimensional plot corresponding to Figure 3.11.

3.5.1.1 Sound pressure in front of a boundary surface

Determination of the sound absorption factor of small specimens and for normal incidence is performed in a so-called *standing wave tube*. This will be treated in detail in section 5.3. In the "classical" method of performing such measurements one needs an expression for the total pressure in front of the specimen surface. At an arbitrary distance x in front of the surface we get

$$p(x,t) = p_i(x,t) + p_r(x,t), \text{ i.e.}$$
$$p(x,t) = \hat{p}_i \left[e^{j(\omega t - kx)} + |R_p| \cdot e^{j(\omega t + kx + \delta)} \right]. \tag{3.71}$$

When carrying out measurements one determines the RMS-value of the sound pressure, by definition given by

$$\tilde{p}(x) = \sqrt{\frac{1}{T} \int_0^T \left[\text{Re}\{p(x,t)\} \right]^2 dt}. \tag{3.72}$$

As before, we shall indicate the RMS-value by using a curly over-bar to distinguish it from the amplitude value, the latter is indicated by a "hat" on top of the symbol. Inserting Equation (3.71) into (3.72) we get

$$\tilde{p}(x) = \frac{\hat{p}_i}{\sqrt{2}} \left[1 + |R_p|^2 + 2|R_p| \cos(2kx + \delta) \right]^{\frac{1}{2}}. \tag{3.73}$$

The pressure will therefore exhibit maximum and minimum values given by the equations

$$(\tilde{p})_{max} = \frac{\hat{p}_i}{\sqrt{2}} \left[1 + |R_p| \right]$$
$$\text{and} \quad (\tilde{p})_{min} = \frac{\hat{p}_i}{\sqrt{2}} \left[1 - |R_p| \right]. \tag{3.74}$$

From the measurements of these pressure values one determines the absolute value of the reflection factor and thereby the absorption factor. (How do we determine the impedance Z_g?)

3.5.2 Oblique sound incidence

We shall extend the above calculations by giving the incident wave (see Figure 3.13) an angle φ with the normal to the surface. We may then rotate the coordinate system and obtain a new x coordinate given as $x' = x \cdot \cos\varphi + y \cdot \sin\varphi$. The sound pressure and the normal component of the particle velocity may then be expressed as

$$p_i(x, y, \varphi) = \hat{p}_i \cdot e^{-jk(x\cos\varphi + y\sin\varphi)} \quad \text{and}$$

$$v_{i,x}(x, y, \varphi) = \frac{\hat{p}_i}{\rho_0 c_0} \cos\varphi \cdot e^{-jk(x\cos\varphi + y\sin\varphi)}. \tag{3.75}$$

As seen, we tacitly infer the time dependence $e^{j\omega t}$. In a similar manner we get for the reflected wave

$$p_r(x, y, \varphi) = \hat{p}_r \cdot e^{jk(x\cos\varphi - y\sin\varphi)} \quad \text{and}$$

$$v_{r,x}(x, y, \varphi) = -\frac{R_p \hat{p}_i}{\rho_0 c_0} \cos\varphi \cdot e^{jk(x\cos\varphi - y\sin\varphi)}. \tag{3.76}$$

In analogy with the use of the Equations (3.66) to (3.69) we now get

$$R_p = \frac{Z_g \cos\varphi - Z_0}{Z_g \cos\varphi + Z_0}. \tag{3.77}$$

Equation (3.73), giving the total sound pressure in front of the surface, will be modified to read

$$\tilde{p}(x, y) = \frac{\hat{p}_i}{\sqrt{2}} \left[1 + |R_p|^2 + 2|R_p| \cos(2kx \cdot \cos\varphi + \delta) \right]^{\frac{1}{2}}. \tag{3.78}$$

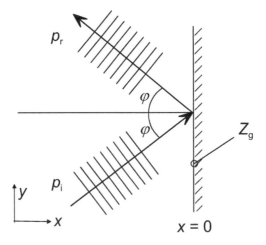

Figure 3.13 Sound incidence at an angle φ. Locally reacting boundary of impedance Z_g.

According to our assumption on local surface reaction, which implies that the impedance Z_g is independent of the angle φ, we may then calculate the *statistical absorption factor* α_{stat}. This is an average value for α over all angles of incidence using the expression

$$\alpha_{\text{stat}} = 2 \int_{0}^{\pi/2} \alpha(\varphi) \sin\varphi \cos\varphi \; \mathrm{d}\varphi = 2 \int_{0}^{\pi/2} \left[1 - \left|R_p\right|^2\right] \sin\varphi \cos\varphi \, \mathrm{d}\varphi. \qquad (3.79)$$

Inserting for R_p according to Equation (3.77) we get

$$\alpha_{\text{stat}} = 8 \cdot \frac{z'}{|z|^2}\left[1 - \frac{z'}{|z|^2} \cdot \ln\left(1 + 2z' + |z|^2\right) + \frac{1}{z''} \cdot \frac{\left(z'\right)^2 - \left(z''\right)^2}{|z|^2} \cdot \mathrm{Arctg}\frac{z''}{1+z'}\right]. \qquad (3.80)$$

The symbol z is the surface impedance normalised by the characteristic impedance Z_0 of the medium, i.e.

$$z = z' + \mathrm{j} \cdot z'' = \mathrm{Re}\left\{\frac{Z_g}{Z_0}\right\} + \mathrm{j} \cdot \mathrm{Im}\left\{\frac{Z_g}{Z_0}\right\}. \qquad (3.81)$$

Figure 3.14 shows the average value α_{stat} as a function of the normalized surface impedance. A comparison with Figure 3.11 generally shows that the statistical absorption coefficient is higher than the normal incidence factor, but also that the absolute maximum is slightly lower; $(\alpha_{\text{stat}})_{\text{max}} \approx 0.95$ at $z' \approx 1.6$.

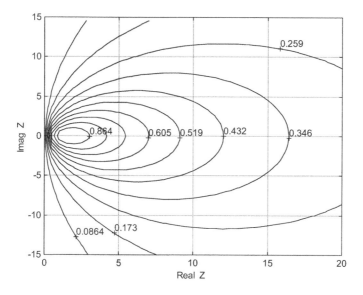

Figure 3.14 Statistical absorption factor as a function of the normalized impedance components, $(Z = Z_g/Z_0)$.

3.5.3 Oblique sound incidence. Boundary between two media

A general treatment of the case of plane wave's incident on a locally reacting surface was given in the previous section. Implicitly, this means that we presuppose the impedance Z_g

being independent of the angle of incidence. We also pointed out that this assumption is reasonably correct for many porous materials. However, one will often encounter cases where one cannot use this assumption but we shall postpone the treatment of calculation models for such cases until later.

Simple models for a homogeneous and isotropic porous material consider it to behave as a fluid; we shall use the term *equivalent fluid*. Such a fluid may be characterized by its propagation coefficient Γ (or by the complex wave number $k' = -$ j·Γ) and its complex characteristic impedance Z_c. Alternative descriptions use the *bulk modulus* K and the *equivalent* (or *effective*) *density* ρ. The relations between these quantities are given by the following expressions

$$Z_c = \sqrt{K\rho} \qquad \text{and} \qquad \Gamma = \text{j} \cdot \omega \sqrt{\frac{\rho}{K}} \qquad \text{with} \qquad K = \frac{-p}{\dfrac{\Delta V}{V}}. \qquad (3.82)$$

The last equation defines the bulk modulus; the ratio between the pressure and the relative change in volume.

As an introduction to these simple models, which we shall treat in more detail in Chapter 5, we shall assume we have an infinitely thick wall of a porous material with a given characteristic impedance $Z_2 = \rho_2 c_2$ and a wave number k_2. The medium of the incident wave is characterized using an index 1 as shown in Figure 3.15. We shall, as before, calculate the reflection coefficient and further examine the conditions necessary for the porous material to behave as locally reacting. The sound pressure for the three partial waves is given by

$$p_{\text{i}} = \hat{p}_{\text{i}} \cdot \text{e}^{-\text{j}(k_1 x \cos\varphi + k_1 y \sin\varphi)},$$
$$p_{\text{r}} = \hat{p}_{\text{r}} \cdot \text{e}^{\text{j}(k_1 x \cos\varphi - k_1 y \sin\varphi)}, \qquad (3.83)$$
$$p_{\text{t}} = \hat{p}_{\text{t}} \cdot \text{e}^{-\text{j}(k_2 x \cos\psi + k_2 y \sin\psi)}.$$

We have as before omitted the time dependence $\text{e}^{\text{j}\omega t}$. Furthermore, the reflection is specular; a condition that immediately will follow from the boundary conditions without being shown in detail here.

Another important law will, however, follow from these equations. The pressure must be equal on both sides of the boundary, i.e. for $x = 0$ we get $p_{\text{i}} + p_{\text{r}} = p_{\text{t}}$. Applying this to Equations (3.83) we obtain *Snell's law*:

$$k_1 \sin\varphi = k_2 \sin\psi, \qquad (3.84)$$

which, by using the sound speeds, may be written

$$\frac{\sin\varphi}{c_1} = \frac{\sin\psi}{c_2} \qquad \text{or} \qquad \cos\psi = \sqrt{1 - \left(\frac{c_2}{c_1}\right)^2 \cdot \sin^2\varphi}. \qquad (3.85)$$

As the two media are in contact with each other, this implies that the normal component of the particle velocity on both sides will be the same. This will give another boundary condition stating that

$$v_i \cos\varphi + v_r \cos\varphi = v_t \cos\psi \qquad \text{or}$$

$$\frac{p_i}{Z_1}\cos\varphi - \frac{p_r}{Z_1}\cos\varphi = \frac{p_t}{Z_2}\cos\psi = \frac{p_i + p_r}{Z_2}\cos\psi. \qquad (3.86)$$

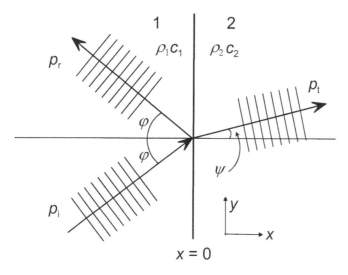

Figure 3.15 Incident wave on a boundary surface between two media having different characteristic impedance. Specular reflection and transmission into medium two.

In this case, the pressure reflection factor will be

$$R_p = \frac{Z_2 \cos\varphi - Z_1 \cos\psi}{Z_2 \cos\varphi + Z_1 \cos\psi}. \qquad (3.87)$$

Comparing with the former Equation (3.77) we find that the surface impedance Z_g now is replaced by $Z_2/\cos\psi$. The requirement that the boundary surface is locally reacting, i.e. $\psi \approx 0$, presupposes that the sound speed in medium two must be much lower than in medium one, which follows from Equation (3.85). Alternatively, we may envisage that the energy losses in medium two are very large; the attenuation of sound waves along the boundary surface is so large that there is, practically speaking, a local reaction only.

3.6 STANDING WAVES. RESONANCE

In section 3.5.1.1, we calculated the total pressure in front of a surface as being the sum of the pressures in the incident plane wave and the reflected (plane) wave, respectively. Letting the reflected wave now hit a second surface, a standing wave may appear between these two surfaces. This may be realized in practice by using a tube with stiff walls and closed at both ends by some kind of lid. The cross sectional dimensions must be much smaller than the wavelength but we shall not put any restrictions on the length L of the tube. For simplicity, we shall assume that the lids closing off the tube are totally reflecting. The surface impedance for these lids is thereby infinite and the particle velocity is zero. Our task is to find an expression for the pressure p in the tube.

As a starting point we assume that there is no sound source inside the tube, which is wholly analogous to the case of free vibration in a mechanical system treated in Chapter 2. Now we have to solve the homogeneous wave equation, the Helmholtz equation (3.7), which in one-dimensional form is given by

$$\frac{d^2 p}{dx^2} + k^2 p = 0. \tag{3.88}$$

The general solution of this equation is

$$p(x) = A\sin(kx) + B\cos(kx), \tag{3.89}$$

where A and B are constants to be determined by the boundary conditions. Having assumed that the particle velocity is zero at each end of the tube, this implies that the gradient dp/dx of the pressure is zero at both tube ends. Setting the coordinate x equal to zero at one end and equal to L at the other end, these conditions imply that the constant A must be equal to zero and that the wave number k may only attain a given set of values. The only possible solutions, the *eigenfunctions* will be

$$p_n(x) = B\cos(k_n x), \tag{3.90}$$

where the wave numbers k_n and the associated natural frequencies are given by

$$k_n = \frac{n \cdot \pi}{L} \quad \text{and} \quad f_n = \frac{n \cdot c_0}{2L}. \tag{3.91}$$

The lowest natural frequency of a tube having a length of one metre is then $f_1 \approx 170$ Hz (20°C). (What is the lowest natural frequency of a similar tube having an open end as an organ pipe? To calculate this one may for simplicity assume that the sound pressure in the open end is equal to zero.)

How does one calculate the sound pressure caused by a given sound source placed at certain point inside the tube? We have to solve the wave equation (3.88) but now having a source term on the right hand side. Assuming a monopole source inside the tube we shall first have to modify Equation (3.2), which is the equation of conservation of mass in the system. We shall now write

$$\frac{\partial \rho}{\partial t} = -\rho_0 \nabla \cdot \boldsymbol{v} + q(\boldsymbol{r}_0, t), \tag{3.92}$$

where q represents the source, a mass flux in an area having coordinates $\boldsymbol{r}_0$. This implies that the sought after right-hand source term of the wave equation will be a time derivative of q. We shall not go into details on how to obtain a solution of the wave equation in this case. Suffice to say that one expresses both the pressure and the source strength by sums of the eigenfunctions (3.90) and then adjusting the coefficients in these sums. Assuming that the source area is very small in comparison with the other dimensions, the pressure in a position x caused by a source placed in position x_0 may be written as

$$p(f, x, x_0) = C \sum_{n=1}^{\infty} \frac{\cos\left(\dfrac{2\pi f_n}{c_0} \cdot x\right) \cdot \cos\left(\dfrac{2\pi f_n}{c_0} \cdot x_0\right)}{f^2 - f_n^2}. \qquad (3.93)$$

The constant C will contain the strength of the source having the frequency f. There are several important comments to be made on this result. First, the pressure will be dominated by the term having a natural frequency nearest to the driving frequency but the response will contain contributions from many terms in the sum.

Second, as we have not introduced any form of energy losses in the tube, the pressure will go to infinity when the driving frequency coincides with any of the natural frequencies. To calculate on a more realistic situation we may formally add a loss term in the denominator. This is carried out by calculating on a situation as depicted in Figure 3.16, where a small loudspeaker is placed at a distance x_0 from the wall in one end and where the sound pressure is measured by a microphone placed at a distance x.

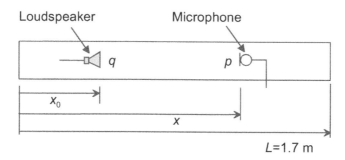

Figure 3.16 A hard-walled tube with a sound source and a receiver (microphone).

The results of the calculations are shown in Figure 3.17. Due to the 1.7 metre length of the tube the first natural frequency will be 100 Hz and the following ones multiples of this frequency. Using one of the source positions we get, as expected, resonances at these frequencies. For the second source position, in the centre of the tube, no resonances shows up at 100 Hz and 300 Hz. (Why is that?)

Last but not least a very important comment should be made on Equation (3.93) when it comes to the symmetry in the expression: the source and the receiver may change places without altering the results. This is an example of the aforementioned acoustical *reciprocity principle*, a principle that is quite general and in many instances very useful in practice. A generalization to three-dimensional sound fields is evident but the principle of exchanging source and receiver does also apply when the system contains structural components, albeit subject to some limitations. We shall return to this theme when treating the subject of sound transmission in Chapter 6 where we deal with the general subject of *vibroacoustic reciprocity* (see section 6.6.1).

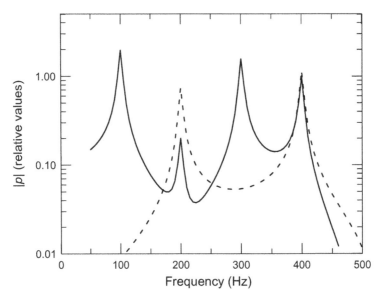

Figure 3.17 Relative sound pressure at a fixed position, $x = 1.2$ m, in the tube shown in Figure 3.16. Solid line – source at $x_0 = 0.5$ m. Dashed line – source at $x_0 = 0.85$ m (centre of tube).

3.7 WAVE TYPES IN SOLID MEDIA

As mentioned in the introduction to this chapter, acoustic waves in solid media are distinctly different from their counterpart in fluids due to shear stresses and shear deformations. This leads to the occurrence of several other wave types besides the compressional or longitudinal treated up to now. We shall mainly give an outline of these wave types but go into some detail on bending or flexural waves, the wave type of particular importance in sound transmission phenomena in buildings.

Two types of wave may exist at the same time in a medium of infinite extent; ideal longitudinal waves, as in fluids, and ideal transverse or shear waves. In the latter, the particle displacement will be normal to the direction of propagation; see below. From the basic equations of elasticity we may show that all wave motion in solids may be seen as a combination of these two "pure" waves but many of these combinations have specific names. It could be mentioned that in a semi-infinite medium surface waves (Rayleigh waves) may occur but these have little relevance in building acoustics. We shall be more interested in which combinations of the two basic waves may exist in structural elements such as beams and plates, an example being the aforementioned bending waves.

3.7.1 Longitudinal waves

Ideal or pure longitudinal waves may only exist in a medium of infinite extent. Practically speaking, this implies that the solid structure must be very large compared with the wavelength. When taking into account the actual dimensions of building elements and the relevant frequency range, displacements normal to the direction of wave propagation will occur, i.e. longitudinal stresses will produce lateral strains on the outer free surfaces. This is called the Poisson contraction phenomenon. The associated wave type is therefore called *quasi-longitudinal* and Figure 3.18 a) may serve as an illustration.

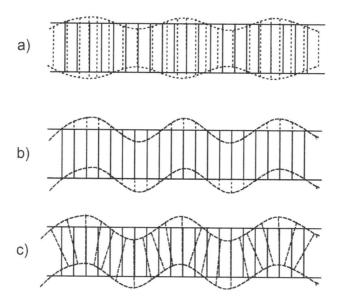

Figure 3.18 Wave types in solids. a) Quasi-longitudinal wave. b) Shear wave. c) Bending wave.

The solid lines in the figure represent elements of the structure at rest whereas the broken lines illustrate the deformations of these elements both in the direction of wave propagation and laterally. Using the particle velocity as the variable, we may show that for free waves (one-dimensional case) the following differential equation applies

$$E' \frac{\partial^2 v_x}{\partial x^2} - \rho \frac{\partial^2 v_x}{\partial t^2} = 0, \tag{3.94}$$

where v_x is the velocity in the x-direction (the direction of propagation) and ρ is the density of the medium. E' is a property of the material which depends on the actual lateral displacements. Taking a plate as an example, we get lateral displacements or contractions in one direction giving

$$E' = \frac{E}{1 - \upsilon^2}, \tag{3.95}$$

where E is the modulus of elasticity (Young's modulus) of the material and υ is the Poisson's ratio. The latter is defined as the ratio of the magnitudes of the lateral strain to the longitudinal strain. As seen from Table 3.1, this ratio varies between 0.2 and 0.35 for common building materials.

It should be noted that using the particle velocity as a variable, which we have done in Equation (3.94) and also further on, is just a choice. A corresponding equation for e.g. the displacement could be used as well. The phase speed of the longitudinal wave, according to the equations above, will be given by

$$c_L = \sqrt{\frac{E}{\rho\left(1-\upsilon^2\right)}}. \tag{3.96}$$

Examples of data for common materials are given in Table 3.1, which we may use to calculate the wave speed for longitudinal waves. It should be noted that wave speed normally found in tabled data applies to pure longitudinal waves, i.e. calculated from the formula $(E/\rho)^{1/2}$. The loss factor given in the table applies to the internal energy losses in the material.

Table 3.1 Examples of material properties.

Material	Density kg/m^3	E-modulus[1] 10^9 Pa	Poisson's ratio	Loss factor $\eta_{int}\cdot10^{-3}$
Steel	7700–7800	190–210	0.28–0.31	~ 0.1
Aluminium	2700	66–72	0.33–034	~ 0.1
Glass	2500	60	-	0.6–2.0
Concrete	2300	32–40	0.15–0.2	4–8
Concrete (lightweight aggregate)	400–600	1.0–2.5	~ 0.2	10–20
Concrete (autoclaved aerated)	1300	3.8 [2]	~ 0.2	10–20
Gypsum plate (plasterboard)	800–900	4.1	~ 0.3	10–15
Chipboard	650–800	3.8	~ 0.2	10–30
Fir, spruce	400–700	7–12	~ 0.4	8–10

[1] Dynamic E-modulus. [2] E-modulus for static pressure.

3.7.2 Shear waves

In a pure shear wave, also referred to as a transverse wave, we only get shear deformations and no change of volume (see b) in Figure 3.18). The particle movements are normal to the direction of wave propagation, and the wave equation for free wave motion will be analogous to Equation (3.94), i.e.

$$G\frac{\partial^2 v_y}{\partial x^2} - \rho\frac{\partial^2 v_y}{\partial t^2} = 0, \tag{3.97}$$

where v_y represents the particle velocity normal to the direction of propagation. The shear modulus is given by

$$G = \frac{E}{2(1+\upsilon)}, \tag{3.98}$$

and for the phase speed c_S we get

$$c_S = \sqrt{\frac{G}{\rho}}. \tag{3.99}$$

Using Equation (3.96), (3.98) and (3.99) we get for a plate

$$\frac{c_S}{c_L} = \sqrt{\frac{1-\upsilon}{2}}. \tag{3.100}$$

Inserting a Poisson's ratio of 0.3 as an example will give a ratio of these wave speeds of approximately 0.59. The wave speed of a shear wave will then always be less than the speed of a longitudinal wave. Lastly, we point to the fact that pure shear waves will, like pure longitudinal waves, generally occur only in bodies where the dimensions are very large compared with the wavelength.

3.7.3 Bending waves (flexural waves)

Bending waves are likely to be excited in bodies or structures where one or two dimensions are becoming small compared to the wavelength at an actual frequency. This implies that this wave type will be dominant in common construction elements, for example beams and plates. This again means that it takes on a central position in building acoustics, also due to these waves being easy to excite. Furthermore, the particle velocity will be normal to the direction of propagation, which also means that it is normal to the surface of a beam or plate (see c) in Figure 3.18). This again implies that there will be an efficient coupling to the surrounding medium (air), which means that the plate or beam potentially could be an efficient sound source. We may easily be aware of this fact by knocking on a thin metal plate.

Our treatment of bending waves will mainly be concerned with simple *thin plate models*, also called *Bernoulli–Euler models*. In these models, one presuppose that the deformation of an element due to bending is much larger than the one caused by shear and, furthermore, the rotation of the element is neglected. A limit for using thin plate models, often referred to in the literature, is that the wavelength of the bending wave must be larger than six times the thickness of the beam or plate. For quite common thicknesses of concrete this may be a limitation and one should then apply *thick plate models* (*Reissner–Mindlin*). We shall limit ourselves to giving some examples of the differences one may encounter by using these models.

The treatment will also, if not pointed out otherwise, be limited to plates of *isotropic* materials, which means that the material properties are independent of direction. One then needs only two quantities, the modulus of elasticity and Poisson's ratio, to describe the linear relationship between forces and displacements. Unfortunately, a large group of building materials exhibit *anisotropy*, the material properties depend on direction. Wooden materials are typical examples where the properties depend on the direction of the fibres. Other examples are composite materials reinforced by fibres. A special type of *anisotropy* is denoted *orthotropic*. An orthotropic plate is a plate where the material properties are symmetric about three mutually perpendicular axes. Well-known examples are corrugated panels often used in industrial buildings; having a waveform or a more sophisticated trapezoidal cross section, the latter normally called cladding. It should be noted, however, that to apply the general theory of orthotropic plates to corrugated panels one has to find the equivalent orthotropic constants for these panels. We shall give an example below (see section 3.7.3.3) and the reason is that further on we shall calculate the sound transmission through such panels.

3.7.3.1 Free vibration of plates. One-dimensional case

The differential equation describing the wave motion is substantially more complicated than for the ones treated above. The reason is that each element of the plate, as sketched in Figure 3.18 c), will be influenced by moments as well as shear forces. We shall not derive the equation, just state that the equation for the particle velocity normal to the plate surface may be written as

$$B \frac{\partial^4 v_y}{\partial x^4} + m \frac{\partial^2 v_y}{\partial t^2} = 0, \tag{3.101}$$

where B and m is the plate bending stiffness per unit length and the mass per unit area, respectively. The same differential equation applies to other quantities such as displacement, angular velocity, shear force and bending moment but we shall use the particle velocity as the characterizing quantity. Assuming a solution of the form

$$v_y = \hat{v}_y \cdot e^{j(\omega t - k_B x)},$$

we get the following expression for the wave number k_B by insertion into Equation (3.101):

$$k_B = \frac{\omega}{c_B} = \sqrt{\omega} \cdot \sqrt[4]{\frac{m}{B}}, \tag{3.102}$$

where the phase speed c_B is given by

$$c_B = \sqrt{\omega} \cdot \sqrt[4]{\frac{B}{m}}. \tag{3.103}$$

As seen from this equation, the medium will be *dispersive* for bending waves, which means that the phase speed will be frequency dependent. A broadband-pulsed signal will therefore change its shape during propagation; the high frequency wave components will outrun the components having a lower frequency. For a homogeneous plate having a thickness h we get from Equation (3.103):

$$c_B = \sqrt{\frac{\pi}{\sqrt{3}} c_L h f} \approx \sqrt{1.8 \cdot c_L h f}, \tag{3.104}$$

where f is the frequency in Hz and where the phase speed c_L for longitudinal waves in the medium is given by Equation (3.96). We arrive at this expression by substituting for the quantities m and B, respectively, using the following formulae

$$m = \rho \cdot h \quad \text{and} \quad B = \frac{E}{1 - \upsilon^2} \cdot I = \frac{E}{1 - \upsilon^2} \cdot \frac{h^3}{12}. \tag{3.105}$$

The quantity I is the cross sectional area moment of inertia of the plate per unit width. As mentioned above, the expressions given for the wave number and phase speed presupposes that the plate is thin, i.e. the wavelength should be larger than six times the plate thickness. Another way of expressing this is by demanding that c_B should be less than $0.3 \cdot c_L$. (How can you show this?) If this condition is fulfilled the error should be less than 10%.

We may if need be, by using results from Mindlin (1951), calculate a corrected phase speed c'_B if the condition above is not fulfilled. The corrected phase speed is given by

$$\frac{1}{c'^3_B} = \frac{1}{c^3_B} + \frac{1}{\gamma^3 \cdot c^3_G}, \qquad (3.106)$$

where γ is a factor depending on Poisson's ratio υ according to Table 3.2.

Table 3.2 Correction table for Equation (3.106).

υ	0.2	0.3	0.4	0.5
γ	0.689	0.841	0.919	0.955

Examples of calculated phase speed for concrete plates, in the thickness range of 50–200 mm, are shown in Figure 3.19. Corresponding data for steel plates, having thickness covering the range of 1–10 mm, are shown in Figure 3.20. Calculated results in both diagrams are performed using thin plate theory as well as thick plate theory (see Equations (3.103) and (3.106)). For the chosen range of plate thickness and frequency range there is practically no difference when it comes to the steel plates. The limit on the thin plate theory will in this case correspond to a phase speed of approximately 1500 m/s. As for the concrete, however, the corresponding limit will approximately be 1000 m/s, which may be seen clearly from the two sets of curves.

Shown in both figures is also the phase speed in air. The point of intersection between this line and the corresponding curves for the different plate thickness, i.e. where c_{air} is equal to c_B, is defining the so-called *critical frequency* f_c. This quantity is of fundamental importance when it comes to sound radiation from plates in bending vibrations (see section 6.3.3).

3.7.3.2 Eigenfunctions and eigenfrequencies (natural frequencies) of plates

In accordance with the general observations in section 2.5.3, we are in a position to describe the vibrations in structural elements, such as beams, plates and shells, by eigenfunctions and corresponding eigenfrequencies. Starting out from these functions we may, in an analogous way as in section 3.6 above, e.g. calculate transfer functions between an input force and a chosen velocity component. This will be a continuance of the calculations on discrete (lumped) mechanical systems given in sections 2.5.1 and 2.5.2.

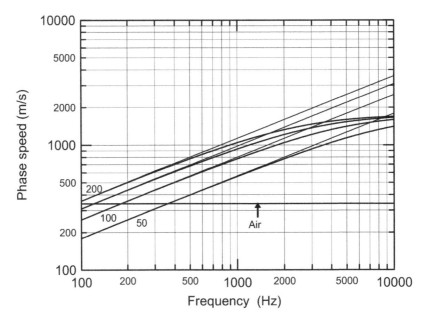

Figure 3.19 Phase speed of bending waves in concrete slabs of thickness 50, 100, 150 and 200 mm. Thin lines – calculated using thin plate theory. Thick lines – calculated using Equation (3.106). Horizontal line – acoustic wave speed in air.

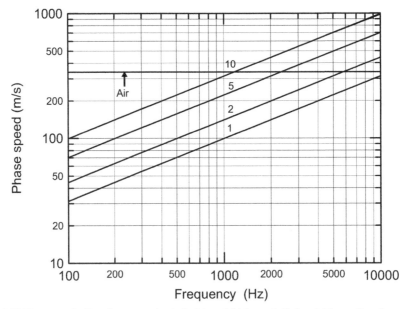

Figure 3.20 Phase speed of bending waves in steel plates of thickness 1, 2, 5 and 10 mm. See also caption of Figure 3.19.

In the literature, e.g. Blevins (1979), we find the eigenfunctions and eigenfrequencies listed for various types of structural element having different shapes and dimensions, and subjected to different boundary conditions. As a rule, exact data are only found for regular-shaped elements having idealized boundary conditions. For more complicated cases one must resort to finite element methods (FEM), but given the versatility of modern FEM software packages this seldom gives practical problems. As an illustration we give the results of an exact calculation on a typical isotropic element; a thin rectangular panel *simply supported* along the edges having length a and b, respectively. This boundary condition implies that the velocity as well as the moment is zero along the edges. We may remark that measurement results of natural frequencies for floors in buildings of monolithic concrete give reasonable agreement with calculations when using this condition.

The eigenfunctions $\Psi_{i,n}(x,z)$ for the plate, placed in the plane x–z, must satisfy the following wave equation (presupposing harmonic time variation $e^{j\omega t}$)

$$B \cdot \nabla^2 \nabla^2 \Psi_{i,n}(x,z) - \omega_{i,n}^2 \cdot m \cdot \Psi_{i,n}(x,z) = 0. \qquad (3.107)$$

Imposing simply supported boundaries gives the solutions

$$\Psi_{i,n}(x,z) = \sin\frac{i\pi}{a}x \cdot \sin\frac{n\pi}{b}z \qquad \text{where} \quad i,n = 1,2,3,\dots \qquad (3.108)$$

The associated eigenfrequencies will be given by

$$f_{i,n} = \frac{\pi}{2}\sqrt{\frac{B}{m}}\left[\left(\frac{i}{a}\right)^2 + \left(\frac{n}{b}\right)^2\right]. \qquad (3.109)$$

For a homogeneous plate this equation may be expressed as

$$f_{i,n} = \frac{\pi}{4\sqrt{3}}c_{\mathrm{L}}h\left[\left(\frac{i}{a}\right)^2 + \left(\frac{n}{b}\right)^2\right]. \qquad (3.110)$$

Each of these eigenfunctions, or each set of indices (i,n), thereby defines a mode of vibration, a *natural mode* or *eigenmode*. Any complex pattern of vibration may then be expressed by a sum of these modes. It should be noted that none of the indices i and n may be equal to zero. The first eigenfrequency or natural frequency of a plate is therefore $f_{1,1}$.

Figure 3.21 gives examples of natural modes of vibration for a plate according to Equation (3.108) and calculated for some of the lowest set of indices. Figure 3.22 gives corresponding examples on natural frequencies for a simply supported 180 mm thick concrete slab. The edges a and b are 4.0 and 6.0 metres, respectively.

3.7.3.3 Eigenfrequencies of orthotropic plates

In contrast to the isotropic plates (or panels), the material properties for orthotropic plates will depend on the direction. These properties are by definition, as also mentioned above, symmetric about three mutually perpendicular axes. We will again assume that the panel is placed in the xz-plane, furthermore that the x- and z-axis are axes of symmetry with corresponding bending stiffness B_x and B_z. In several cases, one may apply the geometric average of the stiffness in these two directions to characterize the panel stiffness but this may also turn out to be completely wrong, e.g. when calculating the natural frequencies.

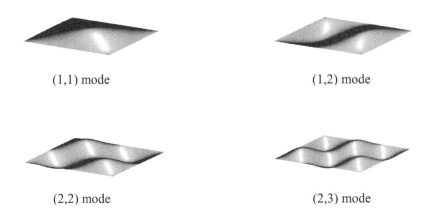

(1,1) mode   (1,2) mode

(2,2) mode   (2,3) mode

Figure 3.21 Rectangular plates in bending motion. Examples on mode shape.

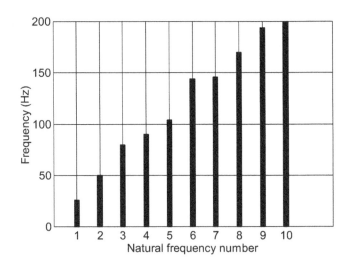

Figure 3.22 Natural frequencies (eigenfrequencies) of a 180 mm thick concrete slab, dimensions 4.0 by 6.0 metres; the frequencies of the first 10 modes of the slab that is simply supported.

We shall characterize the orthotropic plate using the material properties E_x, E_z, υ_x, υ_z and G_{xz}, i.e. by the elasticity modulus and Poisson's ratio for the two directions together with the shear modulus. There are really four independent quantities only as taking symmetry into consideration gives

$$E_x \upsilon_z = E_z \upsilon_x. \tag{3.111}$$

The bending stiffness in the two directions will then be given by

$$B_x = \frac{E_x h^3}{12(1 - \upsilon_x \upsilon_z)} \qquad \text{and} \qquad B_z = \frac{E_z h^3}{12(1 - \upsilon_x \upsilon_z)}. \tag{3.112}$$

Analogous to Equation (3.109), we shall give a formula for the natural frequencies of a simply supported rectangular plate with dimensions a and b. Formulae covering cases with other types of boundary condition may be found in the literature, e.g. Blevins (1979). For the simply supported plate we get

$$f_{i,n} = \frac{\pi}{2\sqrt{m}} \left[\frac{i^4}{a^4} B_x + \frac{n^4}{b^4} B_z + \frac{2 i^2 n^2}{a^2 b^2} B_{xz} \right]^{\frac{1}{2}}, \tag{3.113}$$

where B_{xz} is given by

$$B_{xz} = B_x \nu_z + 2 \cdot \frac{G_{xz} h^3}{12}. \tag{3.114}$$

For the isotropic case, where

$$E_x = E_z = E, \upsilon_x = \upsilon_z = \upsilon, \text{ and } G_{xz} = E / [2(1+\upsilon)],$$

it is easy to show that Equation (3.113) simplifies to Equation (3.109). It was formerly pointed out that the types of orthotropic plate normally found as building components, mainly in industrial buildings, are plates with attached stiffeners or corrugated plates. The latter may have many different shapes; from the "wavy" corrugated type to the more sophisticated having trapezoidal corrugations denoted as cladding. When applying the general theory of orthotropic plates on corrugated plates several assumptions must be fulfilled. We shall not delve into these assumptions, but just point to the fact that for many types equivalent expressions for the stiffness components B_x, B_z and B_{xz} exist in the literature, expressions which one may use to calculate e.g. the natural frequencies.

One example we shall use is the "wavy" type of corrugations; a panel having thickness h and where the "waves" have sinusoidal shape with wavelength L and amplitude H. The total height of the panel is then $2H$. Following Timoshenko and Woinowsky-Krieger (1959)[2] we may write:

---

[2] These equations are also referenced in Blevins (1979), unfortunately, with a misprint in the expression for B_z.

$$B_x = \frac{Eh^3}{12\left(1-\upsilon^2\right)\left(1+\left(\dfrac{\pi H}{L}\right)^2\right)},$$

$$B_z = \frac{EH^2h}{2}\left[1-\frac{0.81}{1+\dfrac{5}{2}\left(\dfrac{H}{L}\right)^2}\right] \qquad (3.115)$$

and $\quad B_{xz} = \dfrac{Eh^3}{12\left(1+\upsilon\right)}\left[1+\left(\dfrac{\pi H}{L}\right)^2\right].$

Example We shall compare the natural frequencies of a flat square plate with the corresponding ones for a wave corrugated plate. For a 1 mm thick steel plate with sides 1 metre long we get $f_{1,1} \approx 4.9$ Hz, this by using Equation (3.109) with E equal to $2.1 \cdot 10^{11}$ Pa, m equal to 7.8 kg/m^2 and υ equal to 0.3. Letting the height of the wave corrugated plate be 20 mm (H equal to 10 mm) and the "wavelength" be equal to 100 mm, we get $f_{1,1} \approx 25.5$ Hz by using Equations (3.113) and (3.115). Proceeding to the (2,2) mode we will have $f_{2,2} \approx 19.7$ Hz for the flat plate and 102 Hz for the corrugated one. Selecting a larger height and/or shorter wavelength for the corrugations will give even larger differences. It should be observed that we have to take into account the fact that the mass per unit area will increase when making the corrugations.

We shall later (see Chapter 6) demonstrate the effect of such corrugations on the sound transmission as compared to a flat plate. However, we shall then employ the more commonly used cladding type of plate, i.e. the type having trapezoidal corrugations. Predicting the bending stiffness in this case, analogous expressions to the ones above must be used. These are given in the literature; see e.g. Hansen (1993) or Buzzi et al. (2003). The latter also cite expressions for L-shaped plates in additions to the trapezoidal ones.[3]

3.7.3.4 Response to force excitation

If the eigenfunctions for a given system are known we may, by analogy with the calculations performed on the air-filled tube in section 3.6, calculate the response to a given mechanical input. Again using a plate as an example we may calculate the transfer function between a force (or moment) in a given point and a given response quantity such as velocity or acceleration in the same or in another point. We shall have to solve the wave Equation (3.107) but now modified by a term on the right-hand side representing the excitation. The response and the relevant transfer function may then be expressed by a sum of the eigenfunctions for the plate.

The measurement technique to determine such transfer functions is either by attaching an electrodynamic vibration exciter to the structure or using a transient excitation with a hammer blow or equivalent. The response quantity is measured at the driving point or at other relevant positions. The latter option gives basic data to determine the resonant vibration modes of the structure, a so-called *modal analysis*. We shall present an example on such transfer functions, using bending waves on a plate. This

---

[3] The quantity b in equation (38) lacks definition. It is the distance from the y-axis to the plate neutral axis.

example will, however, be on an input mobility, not a transfer mobility (the ratio of velocity at one point and the force at another point) as used in modal analysis. The mobility we are using is also calculated, not measured, but this gives us the opportunity to vary the parameters to illustrate the influence of, for example damping.

The object is a 15 mm thick glass plate of rectangular shape, simply supported along the edges of length 1.10 metre and 1.50 metre, respectively. It is driven by a constant force normal to the plate at a point (x,z) equal to $(0.20, 0.50)$. The input mobility, again using a logarithmic scale, is shown in Figure 3.23 for two values of the loss factor η. The highest value used is not a realistic one but is used to illustrate the influence of damping.

The thick horizontal line in the figure gives the mobility for an infinitely large plate of the same material and thickness. It may be surprising that this mobility is independent of frequency and at the same time is a real quantity. It should be noted that this is not generally true; e.g. neither the input mobility at the midpoint of an infinitely long beam nor at the end of a half-infinite beam has these characteristics.

On this horizontal line some marks have been made indicating some of the lowest natural frequencies of the plate, five frequencies altogether. The fourth and the fifth nearly fall together, which indicates that it is not always possible to detect the natural frequencies from such measurement data. The bandwidth of a resonance maximum may be larger than the distance between the natural frequencies. Several natural frequencies may then, depending on the damping, "hide" themselves inside a resonance maximum. It should also be remembered that one does not get any response from a certain mode having a node, i.e. zero displacement, at the driving point.

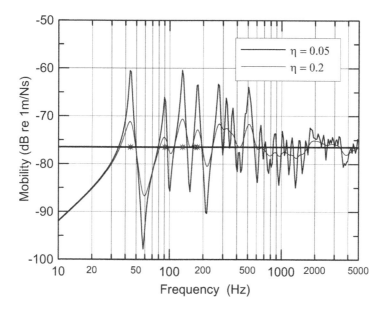

Figure 3.23 Input point mobility of a 15 mm thick glass plate, dimensions 1.10 x 1.50 metre, for two different loss factors. Solid horizontal line – mobility of an infinite size plate, with marks indicating the first five natural frequencies of the finite plate.

3.7.3.5 Modal density for bending waves on plates

Detailed frequency information is normally not required for quantities used in building acoustics. In general, one is interested in average values taken over relatively broad frequency bands such as one-third-octave bands or octave bands. This is true for sound pressure levels as well as for levels of vibration, e.g. one wish to determine an average velocity level for a wall or floor in a building. To ensure that the measurement result has the desired accuracy one has to estimate the required number of measurement points on the structure and this is directly linked to the number of modes having their natural frequencies within the frequency band measured. It is therefore important to estimate the density of natural frequencies in the structure to be measured, the so-called *modal density*.

A suitable procedure for this calculation is to make the modes expressed by their *modal wave numbers* $k_{i,n}$, instead of by their natural frequencies. These wave numbers will be

$$k_{i,n} = \frac{\omega_{i,n}}{c_B} = \frac{2\pi f_{i,n}}{c_B} = \left[\left(\frac{i\pi}{a}\right)^2 + \left(\frac{n\pi}{b}\right)^2 \right]^{\frac{1}{2}} \tag{3.116}$$

or

$$k_{i,n}^2 = \left(\frac{i\pi}{a}\right)^2 + \left(\frac{n\pi}{b}\right)^2 = k_{ix}^2 + k_{nz}^2, \tag{3.117}$$

where k_{ix} and k_{nz} are the wave number components in the x- and z-direction, respectively. All natural frequencies may therefore be plotted in a wave number diagram as shown in Figure 3.24, where each point represents a mode (eigenmode).

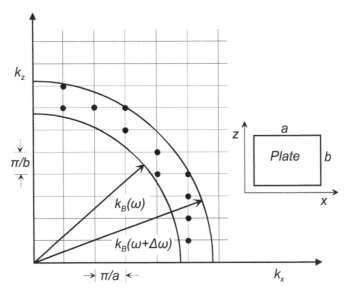

Figure 3.24 Modal wave numbers for bending waves on a rectangular plate of dimensions a and b. The eigenmodes within a frequency band $\Delta\omega$ are indicated.

If we wish to calculate the number of modes inside a given frequency interval $\Delta\omega$ (or Δf), we just count the number of points ΔN inside an area defined by two quarter circular arcs having wave numbers $k_B(\omega)$ and $k_B(\omega+\Delta\omega)$, these being wave numbers corresponding to the lower and upper cut-off frequencies for the given band pass filter. We then get

$$\Delta N = \frac{\pi}{4} \cdot \left[k_B^2 (\omega+\Delta\omega) - k_B^2 (\omega) \right] \cdot \frac{a \cdot b}{\pi^2}. \tag{3.118}$$

In the case of k_B not being too small, an approximate expression for the modal density is

$$n(\omega) = \frac{\Delta N}{\Delta\omega} = \frac{S}{4\pi} \cdot \left[\frac{k_B^2 (\omega+\Delta\omega) - k_B^2 (\omega)}{\Delta\omega} \right] \approx \frac{S}{4\pi} \cdot \frac{\partial(k_B^2)}{\partial\omega} = \frac{S}{4\pi} \cdot \sqrt{\frac{m}{B}} \tag{3.119}$$

or

$$n(f) = \frac{\Delta N}{\Delta f} = \frac{S}{2} \cdot \sqrt{\frac{m}{B}}, \tag{3.120}$$

where S is the plate area. As we can see, the modal density of thin plates is frequency independent. This is again not the case for other types and shapes of structure; see e.g. Blevins (1979).

Example The bandwidth of a third-octave-band filter is approximately $\Delta f \approx 0.23 \cdot f_0$, where f_0 is the centre frequency of the band. Choosing $f_0 = 1000$ Hz and taking the concrete slab used in Figure 3.22 as an example we get, even at this relatively high frequency, $\Delta N \approx 14$. In contrast to this, a room above this concrete slab (taken as the 24 m^2 floor of the room) having a ceiling height 2.5 m will have approximately 4 400 modes inside the same bandwidth! The latter number is calculated using the expression

$$\Delta N_{room} \approx \frac{4\pi V}{c_0^3} \cdot f^2 \cdot \Delta f,$$

where V is the room volume. This expression is derived using an analogous procedure to the one used above taking into account the equation for the natural frequencies of a three-dimensional air-filled space (see section 4.4.1).

3.7.3.6 Internal energy losses in materials. Loss factor for bending waves.

In a previous chapter, when dealing with oscillations in simple mass-spring systems, we introduced the loss factor by way of a complex stiffness. In a similar way we shall, for bending waves in a given structure, define a complex bending stiffness B':

$$B' = B(1 + j \cdot \eta), \tag{3.121}$$

where η is the loss factor. By formal definition, as found in the literature, it is given by the ratio of the mechanical energy E_d dissipated in a period of vibration to the reversible mechanical energy E_m:

$$\eta = \frac{E_d}{2\pi E_m}.$$ (3.122)

Inserting Equation (3.121) into expressions for natural frequencies of an element, e.g. into Equation (3.109), we shall get complex natural frequencies

$$f'_{i,n} = f_{i,n}\sqrt{1+j\eta} \underset{\eta \ll 1}{\approx} f_{i,n}\left(1+j\frac{\eta}{2}\right).$$ (3.123)

This is another formal way of stating that the system has internal energy losses, meaning that the amplitude always has a finite value at resonance.

The loss factor may be determined by measuring either the quality factor Q, the bandwidth Δf for a given mode at resonance or the reverberation time T following an excitation of the system at the given natural frequency f_0. The relations between these quantities are:

$$\eta = \frac{1}{Q} = \frac{\Delta f}{f_0} = \frac{2.2}{T \cdot f_0}.$$ (3.124)

The energy losses of a given element will, however, always be caused by several mechanisms; first, there will be inner losses in the material, where the vibration energy is converted into heat, second, there will be energy radiated as sound. Another important loss mechanism is "leakage" to connected structures, which we may call *edge losses*. The total loss factor may therefore be expressed by a sum of loss factors representing these mechanisms:

$$\eta_{total} = \eta_{internal} + \eta_{radiation} + \eta_{edges}.$$ (3.125)

The crucial questions will then be: 1) which one or which ones of these are the most important in an actual case and 2) how shall we arrive at the data, using either calculations or measurement. The internal losses of metal elements are normally very small; $\eta_{internal}$ is of the order 10^{-3}–10^{-4}. A producer of viscoelastic layers intended for damping of metal panels would certainly be concerned with the question of how much the $\eta_{internal}$ will increase by bounding the layer to the panel. He will then certainly apply a measurement method where the other contributions to the losses are small, i.e. by freely suspending the specimen sample and at the same time ensure that the amount of radiation is small.

For common building constructions composed of materials such as concrete, gypsum etc. we may find that the loss factor due to internal losses is of the order 0.01. Being part of a building construction one normally finds that the edge losses tend to dominate. This implies that, when performing measurements in the field, one can only determine the total loss factor. This is however the important factor when it comes to sound transmission and its estimation and measurement in the field or laboratory. Lastly, we shall therefore give a supplementary expression for the last two terms in Equation (3.125). This will apply to a plate or panel element having a mass per unit area m, an area S and the length l of the edges. The expression, given below, is taken from the standard EN 12354–1.

$$\eta_{\text{total}} = \eta_{\text{internal}} + \frac{\rho_0 c_0}{\pi f \cdot m} \sigma + \frac{c_0}{\pi^2 S \sqrt{f \cdot f_c}} \sum_{k=1}^{4} \ell_k \alpha_k. \tag{3.126}$$

The element has critical frequency f_c (see section 3.7.3.1) and radiation factor σ for free bending waves. The energy losses along the edges ℓ_k are characterized by an absorption factor α_k for bending waves. This factor may in a field situation be in the range 0.05 to 0.5. Further on we shall look into ways of estimating this factor.

3.8 REFERENCES

EN 12354–1: 2000, Building acoustics – Estimation of acoustic performance of buildings from the performance of elements. Part 1: Airborne sound insulation between rooms.

ISO 3744: 1994, Acoustics – Determination of sound power levels of noise sources using sound pressure – Engineering methods in an essentially free field over a reflecting plane.

ISO 3746: 1996, Acoustics – Determination of sound power levels of noise sources using sound pressure – Survey method using an enveloping surface over a reflecting plane.

ISO 9614–2: 1996, Acoustics – Determination of sound power levels of noise sources using sound intensity. Part 2: Measurement by scanning.

ISO 5136: 2003 Acoustics – Determination of sound power radiated into a duct by fans and other air-moving devices – In-duct method.

ISO 80000–8: 2007, Quantities and units. Part 8: Acoustics. [At the stage of ISO/FDIS in 2007.]

Abramowitz, M. and Stegun, I. A. (1970) *Handbook of mathematical functions.* Dover Publications Inc., New York.

Blevins, R. D. (1979) *Formulas for natural frequency and mode shape.* Van Nostrand Reinhold Company, New York.

Buzzi, T., Courné, C., Moulinier, A. and Tisseyre, A. (2003) Prediction of the sound reduction index: A modal approach. *Applied Acoustics*, 64, 793–814.

Hansen, C. H. (1993) Sound transmission loss of corrugated panels. *Noise Control Eng. J.*, 40, 187–197.

Kinsler, L. E., Frey, A. R., Coppens, A. B. and Sanders, J. V. (2000) *Fundamentals of acoustics*, 4th edn. John Wiley & Sons, New York.

Mindlin, R.D. (1951) Influence of rotary inertia and shear on flexural motion of isotropic plates. *J. Appl. Mech.*, 18, 31–38.

Timoshenko, S. P. and Woinowsky-Krieger, S. (1959) *Theory of plates and shells*, 2nd edn. McGraw-Hill, New York.

CHAPTER 4

Room acoustics

4.1 INTRODUCTION

In talking about the concept of room acoustics we shall include all aspects of the behaviour of sound in a room, covering both the physical aspects as well as the subjective effects. In other words, room acoustics deals with measurement and prediction of the sound field resulting from a given distribution of sources as well as how a listener experiences this sound field, i.e. will the listener characterize the room as having "good acoustics"? When designing for a good acoustic environment, which could be everything from introducing some absorbers into an office space to the complete design of a concert hall, one must bear in mind both the physical and the psychological aspects. This implies having knowledge on how the shape of the room, the dimensions and the material properties of the construction influences the sound field. Just as important, however, is a knowledge of the relationship between the physical measurable parameters of this field and the subjective impression for a listener. Finding such objective parameters, either measurable or predictable, which correlate well with the subjective impression of the acoustic quality, is still a subject of research. It goes without saying that the number of suggested parameters is quite large. The reverberation time in a room has been, and still is, an important parameter in any judgement of quality. Another large group of parameters are also based on the impulse responses of the room but here the emphasis is on the relative energy content in given time intervals.

In this chapter, the primary emphasis will be on the physical properties, partly to give a background for the most common measurement methods in room acoustics. Suggested requirements for parameters, other than the reverberation time, will to some extent also be touched on.

4.2 MODELLING OF SOUND FIELDS IN ROOMS. OVERVIEW

In principle, we should be able to calculate the sound field in a room, generated by one or more sources, applying a wave equation of the same type as used earlier in the one-dimensional case (see section 3.6). There we introduced a sound source as a mass flux q, having the dimensions of $\text{kg} \cdot \text{m}^{-3} \cdot \text{s}^{-1}$, in the equation of continuity. In the three-dimensional case, we obtain

$$\nabla^2 p - \frac{1}{c_0^2} \cdot \frac{\partial^2 p}{\partial t^2} + \frac{\partial q}{\partial t} = 0. \tag{4.1}$$

Solving this equation analytically will normally become very difficult except for simple room shapes and simple boundary conditions, e.g. an empty rectangular-shaped room having walls of infinite stiffness. Solutions for such special cases may, however, give some general information on sound fields in rooms. It is therefore useful to discuss some of these cases, which we shall return to in section 4.4.2.

The development of numerical techniques in recent time has been formidable, which include FEM (finite element methods), BEM (boundary element methods) and various other numerical methods for predicting sound propagation in bounded spaces. Using these, accurate solutions may be obtained for complex room shapes and boundary conditions. First and foremost, these techniques are suitable in the lower frequency ranges, i.e. when the ratio between a typical room dimension and the wavelength is not too large. When using a FEM technique a reasonable number of elements per wavelength are of the order three to four. If the typical room dimension is 10 metres one may at 100 Hz perhaps use 1000 elements. However, to calculate with the same accuracy at 1000 Hz one needs 1000000 elements. Depending on the specific computer FEM software, different types of elements are implemented, having some 8 to 20 nodes. At each of these nodes we shall then calculate the sound field quantity in question. In spite of the large capacity of modern computers, the limitations imposed on these calculations should be obvious. It should, however, be stressed that FEM calculations have become very important tools in the area of sound radiation and sound transmission, in particular where a strong coupling between a vibrating structure and the surrounding medium is expected.

A number of other approximate methods have a long history in room acoustics. The reason is that one normally is not interested in a detailed description frequency by frequency. The average value in frequency bands, being either octave or one-third-octave bands, has been more relevant. In the literature one will therefore find methods characterized under headings such as *statistical room acoustics* and *geometrical room acoustics*. The first term implies treating the sound pressure in a room as a stochastic quantity with a certain space variance. The classical *diffuse field* model, also called the *Sabine* model, is an extreme case in this respect. The latter name is a recognition of the American scientist Wallace Clement Sabine (1868–1919) who published his famous article "Reverberation" in the year 1900 containing a formula for the reverberation time in rooms, a formula still being the most used. In a diffuse field model, the space variance of the sound pressure is zero, the energy density is everywhere the same in the room. Such a model may be seen as the acoustic analogue of the classical kinetic gas model.

There is also a long tradition for using geometrical models in acoustics, see e.g. Pierce (1989). For geometrical acoustics in general, also denoted ray acoustics, the concept of *wave front* is central. At a given frequency, a wave front is a surface where the sound pressure everywhere is in phase. As the wave front moves in time, the line described in space by a given point on the surface is called the *ray path*. Generally, it is not necessary to assume that the amplitude is constant over the wave front or that the wave front is a plane surface but in room acoustics this is assumed. Curved paths have no place in geometrical room acoustics; the sound energy propagates along straight ray paths just like light. Inherent in these geometrical models there is no frequency information and the validity of the calculated results is in principle limited to a frequency range where we may assume specular reflections and where diffraction phenomena may be neglected. Such phenomena may, however, be included in these models by certain artifices. We shall deal with them by giving an overview of the principles.

4.2.1 Models for small and large rooms

We have given an overview and some general remarks concerning the different models used to predict the sound field in rooms. We shall proceed by going into more detail on the suitability of these models for given situations. Simple diffuse field models may in practice be quite sufficient predictors given that a certain minimum number of room modes are being excited and participate in the build-up of the sound field. However, there are also a number of other conditions that have to be fulfilled before it is reasonable to assume that a global sound pressure level or a global reverberation time exists. The linear dimensions of the room must not be too different; the absorption material must be reasonably evenly distributed on the room surfaces and the total absorption area must not be too high.

To apply the simple expressions for the reverberation time, given in section 4.5.1.2 below, also presupposes that only the room volume and the total surface area determine the *mean free path* of the sound, i.e. the distance between each reflection. When filling the room with a certain number of scattering objects an "internal" reverberation process may be set up between these objects and the common reverberation time formulae are no longer applicable. We should then bear in mind how to explain the diffusing elements required for laboratories performing standard absorption measurements according to ISO 354. We shall return to this question when treating the subject of scattering.

In conclusion, large discrepancies between the ideal conditions demanded for a diffuse field and the actual room conditions make such models unsuitable. It may be that the linear dimensions are quite different; e.g. the room is "flat" in the sense that the ceiling height is small compared to the length and width of the room (industrial hall, landscaped office etc.) or the room is "long" (a corridor etc.). Absorbing materials or objects may also be unevenly distributed and the room may also contain a number of different types of reflecting and/or scattering object.

The choice of models to use on such "large rooms" is obviously dependent on the intended function for the room, a function that also determines the parameters we shall use to validate the acoustic quality. On industrial premises, e.g. large industrial halls, where a large attenuation between the various noise sources and the workers is aimed at, the decrease in decibels per metre distance may be a suitable parameter to estimate. For rooms having a simple shape, such a parameter could be estimated by an analytical model.

In performance spaces, theatres, auditoria, concert halls etc., the function of the room is to forward the sound to the audience, which implies that a quite different set of parameters, are necessary. Predicting the sound field in such rooms is generally based on methods from geometrical acoustics, partly combined with statistical considerations to include scattering (diffusion) phenomena. Two methods, principally different, are used: the ray-tracing method and the mirror-source method. The former simulates a sound source by emitting a large number of "sound rays", these being evenly distributed over the solid angle covered by the actual sound source. Each ray is followed as it hits the various surfaces in the room, being specularly reflected and radiated having a reduced energy caused by the absorption factor of the surface.

According to the name, the mirror-source method is based on the mirror images of the real source. The sound from a mirror source received at a given point is reflected *once* in the surface of the mirror. These first-order sources are then being mirrored by all room surfaces giving second-order sources and so on. Short descriptions of these two geometrical prediction models are given in section 4.8. Software having implemented these methods is commercially available. Most of them are based on a hybrid method combining the principles outlined above. A number of them have the possibility of simulating simple types of scattering effect.

4.3 ROOM ACOUSTIC PARAMETERS. QUALITY CRITERIA

The parameters used for assessing the acoustic quality of a room obviously depend on its intended use. Whereas the reverberation time and/or the sound level reduction by distance from the source may be sufficient in an industrial hall, a more comprehensive set of parameters must be used in e.g. concert halls. It is acknowledged that the reverberation time has an important role and there is sufficient background experience on how long or short it should be depending on the size of the room and related to the type of the performance room; theatre, room for music performance etc. As for music performance, the type of music will be a vital factor; see e.g. Kuttruff (1999).

A number of other parameters that correlates well with the subjective impression are based on data calculated from measured impulse responses in the room; see ISO 3382. An example is shown in Figure 4.1, a measured impulse response using an MLS technique (see section 1.5.2).

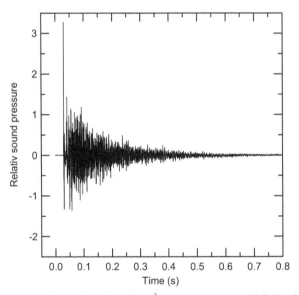

Figure 4.1 A measured impulse response in an 1800 m³ auditorium using a MLS signal (sequence length of order 16 and sampling frequency 25 kHz of which only every second point is shown). After Lundeby et al. (1995).

Irrespective of the intended use of the room, whether for speech or music, it is important to design the room in such a way as to give a balanced set (in time) of the early reflections onto the audience area. Reflections following the direct sound within a time span of approximately 50 milliseconds will contribute to the strength of the direct sound. A listener will not perceive these reflections as a separate part or as an echo, but will if a strong reflection has a longer delay. This phenomenon is called the *precedence effect* or

Haas effect, the latter name in recognition of one of the many researchers on the phenomenon, Haas (1951).

Added to the time arrival of the reflections, it is important for rooms for music performances to know *where* the reflections are coming from. The directional distribution is critical for the listener's feeling of spaciousness of the sound field, i.e. lateral reflections are just as important as reflections from the ceiling. Added to this fact, there has in the last 20 years been a growing awareness that diffuse reflections are also very important, again for rooms for music performances. We shall therefore give some examples of these other objective acoustic parameters used for larger halls, how they are determined and, to a limited extent, on the underlying subjective matter.

4.3.1 Reverberation time

The reverberation time T is defined as the time required for the sound pressure level in a room to decrease by 60 dB from an initial level, i.e. the level before the sound source is stopped. This is not necessarily coincident with a listeners feeling of reverberation and in ISO 3382 one will find that measurement of the *early decay time* (EDT) is recommended as a supplement to the conventional reverberation time. Both parameters are determined from the *decay curve*, EDT from the first 10 dB of decay, and T normally from the 30 dB range between −5 and −35 dB below the initial level. Both quantities are calculated as the time necessary for a 60 dB decay having the rate of decay in the ranges indicated.

Throughout the time a number of methods have been used to determine the decay curves and thereby the reverberation time. A common method is to excite the room by a source emitting band limited stochastic noise, which is turned off after a constant sound pressure level is reached. For historical reasons, we shall mention the so-called level recorders, a level versus time writer, recording directly the sound pressure level decay, where the eye could fit a straight line. Later developments included instruments giving out the decay data digitally, enabling a line fit e.g. by the method of least squares.

Modern methods based on deterministic signals such as MLS or SS, however, are superior in the dynamic range achieved in the measurements and may well measure over a decay range of 60 dB or more. It may be shown that the decay curve is obtained by a "backward" or reversed time integration of impulse responses as the one shown in Figure 4.1. Normally as we are interested in the reverberation as a function of frequency, the impulse response is filtered in octave or one-third-octave bands before performing this integration. The decay as a function of time is then given by

$$E(t) = \int_{t}^{\infty} p^2(\tau)\,d\tau = \int_{\infty}^{t} p^2(\tau)\,d(-\tau), \qquad (4.2)$$

where p is the impulse response. Certainly, this equation was also utilized when analogue measuring equipment was used by splitting the integral into two parts as follows

$$E(t) = \int_{t}^{\infty} p^2(\tau)\,d\tau = \int_{0}^{\infty} p^2(\tau)\,d\tau - \int_{0}^{t} p^2(\tau)\,d\tau. \qquad (4.3)$$

The upper limit of the integration poses a problem as the background noise unrelated to the source signal will be integrated as well. Different techniques are suggested to minimize the influence of background noise. One method is to estimate the background noise from the later part of the impulse response, thereafter compensating for the noise by assuming that the energy decays exponentially with the same decay rate as the actual one at a level 10–15 dB above the background level. Such a technique (see Lundeby et al. (1995)) is used calculating the decay curves shown in Figure 4.2. The impulse response shown in Figure 4.1 is filtered by a one-third-octave band of centre frequency 1000 Hz and the decay curves are calculated with and without being compensated for background noise. In one set of curves, the level of the background is equal to the one present at the time of measurement. In the second set, the background noise is artificially increased to show that also in this case one will obtain a decay curve having an acceptable dynamic range. Ideally, all the solid curves should be coincident but this will only be the case if the decay rate is everywhere the same.

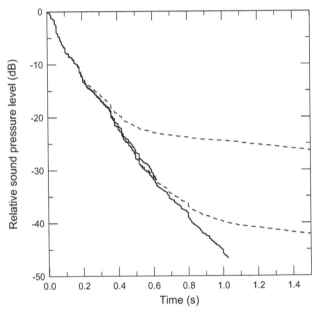

Figure 4.2 Decay curves based on filtering, one-third-octave band 1000 Hz and reverse time integration of the impulse response shown in Figure 4.1. Solid curves – integration with background noise compensation. Dashed curves – integration of the total impulse response. One set of curves is using an artificially added noise. After Vigran et al. (1995).

4.3.2 Other parameters based on the impulse response

A large number of parameters suggested in the literature and applied over the years are listed and commented on in ISO 3382. These are all derived from measured impulse responses, and we shall present a selection of these measures.

The balance between the early and late arriving sound energy, which concerns the balance between the clarity (or distinctness) and the feeling of reverberation, is important for music as well as for speech. Several parameters are suggested to cover this matter in room acoustics. The simplest ones deal with the ratio of the total sound energy received in the first 50 or 80 milliseconds to the rest of the energy received. We have an *early-to-late index* C_{te} defined by

$$C_{t_e} = 10 \cdot \lg \left(\int_0^{t_e} p^2(t)\,dt \middle/ \int_{t_e}^{\infty} p^2(t)\,dt \right), \tag{4.4}$$

where t_e is 50 ms for speech and 80 ms for music. A recommended value for this parameter is 0 dB.

An early variant of this parameter was D_{50}, which is denoted *definition* in line with the original German notion of *Deutlichkeit*. The difference from the above is that, instead of the late energy, one is using the total energy received. Hence

$$D_{50} = \int_0^{50\text{ms}} p^2(t)\,dt \middle/ \int_0^{\infty} p^2(t)\,dt. \tag{4.5}$$

The relationship between C_{50} and D_{50} is then given by

$$C_{50} = 10 \cdot \lg \left(\frac{D_{50}}{1 - C_{50}} \right), \tag{4.6}$$

making it unnecessary to measure both parameters.

By way of introduction, we pointed out that the direction of sound incidence was important for the feeling of spaciousness. Of special importance are the lateral reflections, which also contribute to an impression of widening a source or a source area. Several early lateral energy measures are proposed, one is the *lateral energy fraction LF* based on measured impulse responses obtained from an omni-directional and a figure-of-eight pattern microphones. It is defined as

$$LF = \int_{5\text{ms}}^{80\text{ms}} p_L^2(t)\,dt \middle/ \int_0^{80\text{ms}} p^2(t)\,dt, \tag{4.7}$$

where p_L is the sound pressure obtained with the figure-of-eight microphone. This microphone is intended to be directed in such a way that it responds predominantly to sound arriving from the lateral directions and is not significantly influenced by the direct sound.

Because the directivity of a figure-of-eight microphone essentially has a cosine pattern and the pressure is squared, the resulting contribution from a given reflection will vary with the square of the cosine of the angle between the reflection relative to the axis of maximum sensitivity of the microphone. An alternative parameter is *LFC*, where the contributions will be a function of the cosine to this angle. This parameter, which is believed to be subjectively more accurate, is defined by

$$LFC = \int_{5\text{ms}}^{80\text{ms}} |p_L(t) \cdot p(t)|\,dt \middle/ \int_0^{80\text{ms}} p^2(t)\,dt. \tag{4.8}$$

In addition to the parameters given above, there are others related to our binaural hearing, based on measurements using an artificial or dummy head. These so-called *inter-aural cross correlation* measures are correlated to the subjective quality of "spatial impression".

4.4 WAVE THEORETICAL MODELS

Obtaining analytical solutions to the wave equation (4.1) are difficult except in cases where the room has a simple shape and simple boundary conditions. In section 3.6, we arrived at a solution for the sound field in a simple one-dimensional case: a tube closed in both ends and with stiff walls where we assumed that the particle velocity everywhere was equal to zero. We may easily generalize these results to the three-dimensional case if we assume that the room has a rectangular shape with dimensions L_x, L_y and L_z. We shall use this as an example to illustrate some important properties of sound fields in rooms; how the impulse response will depend on e.g. the room dimensions and furthermore, how we may predict the impulse responses.

For a free wave field we shall have to solve the wave equation without the source term. Assuming harmonic time dependence, we get the Helmholtz equation for the sound pressure in three-dimensional form

$$\nabla^2 p + k^2 p = 0, \tag{4.9}$$

where k is the wave number. Initially, we shall assume that all boundary surfaces are infinitely stiff and there are no other energy losses in the room. The eigenfunctions for the pressure will then be given by

$$p_{n_x n_y n_z}(x,y,z) = C \cdot \cos(k_x x) \cdot \cos(k_y y) \cdot \cos(k_z z), \tag{4.10}$$

where C is a constant and where the eigenvalues for the wave number is given by

$$k_n^2 = k_x^2 + k_y^2 + k_z^2 = \left(\frac{n_x \pi}{L_x}\right)^2 + \left(\frac{n_y \pi}{L_y}\right)^2 + \left(\frac{n_z \pi}{L_z}\right)^2. \tag{4.11}$$

The corresponding eigenfrequencies are given by

$$f_{n_x n_y n_z} = \frac{c_0}{2}\left[\left(\frac{n_x}{L_x}\right)^2 + \left(\frac{n_y}{L_y}\right)^2 + \left(\frac{n_z}{L_z}\right)^2\right]^{\frac{1}{2}}. \tag{4.12}$$

To each of these eigenfunctions or normal modes there is a set of numbers, a set of indices. Equation (4.10) then represents a three-dimensional standing wave if we multiply with the time-dependent factor exp($j\omega t$). In the literature special names are used for the wave forms associated with these sets of indices. We have an *axial mode* when two of the indices are equal to zero, a *tangential mode* when just one of the indices is zero, and finally, an *oblique mode* when all indices are different from zero. (Can you tell the direction of the wave in the room in these three cases?)

For the case of the one-dimensional standing wave, we named the points where the sound pressure was zero as *nodal points*. By analogy, here we shall have *nodal planes* if one or more of these indices is zero and the indices will indicate the number of such planes normal to the x-, y- and z-axis, respectively. That the nodal points have the form of a plane is a special case due to the example we have chosen, the rectangular room. For other shapes we shall have other types of geometric surface; we shall call them *nodal surfaces*.

4.4.1 The density of eigenfrequencies (modal density)

Concerning measurements in building acoustics, such as sound insulation, sound absorption, sound power etc. the eigenfrequencies per se are not particularly important. The relative density, i.e. the number of eigenfrequencies within a given bandwidth, is, however, of crucial importance for measurement accuracy. By analogy to the calculation of the modal density for a plate (see section 3.7.3.5), we may develop a wave number diagram having the shape as the octant of a sphere. Summing up the number of "points" or eigenfrequencies N below a given frequency f, we arrive at the following approximate expression

$$N \approx \frac{4\pi f^3}{3c_0^3} \cdot V + \frac{\pi f^2}{4c_0^2} \cdot S + \frac{Lf}{8c_0},$$ (4.13)

where V, S and L are the room volume, the total surface area of the room and the total length of the edges, respectively. Differentiating this expression with respect to frequency we arrive at the following approximate expression for the modal density

$$\frac{\Delta N}{\Delta f} \approx \frac{4\pi f^2}{c_0^3} \cdot V + \frac{\pi f}{2c_0^2} \cdot S + \frac{L}{8c_0}.$$ (4.14)

As seen, the first term will be the dominant one at higher frequencies, and in the literature one often finds this term alone. This certainly has the advantage of requiring the room volume only, but this practice may introduce large errors at low frequencies.

Example An ordinary sitting room in a dwelling with dimensions $L_x \cdot L_y \cdot L_z$ equal to $6.2 \cdot 4.1 \cdot 2.5$ metres, gives us a floor area of 25.4 m^2 and a volume of 63.6 m^3. Choosing a frequency of 100 Hz, Equation (4.14) gives us $\Delta N/\Delta f$ equal to 0.361. If we measure using one-third-octave bands filters, at centre frequency 100 Hz we get a bandwidth $\Delta f \approx 0.23 \cdot 100 = 23$ Hz. We will then get $23 \cdot 0.361 \approx 8$ eigenfrequencies inside this band, which compares well with an exact calculation giving seven eigenfrequencies. If we just use the first term we will get five eigenfrequencies. However, going up in frequency the first term will become dominant. Keeping a fixed bandwidth of 23 Hz and moving up to 1000 Hz, we expect to find approximately 500 eigenfrequencies (the first term alone gives 470). Using a one-third-octave filter we arrive at approximately 5000 eigenfrequencies inside the band.

4.4.2 Sound pressure in a room using a monopole source

We shall proceed by calculating the sound field in a room of rectangular shape where we have placed a sound source in a given position. This is again a generalization of the one-dimensional case of a tube with a sound source (see section 3.6). We shall assume that the source is a monopole, pulsating harmonically in time. The task is then to solve the Helmholtz equation (4.9) but now modified with a source term on the right side of the equation. We shall characterize the monopole source by its volume velocity or source strength Q having unit m³/s, i.e. not by the mass q as in Equation (4.1). The pressure root-mean-square-value in a given point (x,y,z) caused by the source in a position (x_0,y_0,z_0) may be written

$$\tilde{p}(x,y,z) = \rho_0 c_0^2 \tilde{Q} \sum_{n_x=0}^{\infty} \sum_{n_y=0}^{\infty} \sum_{n_z=0}^{\infty} \frac{\omega \cdot \Psi_{n_x n_y n_z}(x,y,z) \cdot \Psi_{n_x n_y n_z}(x_0,y_0,z_0)}{V_{n_x n_y n_z}\left(\omega^2 - \omega_{n_x n_y n_z}^2\right)}. \quad (4.15)$$

The quantity ω is the angular frequency of the source, and $\omega_{n_x n_y n_z}$ are the eigenfrequencies according to Equation (4.12). The Ψ-functions are the corresponding eigenfunctions:

$$\Psi_{n_x n_y n_z}(x,y,z) = \cos\frac{n_x \pi x}{L_x} \cdot \cos\frac{n_y \pi y}{L_y} \cdot \cos\frac{n_z \pi z}{L_z}. \quad (4.16)$$

$V_{n_x n_y n_z}$ is a normalizing factor, depending on the modal numbers, given by

$$V_{n_x n_y n_z} = V \cdot \varepsilon_{n_x} \cdot \varepsilon_{n_y} \cdot \varepsilon_{n_z}, \quad \text{where} \quad \varepsilon_n = 1 \text{ for } n = 0$$

$$\text{and} \quad \varepsilon_n = \frac{1}{2} \text{ for } n \geq 1. \quad (4.17)$$

The equations are derived assuming no energy losses in the room. However, as shown earlier in section 3.7.3.6, we may introduce small losses by complex eigenfunctions. We shall write

$$\bar{\omega}_{n_x n_y n_z} = \omega_{n_x n_y n_z} \sqrt{1 + j \cdot \eta} = \omega_{n_x n_y n_z} \sqrt{1 + j \cdot \frac{4.4 \cdot \pi}{\omega_{n_x n_y n_z} \cdot T}}, \quad (4.18)$$

where η is the loss factor and T the corresponding reverberation time. As an example of the use of Equation (4.15), we shall calculate the pressure at a given position in the same room as used in the example in section 4.4.1. We shall make the reverberation time 1.0 seconds independent of frequency.

The pressure response is shown in Figure 4.3 represented by the transfer function $p/(Q \cdot \omega)$ on a logarithmic scale for a frequency range up to 1000 Hz. This implies that we have related the pressure to the volume acceleration of the monopole source, both given by their root-mean-square-values. Also shown in the diagram are the lowest 10 eigenfrequencies. It will appear that only the very low frequency resonances may be identified. In the higher frequency range we find that the response is made up by contributions from many modes.

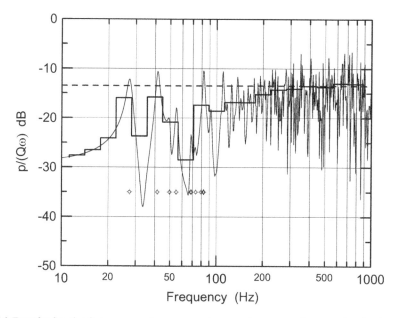

Figure 4.3 Transfer function between sound pressure and monopole source volume acceleration in a room of dimensions 6.2 x 4.1 x 2.5 metres and reverberation time 1.0 seconds. Source position (1.7, 1.0, 1.5), receiver position (3.5, 2.5, 1.5). Thick solid curve – analysis in one-third-octave bands. Dashed line – diffuse-field model. The points show calculated resonance frequencies.

The response is also shown resulting from an analysis in one-third-octave bands, a normal procedure when performing measurements in buildings. It is then of interest to calculate the result if one is using a simple diffuse field model for this case (see section 4.5.1 below). Assuming that the pressure at the receiver position is not affected by the direct field from the source, we may use the simple relationship between the source power W and the average sound pressure in the room stating that

$$W = \frac{\tilde{p}^2}{4\rho_0 c_0} \cdot A = \frac{\tilde{p}^2}{4\rho_0 c_0} \cdot \frac{55.3 \cdot V}{c_0 T}, \qquad (4.19)$$

where A is the total absorbing area in the room. A monopole source freely suspended in the room will radiate a power

$$W_{\text{monopole}} = \frac{\rho_0 c_0 k^2 \tilde{Q}^2}{4\pi}. \qquad (4.20)$$

Equating these powers, we obtain

$$\frac{\tilde{p}}{\tilde{Q}\omega} = \rho_0 \sqrt{\frac{c_0 T}{55.3\pi V}}. \qquad (4.21)$$

The result is shown by the dashed line in Figure 4.3. We see that there is a good fit between this result and the frequency averaged data in the frequency range above 200 Hz. However, it must be noted that we have performed a calculation just for one receiver position. Determining the emitted power from a source in a standard reverberation room test (see ISO 3741) the squared sound pressure is space averaged by using a number of microphone positions. It is interesting to note that this standard requires a minimum room volume of 70 m^3 (the volume in our example is approximately 64 m^3) permitting measurements upwards from 200 Hz.

4.4.3 Impulse responses and transfer functions

The common measurement procedure today is to determine pertinent impulse responses, hereby using these to calculate reverberation time, other room acoustic measures and transfer functions if required. In the preceding section, we calculated the transfer function between the sound pressure at a given position in a room and the volume acceleration of a source at another position. Vice versa, by an inverse Fourier transform of the transfer function we shall arrive at the impulse response, from which we may calculate the reverberation time and check that it is correct. The latter means that it is 1.0 second independent of frequency, as presupposed when calculating the transfer function shown in Figure 4.3.

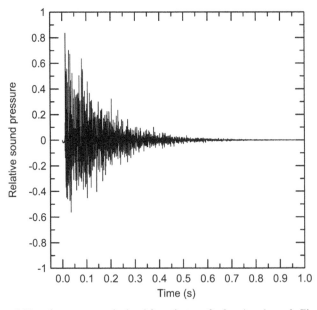

Figure 4.4 Impulse response calculated from the transfer function shown in Figure 4.3.

The unfiltered impulse response (for the frequency range up to 1000 Hz) corresponding to the transfer function in Figure 4.3 is shown in Figure 4.4. It should be noted that when calculating the inverse transform one must ensure that the result, the impulse response, turns out to be a purely real quantity, which implies a meticulous treatment of the real and imaginary part of the transfer function.

The unfiltered impulse response may now be filtered in either octave or one-third-octave bands to arrive at the reverberation time in these bands. This is carried out using octave bands with centre frequencies 125, 250 and 500 Hz and the decay curves are shown in Figure 4.5. Fitting straight lines to these curves, one will find that the time for the sound pressure level to decrease 60 dB is 1 second, which was input to the calculations using Equation (4.15). For simplicity, the decay curves are not calculated using the integration procedure given by Equation (4.2) but by a running short-time (50 milliseconds) integration of the squared response. In fact, such a procedure simulates the working of the old level recorders.

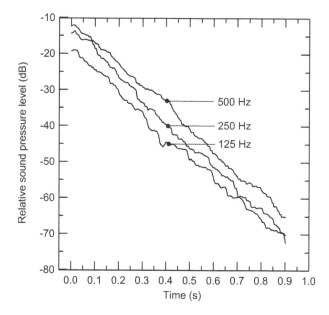

Figure 4.5 Decay curves in octave bands with centre frequencies 125, 250 and 500 Hz, calculated from the impulse response shown in Figure 4.4.

To conclude on this topic, we shall present examples of transfer functions based on impulse responses obtained in a real room like the one shown in Figure 4.1. The purpose is, for one thing, to show that transfer functions obtained in real rooms have the character as calculated and depicted in Figure 4.3. We shall use transfer functions based on impulse responses measured in the same auditorium as the one used for measuring the impulse response in Figure 4.1. The result is shown in Figure 4.6 where the sound pressure level (arbitrary reference) is given for the frequency range 100–200 Hz. One of these curves corresponds to the impulse response shown in Figure 4.1, for the other two curves the axis of the loudspeaker source is rotated 30° and 60°, respectively, from the horizontal plane. It goes without saying that the results exhibit the expected deterministic behaviour depending, among other factors, on the physical dimensions of the room.

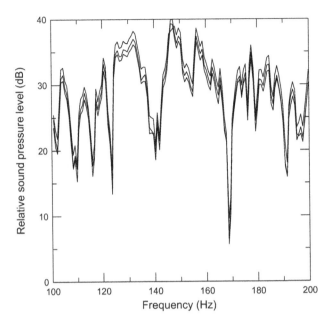

Figure 4.6 Some examples of transfer functions measured in an auditorium of volume 1800 m³. Measurements by varying the direction of the loudspeaker axis.

4.5 STATISTICAL MODELS. DIFFUSE-FIELD MODELS

We demonstrated in section 4.4.2 that, rising to sufficiently high frequencies, one cannot link the various maxima in the transfer functions to the individual eigenfrequencies. These higher frequency maxima are the result of many, simultaneously excited modes adding up in phase. Correspondingly, minima in the response are the results of many modes having amplitudes and phase relationship resulting in a very small vector when added. It is also very important to realize that the general features of these transfer functions such as the distribution of minima, the level difference between minima and maxima, the phase change over a given frequency range etc. is not specifically dependent on the room or the relative position of the source and receiver. A "flat" frequency response curve, which is the aim when designing microphones and loudspeakers, will never be obtained in a room.

At sufficiently high frequencies, however, we may express the abovementioned variables by statistical means. Specifically, we shall be able to do this when the distance between the eigenfrequencies becomes less than the bandwidth of the resonances. The so-called *Schroeder cut-off frequency* f_S, given by

$$f_S = 2000\sqrt{\frac{T}{V}},\tag{4.22}$$

where V and T are the volume (m³) and reverberation time (s), respectively, may be used as a frequency limit above which a statistical treatment is feasible. This corresponds to a frequency where we will find approximately three eigenfrequencies within the bandwidth of a resonance. The formula may be understood from the following facts: the resonance bandwidth is inversely proportional to the reverberation time and the separation between the eigenfrequencies is inversely proportional to the room volume. For the example used in Figure 4.3, we arrive at a cut-off frequency of approximately 250 Hz.

In building acoustics, however, we are not normally interested in a statistical description of pure tone responses for rooms. We shall look for responses averaged over frequency bands, octave or one-third-octave bands and broadband excitation sources are used. This leads to a treatment where we are looking at the energy or the energy density as the primary acoustic variable, which allows us to "forget about" the wave nature of the field as long as we keep away from the low frequency range. In this relation, it is pertinent to start by presenting a model that properly may be denoted the classical diffuse field model. It will appear that the formulae derived from this model are implemented in a number of measurement procedures both for laboratory and field use, in spite of their presumptions of an ideal diffuse field. An ideal diffuse field should imply that the energy density is everywhere the same in the room but, actually, acousticians have agreed neither on the definition nor on a measurements method for this concept. A couple of suggestions for a definition:

- In a diffuse field the probability of energy transport is the same in all directions and the energy angle of incidence on the room boundaries is random.
- A diffuse sound field contains a superposition of an infinite number of plane, progressive waves making all directions of propagation equally probable and their phase relationship are random at all room positions.

Both definitions, and a number of others, should be conceptually adequate but offer little help as to the design of a measurement method. We shall not delve into the various diffusivity measures being suggested, of which none has been generally accepted. In practice, when the international standards on laboratory measurements are concerned, procedures on improving the diffusivity are specified together with qualification procedures to be fulfilled before making the laboratory fit for a certain task. As for measurements in situ one is certainly forced to accept the existing situation.

In a number of standard measurement tasks in building acoustics, determination of sound absorption, sound insulation or source acoustic power, the primary tasks is to determine a time and space averaged squared sound pressure in addition to the reverberation time. In several cases, pressure measurements may be substituted by intensity measurements but still averaging procedures in time and over closed surfaces must be applied. Concerning the measurement accuracy of the averaged (squared) sound pressure and the reverberation time, this may be predicted using statistical models for the sound field. We shall return to this topic after treating the classical model for a diffuse sound field.

4.5.1 Classical diffuse-field model

For the energy balance in a room where a source is emitting a given power W (see Figure 4.7), a simple differential equation may be set up. This power is either "picked up", i.e. absorbed, by the boundary surfaces or other objects in the room or contributes to the build-up of the sound energy density. The boundary surfaces certainly include all absorbers which may be mounted there. We may write

$$W = \sum_j \left(W_{\mathrm{abs}}\right)_j + V \cdot \frac{\mathrm{d}w}{\mathrm{d}t}, \qquad (4.23)$$

where V is the room volume and w is the *energy density* (J/m^3) in the room. We shall, for simplicity, initially assume that the room boundaries are the only absorbing surfaces, thereby relating the first term to the absorption factors α_j of these surface areas S_j. Hence

$$\alpha_j = \frac{(W_{\text{abs}})_j}{(W_{\text{i}})_j} = \frac{(W_{\text{abs}})_j}{I_{\text{b}} \cdot S_j},$$
(4.24)

where W_{i} is the power incident on all boundaries (walls, floor and ceiling) and I_{b} is the corresponding sound intensity.

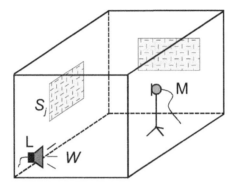

Figure 4.7 Room with a sound source, emitting a power W.

Having assumed that the energy density is everywhere the same implies that the latter quantities are independent of the position on the boundary. Equation (4.23) may therefore be written as

$$W = I_{\text{b}} \cdot \sum_j \alpha_j S_j + V \cdot \frac{\mathrm{d}w}{\mathrm{d}t} = I_{\text{b}} \cdot A + V \cdot \frac{\mathrm{d}w}{\mathrm{d}t},$$
(4.25)

where A (m^2) is the total absorbing area of the room. It remains to find the relationship between the energy density w and the intensity I_{b}. It should be noted that the total sound intensity at any position in the room is ideally equal to zero because the energy transport is the same in all directions but certainly, we may associate an effective intensity with the energy transport in a given direction. The idea is then to calculate the part of the energy contained in a small element of volume that per unit time impinges on a small boundary surface element, thereafter integrating the contributions from the whole volume. We shall skip the details in this calculation, which results in

$$I_{\text{b}} = \frac{w \cdot c_0}{4} = \frac{\frac{\tilde{p}^2}{\rho_0 c_0^2} \cdot c_0}{4} = \frac{\tilde{p}^2}{4 \rho_0 c_0}.$$
(4.26)

Additionally, we have introduced the relationship between the sound energy density and the sound pressure in a plane progressive wave, this due to our assumption that the sound field in the sound is a superposition of plane waves. As seen from the formula, the intensity at the boundaries differs only by the constant 4, different from the corresponding one in a plane progressive wave. Introducing this result into Equation (4.25) we get

$$W = \frac{\tilde{p}^2}{4\rho_0 c_0} \cdot A + \frac{V}{\rho_0 c_0^2} \cdot \frac{\mathrm{d}\left(\tilde{p}^2\right)}{\mathrm{d}t}. \tag{4.27}$$

Obviously, the pressure root-mean-square value here must be interpreted as a short-time averaged variable, i.e. the averaging must be performed over a time interval much less than the reverberation time. The general solution of this equation is given by

$$\tilde{p}^2 = \frac{4\rho_0 c_0}{A} \cdot W + K \cdot \mathrm{e}^{-\frac{Ac_0}{4V}t}. \tag{4.28}$$

The constant K is determined by the initial conditions. We shall look into two special cases, applying this solution.

4.5.1.1 The build-up of the sound field. Sound power determination

We now assume that the sound pressure is zero when the source is turned on, $(\tilde{p} = 0$ at $t = 0)$, which gives

$$K = -\frac{4\rho_0 c_0}{A} \cdot V \quad \text{and}$$

$$\tilde{p}^2 = \frac{4\rho_0 c_0}{A} W \left(1 - \mathrm{e}^{-\frac{Ac_0}{4V}t}\right). \tag{4.29}$$

The sound will then build up arriving at a stationary value when the time t goes to infinity. The RMS-value of the sound pressure becomes

$$\tilde{p}^2_{t\to\infty} = \frac{4\rho_0 c_0}{A} W. \tag{4.30}$$

The equation then gives us the possibility of determining the sound power emitted by a source by way of measuring the mean square pressure in a room having a known total absorbing area. For laboratories this type of room is called a *reverberation room* and procedures for such measurements are found in international standards (see e.g. ISO 3741).

A couple of important points concerning such measurements must be mentioned. As pointed out above, one has to determine the time and space averaged value of the sound pressure squared. This is accomplished either by measurements using a microphone (or an array of microphones) at a number of fixed positions in the room or by a microphone moved through a fixed path in the room (line, circle etc.). One must, however, avoid positions near to the boundaries where the sound pressure is systematically higher than in the inner parts of the room. Waterhouse (1955) has shown that the sound pressure level at a wall, at an edge and at a corner, respectively, will be 3,

6 and 9 dB higher than the average level in the room. This is also easily demonstrated by direct measurements. Restricting the determination of the average sound pressure level to the inner part of a room, normally half a wavelength away from the boundaries, implies that we are "losing" a part of the sound energy. One therefore finds that the standards include a frequency-dependent correction term, the so-called *Waterhouse correction* to compensate for this effect and the power is then calculated from

$$W = \frac{\tilde{p}_\infty^2}{4\rho_0 c_0} A \left(1 + \frac{Sc_0}{8Vf} \right), \tag{4.31}$$

where S is the total surface area of the room. In addition, the standard ISO 3741 includes some minor corrections for the barometric pressure and temperature and furthermore, the absorption area A is substituted by the so-called *room constant R* where

$$R = \frac{A}{1 - \dfrac{A}{S}} = \frac{A}{1 - \bar{\alpha}}, \tag{4.32}$$

and where $\bar{\alpha}$ is the mean absorption factor of the room boundaries. Normally, the mean absorption factor is required to be small for laboratory reverberation rooms making this correction also small. However, in the high frequency range (above 8–10 kHz) this may not be the case, especially due to air absorption (see section 4.5.1.3).

4.5.1.2 Reverberation time

Turning off the sound source when the stationary condition is reached, i.e. setting $\tilde{p}^2(t) = \tilde{p}_\infty^2$ at time $t = 0$, and $W = 0$ for $t > 0$, we get

$$\tilde{p}^2(t) = \tilde{p}_\infty^2 \cdot e^{-\frac{Ac_0}{4V} \cdot t}. \tag{4.33}$$

As the reverberation time T is defined by the time elapsed for the sound pressure level to decrease by 60 dB, or equivalent, that the sound energy density has decreased by a factor 10^{-6}, we write

$$\frac{\tilde{p}^2(T)}{\tilde{p}_\infty^2} = 10^{-6} = e^{-\frac{Ac_0}{4V} \cdot T}, \tag{4.34}$$

which gives us the reverberation time, commonly denoted T_{60}, as

$$T_{60} = \ln\left(10^6\right) \cdot \frac{4V}{c_0 A} \approx \frac{55.26}{c_0} \cdot \frac{V}{A}. \tag{4.35}$$

This is the famous reverberation time formula by Sabine, which is the most commonly used in practice in spite of its simplicity and the assumptions lying behind its derivation. Obviously, it cannot be applied for rooms having a very high absorption area. Setting the absorption factor equal to 1.0 for all surfaces, we still get a finite reverberation time whereas it is obvious that we shall get no reverberation at all. Other formulae have been developed taking account of the fact that the reverberation is not a continuous process but involves a stepwise reduction of the wave energy when hitting the boundary surfaces. We shall not go into detail but just refer to a couple of these formulae. The first one is denoted *Eyring's formula* (see Eyring (1930)), which may be expressed as

$$T_{Ey} = \frac{55.26}{c_0} \cdot \frac{V}{-S \cdot \ln(1-\overline{\alpha})},$$ (4.36)

where $\overline{\alpha}$ as before is the average absorption factor of the room boundaries, i.e.

$$\overline{\alpha} = \frac{1}{S} \sum_i \alpha_i S_i.$$ (4.37)

The formula is obviously correct for the case of totally absorbing surfaces as we then get T_{Ey} equal to zero. For the case of $\overline{\alpha} \ll 1$, the formula will be identical to the one by Sabine.

Still another is the *Millington–Sette formula* (Millington (1932) and Sette (1933)), where one does not form the average of the absorption factors as above but is using the average of the so-called absorption exponents $\alpha' = -\ln(1-\alpha)$. This leads to

$$T_{MS} = \frac{55.26}{c_0} \cdot \frac{V}{-\sum_i S_i \ln(1-\alpha_i)}.$$ (4.38)

One drawback of this formula is that the reverberation time will be zero if a certain subsurface has an absorption factor equal to 1.0. In practice, the absorption factors α_i have to be interpreted as an average factor for e.g. a whole wall. It is claimed (see e.g. Dance and Shield (2000)) that when modelling the sound field in rooms having strongly absorbing surfaces this formula gives a better fit to measurement data than the formulae of Sabine and Eyring.

Sabine's formula is however widely used, also by the standard measurement procedure for determining the absorption area and absorption factors of absorbers of all types (see ISO 354). By the determination of absorption factors one measures the reverberation time before and after introduction of the test specimen, here assumed to be a plane surface of area S_t, into the room. The absorption factor is then given by

$$\alpha_{Sa} = \frac{55.26 \cdot V}{c_0 S_t} \left(\frac{1}{T} - \frac{1}{T_0} \right).$$ (4.39)

T_0 and T are the reverberation times without and with the test specimen present, respectively. One thereby neglects the absorption of the room surface covered by the test specimen but this surface is assumed to be a hard surface, normally concrete, having negligible absorption. We shall return to this measurement procedure in the following chapter.

To conclude this section, we mention that various extensions of the simple reverberation time formulae have been proposed, in particular to cover situations where the absorption is strongly non-uniformly distributed in the room. A review of these formulae may be found in Ducourneau and Planeau (2003), who performed an experimental investigation in two different rooms comparing, altogether, seven different formulae. However, this number includes the three formulae presented above.

Here, we shall present just one example of the formulae particularly developed for covering the aspect of non-uniformity, a formula given by Arau-Puchades (1988). It applies strictly to rectangular rooms only and may be considered as a product sum of Eyring's formula defined for the room surfaces in the three main axis directions, *X, Y* and *Z*, each term weighted by the relative area in these directions. It may be expressed as

$$T_{AP} = \left[q \cdot \frac{V}{-S \ln(1 - \bar{\alpha}_X)} \right]^{\frac{S_X}{S}} \cdot \left[q \cdot \frac{V}{-S \ln(1 - \bar{\alpha}_Y)} \right]^{\frac{S_Y}{S}} \cdot \left[q \cdot \frac{V}{-S \ln(1 - \bar{\alpha}_Z)} \right]^{\frac{S_Z}{S}}, \quad (4.40)$$

where *q* is the factor $55.26/c_0$. Using this formula one may e.g. assign the area S_X to the ceiling and the floor having average absorption factor $\bar{\alpha}_X$, the two sets of sidewalls to the corresponding surface areas and absorption coefficients with indices *Y* and *Z*. It will appear that this formula will predict quite longer reverberation times than predicted by the simple Eyring's formula in case of low absorption on the largest surfaces of the room.

4.5.1.3 The influence of air absorption

In the derivation of the formulae above we assumed that all energy losses were taking place at the boundaries of the room. This is only partly correct as one in larger rooms and/or at high frequencies one may have a significant contribution to the absorption caused by energy dissipation mechanisms in the air itself. This is partly caused by thermal and viscous phenomena but for sound propagation through air by far the most important effect is due to *relaxation* phenomena. This is related to exchange of vibration energy between the sound wave and the oxygen and nitrogen molecules; the molecules extract energy from the passing wave but release the energy after some delay. This delayed process leads to hysteretic energy losses, an excess attenuation of the wave added to other energy losses.

The relaxation process is critically dependent on the presence of water molecules, which implies that the excess attenuation, also strongly dependent on frequency, is a function of relative humidity and temperature. Numerical expressions are available (see ISO 9613–1) to calculate the attenuation coefficient, which include both the "classic" thermal/viscous part besides the one due to relaxation. The standard gives data that are given the title atmospheric absorption, as attenuation coefficient α in decibels per metre. This is convenient due to the common use of such data in predicting outdoor sound propagation. For applications in room acoustics, we shall, however, make use of the *power attenuation coefficient* with the symbol *m*, at the same time reserving the symbol α for the absorption factor. The conversion between these quantities is, as shown earlier, simple as we find

$$\alpha = \text{Attenuation} \, (\text{dB/m}) = 10 \cdot \lg(e) \cdot m \approx 4.343 \cdot m. \quad (4.41)$$

Examples on data are shown in Figure 4.8, where the power attenuation coefficient *m* is given as a function of relative humidity at 20° Celsius, the frequency being the parameter.

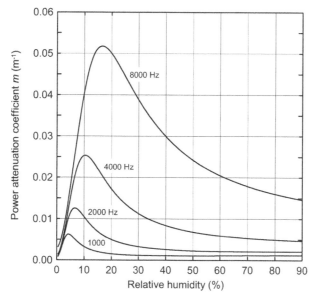

Figure 4.8 Power attenuation coefficient *m* for atmospheric absorption at 20° Celsius. Calculated from ISO 9613–1.

This atmospheric or air absorption brings about a modification of the total absorption area of a room by an added term $4mV$, where V is the volume of the room. Instead of Equation (4.35) we get

$$T_{60} = \frac{55.26}{c_0} \cdot \frac{V}{A_s + 4mV}, \qquad (4.42)$$

where A_s represent the total absorption area in the room exclusive of the air absorption. This added term may certainly also be included in other expressions for the reverberation time by modifying the denominator in the Equations (4.36) and (4.38). (How should we include the air absorption into Equation (4.40)?). Certainly, the air absorption will be important in large rooms. However, at a relative humidity in the range 20–30 %, which is not unusual at certain times of the year in some countries, one will find that the reverberation time at frequencies above 6–8 kHz, even for moderate sized rooms, will be considerably influenced by air absorption.

<u>Example</u> In a room of volume 100 m^3 one measures a reverberation time of 0.5 seconds in the one-third-octave band with centre frequency 8000 Hz. The relative humidity is 20 %. Using Figure 4.8 we find that m is equal to 0.05 m^{-1} at the frequency 8000 Hz. (The figure applies to single frequencies but we shall use it to represent the corresponding frequency band.) The air absorption alone then gives an absorption area of 20 m^2. Applying Equation (4.35) we find the total absorption area A of the room is approximately 32.5 m^2. More than half of this absorption area is then due to air absorption. Without this contribution, the reverberation time would be well over one second.

Evidently, the air absorption may have important implications on the reverberation time but also on sound pressure levels in rooms at sufficiently high frequencies. We referred in section 4.5.1.1 above to the standard ISO 3741 on sound power determination in a reverberation room, where a correction factor $(1 - \bar{\alpha})$ was applied to the absorption area (see Equation (4.32)). Vorländer (1995) has shown that this correction factor is an approximation of the general term $\exp(A/S)$, where the absorption area is given by

$$A = -S \cdot \ln(1 - \bar{\alpha}) + 4mV. \qquad (4.43)$$

If m equals zero, we certainly arrive at the correction term in Equation (4.32) again as

$$\mathrm{e}^{-\frac{A}{S}} = \left[\mathrm{e}^{\frac{-S\ln(1-\bar{\alpha})+4mV}{S}} \right]_{m=0} = 1 - \bar{\alpha}. \qquad (4.44)$$

Using this general correction, Vorländer (1995) obtains a very good fit, even up to 20 kHz, between the sound powers of a reference sound source determined in a reverberation room as compared with a free field determination.

4.5.1.4 Sound field composing direct and diffuse field

When deriving Equation (4.28), we assumed that the sound field was an ideal diffuse one; the energy density was everywhere the same in the room. It is obvious, however, that the source must represent a discontinuity; even in a room having a very long reverberation time there must exists a direct sound field in the neighbourhood of the source. We shall have to distinguish between the source near field, where the sound pressure may vary in a very complicated manner depending on the type of source, and the far field where the sound pressure decreases regularly with the distance from the source (see the discussion on sound sources in Chapter 3).

Assuming a position in the far field, we may apply the formula describing the relationship between the source sound power and the pressure squared in an ideal spherical (or plane) wave field:

$$W = \oint \frac{\tilde{p}^2}{\rho_0 c_0} \cdot \mathrm{d}S. \qquad (4.45)$$

Initially, we shall assume that the source is a monopole, hence

$$\tilde{p}^2 = W \rho_0 c_0 \frac{1}{4\pi r^2}. \qquad (4.46)$$

For other types of source, we may introduce a *directivity factor* D_θ, thus write

$$\tilde{p}^2 = W \rho_0 c_0 \frac{D_\theta}{4\pi r^2}, \qquad (4.47)$$

where r is the distance from the source. The index θ on the directivity indicates that the latter generally depends on a properly defined angle. Combining this expression with the simple one giving the pressure in a diffuse field, Equation (4.30), we arrive at the following expression for the total sound field:

$$\tilde{p}_{\text{tot}}^2 = W \rho_0 c_0 \left(\frac{D_\theta}{4\pi r^2} + \frac{4}{A} \right). \tag{4.48}$$

Expressed by the corresponding levels using standardized reference values for sound pressure and sound power, we may write

$$L_p = L_W + 10 \cdot \lg \left(\frac{D_\theta}{4\pi r^2} + \frac{4}{A} \right). \tag{4.49}$$

For simplicity, we have given the characteristic impedance $\rho_0 c_0$ the value 400 Pa·s/m. The difference between the sound pressure level and the sound power level is shown in Figure 4.9 as a function of the relative distance $r/(D_\theta)^{1/2}$. The parameter on the curves is the absorption area A. The dashed curve indicates the relative level of the direct field.

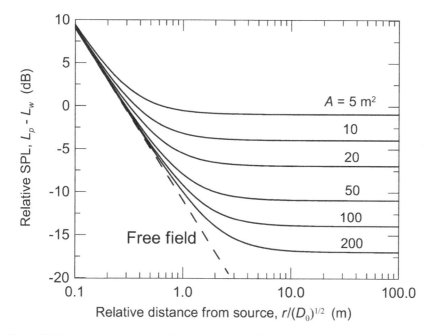

Figure 4.9 Sound pressure level as a function of relative distance from a source of sound power level L_W. The parameter is the room total absorption area A. The dashed line indicates the free field level.

The distance r_H from the source, where the contributions from the direct field and the diffuse field are equal, is called the *hall radius* or also *room radius* for the case where the directivity factor is equal to 1.0:

$$r_H = \sqrt{\frac{A}{16\pi}} = \sqrt{\frac{55.26}{16\pi c_0} \cdot \frac{V}{T}}. \tag{4.50}$$

Example In Figure 4.9 we calculated the difference between the sound pressure and the power levels for the cases where the absorption area varies between 5 and 200 m². Correspondingly, the room radius will vary between 0.32 and 1.99 metres.

4.5.2 Measurements of sound pressure levels and reverberation time

As pointed out in the introduction to section 4.5, the formulae derived using simple diffuse models are used in a number of measurement tasks both in the laboratory and in the field. Quantities such as sound pressure squared and reverberation time are considered, subject to certain presumptions, as global measures but in the sense of being average values with a space variance. We shall therefore have means to estimate this variance to be able to predict the uncertainty in the end results, results obtained by sampling the sound field in the room at a number of microphone positions.

Instead of sampling the sound field in a number of fixed positions, one may use a microphone moving continuously through the room. As the pressure is strongly correlated at adjacent positions, positions within some half a wavelength apart, implies that no new information is gained from close lying positions. The length of the path covered by such a microphone must therefore be carefully chosen by keeping this in mind. We shall return to this question later on, first, treating the case of using discrete sampling of the sound field to determine the average sound pressure squared and the reverberation time.

One may use several quantities to characterize the measurement uncertainty. It should also be noted that the expressions for the variance (or standard deviation) may be given as a relative value or not, which means that they are stated relative to the mean value or not. The relative variance of an actual quantity x shall be defined as

$$\sigma_r^2(x) = \frac{E\left\{\left(x - E\{x\}\right)^2\right\}}{E\{x\}^2}, \tag{4.51}$$

where the index r indicates a relative value and $E\{\ldots\}$ the expectation value. The square root of this expression is denoted the relative, sometimes the normalized, standard deviation. The symbol s is commonly used to indicate the standard deviation, indicating that practical calculations comprises a limited selection of data enabling us just to estimate the underlying expectation value.

4.5.2.1 Sound pressure level variance

An early effort to predict the space variance of the squared sound pressure is due to Lubman (1974), working on the determination of sound power level of sources in a reverberation room. At frequencies above the Schroeder cut-off frequency f_S (see Equation (4.22)) he found a relative variance of 1.0 for pure tone sources assuming that p^2 was exponentially distributed. The corresponding standard deviation $s(L_p)$ of the sound pressure level is then approximately equal to 5.6 dB, which implies that the 95% confidence interval will be as large as 22 dB. It should not come as a surprise that sound power level determination of pure tone sources present special problems in order to arrive at a reasonably correct space averaged value. Sources having a larger bandwidth will tend to "smear out" these space variations, thereby making the measurement task considerably easier. We shall present expressions below taking the bandwidth into account.

The Schroeder cut-off frequency represents an important division in the prediction of the variance. A satisfactory theory does not exist which covers the frequency range below this cut-off frequency. However, we shall present an estimate also for this range, a range where investigations are best conducted by FEM modelling. As for the frequency range above f_S, statistical models will have limited validity if the absorption becomes so large that the direct field is significant, which may happen at sufficiently high frequencies.

Lubman (1974) presented the following expressions for the relative variance:
For the range given by $0.2 \cdot f_S \leq f \leq 0.5 \cdot f_S$ he got

$$\sigma_r^2\left(p^2\right) = \frac{1}{1 + \dfrac{\Delta N}{\pi}}, \tag{4.52}$$

where ΔN is the number of natural modes inside the frequency band Δf (see Equation (4.14)). As for the range $f \geq f_S$ he found

$$\sigma_r^2\left(p^2\right) = \frac{1}{1 + \dfrac{\Delta f \cdot T}{6.9}}, \tag{4.53}$$

where T is the usual reverberation time. It should be noted that both expressions presuppose that the product $\Delta f \cdot T$ is numerically equal or larger than 20.

Normally, one is looking for the corresponding standard deviation $s(L_p)$ in the sound pressure level. However, to calculate this one needs to know the probability distribution of p^2. If the relative variance is less than approximately 0.5 we may make an estimate based on transforming the sound pressure level in the following manner:

$$L_p = 10 \cdot \lg\left(\frac{p^2}{p_0^2}\right) \tag{4.54}$$

into $\quad L_p = 10 \cdot \lg(e) \cdot \ln(p^2) - 10 \cdot \lg\left(p_0^2\right) \approx 4.34 \cdot \ln(p^2) - 10 \cdot \lg\left(p_0^2\right).$

Differentiating the last expression with regards to p^2, we get

$$\frac{d\left(L_p\right)}{d\left(p^2\right)} = 4.34 \cdot \frac{1}{p^2} \quad \text{or} \quad s\left(L_p\right) = 4.34 \cdot \frac{s\left(p^2\right)}{p^2} = 4.34 \cdot \sqrt{\sigma_r^2\left(p^2\right)}. \tag{4.55}$$

Up until now we have concentrated on the spatial variance. In measurements on stochastic signals there will also be a corresponding relative time variance given by

$$\sigma_t^2\left(p^2\right) = \frac{1}{\Delta f \cdot T_i}, \tag{4.56}$$

where T_i is the measuring or integration time used to determine p^2 in a given microphone position. Certainly, we are able to make this time variance arbitrarily small by extending the measuring time but there is, of course, a trade-off here. In practice, one normally chooses a measurement time making the time variance some one-tenth of the expected spatial variance.

If the task is to determine the average stationary sound pressure level in a room set up by a given source, we may choose a number M of microphone positions. Assuming that the sound pressures at these positions are uncorrelated, i.e. the positions are some half a wavelength apart, we may estimate the relative variance in the mean value by the following equation

$$\sigma_r^2\left(\overline{p^2}\right) = \frac{\sigma_r^2\left(p^2\right) + \sigma_t^2\left(p^2\right)}{M}, \qquad (4.57)$$

where we may insert the actual contributions to the variance from the Equations (4.52), (4.53) and (4.56).

The spatial variance expressions given above were developed in connection with the problem of sound power determination in reverberation rooms, i.e. a typical laboratory set-up in hard-walled rooms. They may, however, also be applied to field measurement such as sound insulation between dwellings, from which we shall give some examples taken from a NORDTEST report (see Olesen (1992)). The main content of this report may now be found in the standard ISO 140 Part 14.

In this report, however, some modifications are introduced in the above expressions when calculating the standard deviation $s(L_p)$. In Equation (4.52) the factor π is substituted by the number 8.5, which is claimed to give a better fit to experimental data. Furthermore, an additional term is introduced into Equation (4.53) allowing for a possible influence of the direct field from the source. Figures 4.10 and 4.11 show the results; the measured and the predicted standard deviation of the sound pressure level in two rooms having widely different volumes. Taking the valid range of the theoretical expressions into account, the fit between measured and predicted data are reasonably good. As for the smallest sized room, the expressions are not valid below approximately 150 Hz. For the larger room, there are also some discrepancies in the higher frequency range, most probably due to a relatively high and unevenly distributed absorption (carpeted floor). All results are based on measurements using five microphone positions for each of the two source positions used.

Apart from the determination of sound power of sources in reverberation rooms and the determination of sound insulation, great effort has been put into finding accurate methods for determination of sound pressure levels from service equipment in buildings. Service equipment noise normally involves low frequency components and small rooms makes a correct sampling of the room important, this is so even if legal requirements are commonly specified by the overall A- or C-weighted sound pressure levels. It has been shown (see e.g. Simmons (1997)) that combining a few microphone positions in the room with a corner position, the corner having the highest C-weighted sound pressure level, is an efficient procedure both with respect to the correct average value (less bias error) and to the reproducibility. This procedure has been adopted by the international standard ISO 16032.

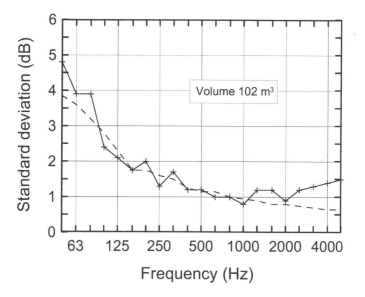

Figure 4.10 Spatial distribution of sound pressure level. Furnished living room with carpet, volume 102 m³. Solid curve – measured standard deviation. Dashed curve – predicted standard deviation. After Olesen (1992).

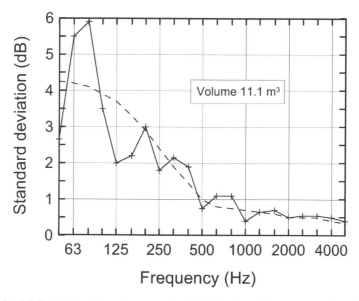

Figure 4.11 Spatial distribution of sound pressure level. Toilet with hard room boundaries, volume 11.1 m³. Solid curve – measured standard deviation. Dashed curve – predicted standard deviation. After Olesen (1992).

4.5.2.2 Reverberation time variance

Measurements of sound decay and reverberation time in rooms are performed either by using a method based on an interrupted noise signal or by a method based on the integrated impulse response, specifically by

- exciting the room using a stochastic noise signal, usually filtered in octave or one-third-octave bands, and recording the sound pressure level after turning off the source, i.e. the method outlined when deriving the reverberation time formula in section 4.5.1.2
- measuring the impulse response, using either a maximum length sequence signal (MLS signal) or a swept sine signal (SS signal), which again is filtered in octave or one-third-octave bands, thereafter applying the method given in section 4.3.1.

As for the first method concerned one will, due to the stochastic noise of the signal, observe variations in the results when repeating the measurement. This will be the case even if both source and microphone positions are exactly the same. The reason is that the stochastic signal is stopped at an arbitrary time making the room excited by different "members" of the ensemble of noise signals produced by the source. It makes no difference if the stochastic signal in fact is pseudo stochastic, i.e. periodically repeats itself, as the source normally is not stopped coincident with this period. The variance due to the variation in the reverberation time measured at a given position we shall call an ensemble variance. This quantity $\sigma_e^2(T)$ is therefore an analogue of the time variance $\sigma_t^2(p^2)$ by a sound pressure measurement (see Equation (4.56)).

By measuring the reverberation time using M microphone positions, repeating each measurement N times in each position, the relative variance in the average reverberation will be given by

$$\sigma_r^2\left(\overline{T}\right) = \frac{\sigma_r^2(T) + \dfrac{\sigma_e^2}{N}}{M}, \tag{4.58}$$

where the first term is the variance due to the spatial variation. It should be noted that the last term will be zero when using an impulse response technique as the excitation signal will be deterministic in this case. This does not, however, imply that systematic errors cannot occur in this case if the system is not *time invariant*, e.g. due to temperature changes etc. during the measurement. The SS technique is less prone to such errors than the MLS technique.

Returning to the method of using interrupted noise, Davy et al. (1979) developed theoretical expressions for the two contributions to the variance, applicable to frequencies above the Schroeder frequency f_S. In effect, they calculated the variance of the slope of the decay curves but the results may easily be transformed to apply to the corresponding reverberation time. As expected, these expressions are functions of the filter bandwidth and reverberation time but also depends on the time constant (or "internal reverberation time") of the measuring apparatus together with the dynamic range available. It has to be remembered that at the time when this work was performed the equipment available was of analogue type such as the level recorder. We shall therefore just give an example applicable for one-third-octave measurements, using a dynamic range of 30 dB and a RC detector (exponential averaging). The time constant of this detector is assumed to be one-quarter of the equivalent time constant for the room. The relative variance of the mean reverberation time may then be written (Vigran (1980)) as

$$\sigma_r^2\left(\overline{T}\right)=\frac{1.31+\dfrac{1.94}{N}}{f_0\overline{T}M},\qquad(4.59)$$

where f_0 is the centre frequency in the one-third-octave band. This expression is also used in the report by Olesen (1992) comparing with measurement results obtained in a small laboratory room of volume 65 m³, having an almost frequency independent reverberation time of two seconds. The numbers N and M of source and microphone positions were two and six, respectively. The result is shown in Figure 4.12, given by the reverberation time standard deviation, i.e. by the expression

$$s(T)=\sigma_r\left(\overline{T}\right)\cdot\overline{T}\cdot\sqrt{M}\ ,$$

and as seen, the fit between measured and predicted results is quite good.

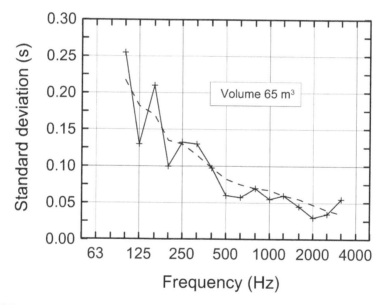

Figure 4.12 Reverberation time standard deviation in a laboratory room of volume 65 m³. The reverberation time is approximately frequency independent (2 seconds). Solid curve – measured. Dashed curve – predicted. After Olesen (1992).

4.5.2.3 Procedures for measurements in stationary sound fields

As is apparent from the discussions above, a number of the standard measurement tasks in building acoustics; e.g. sound insulation, sound absorption and noise measurements, are based on determination of the spatial averaged sound pressure squared and the reverberation time. In the following, we shall use the sound pressure as an example.

We shall further assume that measurements are performed on band-limited stochastic noise. This may comprise measurements on a broadband source of unknown sound power where we apply filtering in octave or one-third-octave bands for the analysis; e.g. in a sound power determination in a reverberation room. In other cases, we shall set up a sound field in a room with a loudspeaker driven by a narrowband signal. In the latter case, we may alternatively measure the impulse response (between the loudspeaker source signal and the signal from the microphone) using MLS or another deterministic signal. The latter procedure is certainly superior when the task is to determine *differences* in the squared sound pressures, e.g. when determining the airborne sound insulation between two rooms.

Regarding a spatial averaged value as a reasonably global one for the room presupposes that the room dimensions are of the same order of magnitude. This means that in those rooms where the dimensions are too different, a corridor, an open plan office or school, a factory hall etc., one will never, using a single source, find areas where the sound pressure level is constant (in the statistical sense of the word). We will experience a systematic variation; the sound pressure level will decrease more or less rapidly with the distance from the source depending on the room shape, the absorption and the presence of scattering objects. We shall return to this subject in section 4.9.

In most measurements standards, the required end result is the mean sound pressure level and quantities derived from it. The underlying quantity, however, is the mean squared pressure. In principle, we may proceed in two ways: We may sample the sound field in a number M of microphone positions, which we in fact assumed when deriving the expressions above, thereby calculating the mean sound pressure using the formula

$$\bar{L}_p = 10 \cdot \lg \left[\frac{\sum_{i=1}^{M} p_i^2}{M p_0^2} \right] \quad \text{(dB)}, \tag{4.60}$$

where p_i^2 denotes the time averaged squared pressure in position i. Alternatively, we may use a microphone moving along a certain path in the room, performing a continuous averaging process in time and space. We will then write

$$\bar{L}_p = 10 \cdot \lg \left[\frac{\dfrac{1}{T_{\text{path}}} \displaystyle\int_0^{T_{\text{path}}} p^2(t)\, dt}{p_0^2} \right] \quad \text{(dB)}, \tag{4.61}$$

where T_{path} is the time used for the complete path.

How do we compare these two methods as to the measuring accuracy? If a given length of the path could be attributed to a certain equivalent number M_{eq} of discrete positions we could apply the equations given in section 4.5.2.1 directly for the calculation of the standard deviation according to Equation (4.57). The time averaging term σ_t should not give any problem as the total measuring time is $T_{\text{path}} = T_i \cdot M$, but how long should the path be to correspond to M positions spaced at a distance ensuring uncorrelated sampling? This may be calculated for frequencies above the Schroeder frequency f_S and for a circular path, which is the most practical one, we approximately (perhaps not particularly surprising) get

$$M_{eq}(\text{circular path}) \approx \frac{4\pi r}{\lambda} = \frac{4\pi r f_0}{c_0}. \qquad (4.62)$$

The quantities r and f_0 are the path radius and the centre frequency in the actual frequency band, respectively. A microphone path corresponding to three discrete microphone positions at 100 Hz should, therefore, have a radius of approximately 0.8 metres.

Accurate estimates of the measurement accuracy at frequencies below f_S are difficult to attain, but there are guidelines in measurement standards to improve the accuracy (see below). As seen from Equation (4.52), the number of modes excited is vital, and exciting the room by band-limited noise will certainly excite most modes inside the frequency band. However, we have seen that a source cannot excite a mode having a node at the source position. This is one reason for the requirements in standards to use several source positions, which is particularly important when measuring at low frequencies. It should not come as a surprise that some laboratories are, using not only a moving microphone but also a moving source.

Eventually, at sufficiently low frequencies, the number of modes will be too small to realistically speak of a space averaged value of the squared pressure. The exception is when the frequency gets so low that there will only be a homogeneous pressure field in the room, i.e. when going below the first eigenmode for the room.

Guidelines and help on these questions are given in national and/or international standards. These give guidance and requirements as to the choice of measuring positions and source positioning; the number of these depending i.a. on frequency and room volume, the distance of microphone positions from the source and from the room boundaries etc. Information is also given on the measurement uncertainty of the procedure or method. Concerning the latter, one will find the concepts of *repeatability* and *reproducibility* standard deviation. The former implies the standard deviation obtained when repeating a given procedure within a short time interval and under identical conditions (same laboratory, same operator, same measuring equipment). Otherwise, when these conditions are unequal, we have reproducibility conditions. The standard deviation of reproducibility therefore includes the standard deviation of repeatability. Data for reproducibility are usually established by round robin experiments by a number of participating laboratories.

To conclude, one will find the necessary instructions in the relevant standards to perform most measurement tasks. The purpose of dealing in some detail with the basis for these measurements are twofold: to give some understanding of the formulations, found in these standards, at the same time give some assistance when presented with a measurement task not covered by any standard.

4.6 GEOMETRICAL MODELS

A number of computer software programs, of which many are commercially available, are developed to predict sound propagation in large rooms, e.g. concert halls or large factory spaces. We shall not present any overview of the various programs or deal with specific published work where these programs are used but limit ourselves to give an outline of the principles behind the models. The majority of prediction models used for large rooms are based on geometrical acoustics, partly combined with statistical concepts to include scattering effects. Judged by the concepts found in the literature dealing with these prediction models, there may be some confusion as to the number of basic methods used. In effect, there are only two basic methods, the *ray-tracing method* and the *image-*

source method. The models implemented in software programs are, however, given special names depending on the specific algorithm used and furthermore, there exist *hybrid* types combining principles from ray-tracing and image-source modelling. A review on computer modelling of sound fields is given a journal special issue (see Naylor (1993)).

4.6.1 Ray-tracing models

A pioneering work on computer modelling using the ray-tracing method is from Krokstad et al. (1968). Calculation involving ray tracing is based on simulating a point source emitting a large number of "rays" evenly distributed per unit solid angle. Each ray then represents a given solid angle part of the spherical wave emitted from the source. The rays are "followed" on their way through the room, either through a sufficiently long time span or until they hit a surface defined as totally absorbing (see Figure 4.13). The seating area in, for example, a concert hall, is a surface of the latter type. What is a "sufficiently" long time if such a surface does not exist? Pragmatically, one may choose the time according to the energy left in the ray after a certain time interval but there are also implementations where the last surface point hit is defined as a new source, in its turn emitting the rest energy of the ray, contributing to the reverberant energy in the room.

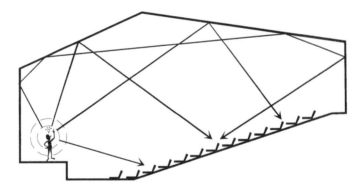

Figure 4.13 The principle of ray tracing.

A major problem using ray tracing is that a ray, per definition, has no extent, i.e. in practice it almost never hits a receiving point. This implies that the detectors ("microphones"), which shall record the rays hitting a given surface and thereby the magnitude and direction of the intensity, must be quite large. One may apply spherical microphones having a diameter in the range of one metre. Certainly, applying a very large number of rays, one may reduce the diameter but there is also the question of calculating time. There are alternative measures, such as using a beam having the shape of a cone of pyramid, but in effect, these are models of a hybrid type (see below).

One will also encounter the notion of "sound particle" instead of the ray and thereby the concept of sound particle tracing (see e.g. Stephenson (1990)). The algorithm to calculate the trajectories is the same; the sound particles or *phonons* propagate along rays. The differences are found on the receiving side; i.e. how the detectors are arranged and how the energy is calculated. In principle, however, it is still a ray-tracing method.

4.6.2 Image-source models

Image-source (or mirror-source) modelling is based on regarding all reflections from the boundary surfaces as sound contributions from images of the real source(s). The strength of this type of modelling, when carried out rigorously, is that it covers all transmission paths between source and receiver. It may give the impulse responses correct inside the framework of geometrical acoustics.

It is relatively simple mathematically to find all these mirror sources. The main problem is that except for rooms of very simple shapes, most of these sources are either not visible in a given receiver position or may be invisible in any part of, for example, the audience area. This means that a number of reflections are not physically valid. To separate out the "valid" image sources is a time-consuming task when coming to the higher order reflections. We may illustrate this by calculating the number of image sources of the order N in a room having M surfaces, which is given by $M(M - 1)^{N-1}$. In a room having e.g. M equal to 12, we get approximately 16 000 image sources of the forth order, approximately 175 000 of the fifth order and so on. Except for rooms having a very simple shape, e.g. rectangular ones, maybe only a few hundred of these sources are valid. As in the case of ray tracing the question arises on when to stop the calculations. "Adding on" to the results using statistical arguments are common having carried out calculations correctly up to a given order of reflections.

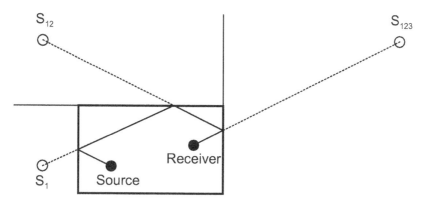

Figure 4.14 Example on trajectory between a receiver and a third-order image source.

Sketches which illustrate some of the aspects discussed above are shown in Figures 4.14 and 4.15. The first one shows, in a cross section (horizontally or vertically) through a room of rectangular shape (parallelepiped), an example of the trajectory between a receiver and a third-order image source. Figure 4.15 gives an example on a first-order image source S_1 (mirrored in wall W), which is not visible in any of the possible receiver positions R within the indicated sector, a sector given by the solid angle defined by the wall surface as seen from the image source.

Finding the image-source positions is in many cases quite easy where regular room shapes are concerned and one may also find analytical expressions as to the sound propagation. An example that we shall also use later on (see section 4.8) is sketched in Figure 4.16, which shows a vertical section of a long "flat" room. Here we shall assume that the ceiling height is much smaller than the other dimensions of the room; i.e. we shall neglect the influence of the sidewalls.

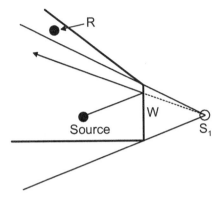

Figure 4.15 Example on image source not being visible in receiver positions *R*.

We shall put a source midway between the floor and the ceiling, initially assuming that the absorption factor α is the same for these boundary surfaces. The energy density *w* at a receiver position may then be expressed by

$$w = \frac{W}{4\pi c_0}\left[\frac{1}{r^2} + 2\sum_{n=1}^{\infty}\frac{(1-\alpha)^n}{r_n^2}\right], \tag{4.63}$$

where *W* is the source sound power, *r* and r_n are the distances between the source and the receiver and between the receiver and the image source with index *n*, respectively.

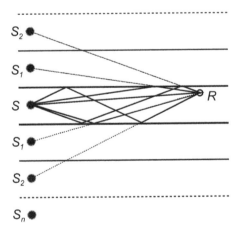

Figure 4.16 Image sources in a "flat" room.

4.6.3 Hybrid models

A number of the computer programs for room acoustic predictions are based on models that we may characterize as being hybrid; they comprise elements from ray-tracing methods as well as from image-source methods. An important aspect when developing such programs is to reduce the computing time.

A common practice is initially by finding available image sources by following ray trajectories, thereby noting the points on the boundaries hit by these rays. Thereafter, one is testing whether these reflection sequences will contribute to the energy in a given receiver position in the same manner as when using a pure image-source method. One makes use of a beam, either in the form of a cone or a pyramid, where the ray itself represents the axis. At each reflection, the highest point in the beam will represent an image source. This approach makes it possible to work with receivers represented by points, not as a large sphere necessary in a pure ray-tracing model. Certainly, the approach is not without its problems. The number of beams is certainly finite, making it possible to find only a limited number of image sources. Another problem is that the ray direction following a reflection is solely determined by the axis of the beam, which implies that the beam is not split up when it hits two or more surfaces. This makes it possible for some image sources to "illuminate" and thereby contribute to the energy in receiver points that in effect are not visible. And, vice versa, some image sources may not illuminate receivers that in fact should be visible. For a closer description of the procedure, see for example, Lewers (1993).

The necessary finite number of rays or beams will impose a limit on the accuracy of the calculated impulse response. One therefore has to apply other methods to add a reverberant "tail" to the response. This is coupled to the aspect of adding some diffuse reflections to the response. Obviously, scattering phenomena have strictly no place in geometrical acoustics but certainly being present in real rooms due to surface irregularities and objects filling the room. A strong element of diffuse reflections is also important in performance spaces such as concert halls etc., making it necessary by some artifice also to implement this aspect in the prediction models, mainly by some statistical type of reasoning.

4.7 SCATTERING OF SOUND ENERGY

With the concept of diffraction, it is generally understood that changes are taking place in the direction of sound propagation, thereby including both the concept of reflection and scattering. As to the former, one assumes that the dimensions of the reflecting surface are large as compared with the wavelength, the reflection is considered to be specular. The word scattering is commonly used when the dimensions of the surface or object hit are comparable or less than the wavelength. As pointed out above, scattering has strictly no place in geometrical acoustics. By e.g. ray-tracing modelling there is certainly no impediment for not making the reflection specular; the ray may be reflected in a random direction, however a physical reason for allowing such a diffuse reflection must exist.

Several hybrids models (see e.g. Heinz (1993); Naylor (1993b)) combine a strict calculation using specular reflections together with the addition of a certain number of such diffuse reflections. When modelling the sound field in large assembly halls, concert halls etc. one might say that the inclusion of diffuse energy is justified by the necessary partially detailed description of the room. In addition, scattering phenomena certainly exist when increasing the frequency and the wavelength is becoming comparable to the size of objects. The energy in the incident wave will be redistributed with a directional distribution depending both on the shape of the object and on the ratio of wavelength to object dimensions.

Since 20 years ago, there has been a growing awareness that diffuse reflections are very important, especially for rooms for music performances. It is realized that an important contribution to the fame of some older concert halls, e.g. the Grosser Musikvereinsaal in Vienna, is the diffuseness provided by numerous surface irregularities: various types of surface decoration, columns, balconies etc. Following the work of Schroeder (1975, 1979) on the design of artificial diffusing elements based on number theory, a range of commercial as well as non-commercial diffusing elements are now in use in rooms for music production and reproduction. A comprehensible treatise may be found in Cox and D'Antonio (2004). Here we shall just give a short overview on these types of diffuser element. In this connection a series of measurement methods are developed to characterize the acoustic properties of such elements both in ISO (ISO 17497) and in AES (Audio Engineering Society).

4.7.1 Artificial diffusing elements

The sound scattering properties of solid bodies and surfaces is of great interest in many areas of acoustics and the distribution of the scattered energy around structures of various shapes for a given incident wave is well known. Such distributions are normally given in the form of a directivity pattern for the scattered wave. In room acoustic modelling, however, one is in most cases not interested in such a detailed pattern. A surface property of major interest is the total amount of non-specularly reflected sound energy in relation to the total reflected energy. In ISO 17497 Part 1 a quantity named the *scattering coefficient* s is defined,[1] as one minus the ratio of the specularly reflected acoustic energy E_{spec} to the total reflected acoustic energy E_{total}:

$$s = 1 - \frac{E_{spec}}{E_{total}}. \tag{4.64}$$

Theoretically, this quantity can take on values between zero and one, where zero means a totally specular reflecting surface and one means a totally scattering surface. Being measured in a reverberation room as a random incidence quantity in one-third-octave or octave bands, it represents a direct analogue to the statistical absorption factor.

The main purpose of the artificially diffusing elements is certainly to reduce the specularly reflected energy. However, from the point of view of the producers of such elements one would like to have a corresponding measure characterizing the uniformity of the reflected sound, in the same way as characterizing radiated sound from sources, e.g. loudspeakers. There seems as yet no universal agreement concerning such a *diffusion coefficient* (or *factor*) to characterize these so-called diffusers but there is ongoing work e.g. inside ISO. The problem is to arrive at a single number measure characterizing the scattering directivity pattern.

These artificial types of diffuser element constitute a hard surface with grooves or protrusions of various shapes. The surface irregularity used may be one-dimensional or two-dimensional, according to the task of making a diffuser working in one or two

---

[1] Having the unit of 1, it should have been termed *scattering factor*.

planes. We shall confine ourselves to the first type, as the extension to two dimensions is reasonably straightforward, conceptually at least.

Schroeder (1975) began his work on what we may term mathematical diffusers by investigating the scattering from surfaces shaped in the form of a maximum length sequence (MLS). We showed in section 1.5.2 the particular Fourier properties of these sequences giving a completely flat power spectrum. Then, quoting Schroeder: "Thus, because of the relation between the Fourier transform and the directivity pattern, a wall with reflection coefficients alternating between +1 and −1, would scatter an incident plane wave evenly (except for a dip in the specular direction which corresponds to the DC component in the spectrum)." The "MLS wall" was realized as a hard wall with "grooves" or wells a quarter of a wavelength deep in the area where a reflection factor of −1 was called for. In practice, such diffusers work, however, over a rather limited frequency range, approximately one octave. There are means of increasing the workable bandwidth, as recent research shows, but this implies adding active components to the diffuser (see Cox et al. (2006)).

However, there are other periodic sequences having useful Fourier properties, which make them excellently suited for modelling diffusing elements having a much broader bandwidth than the MLS. These are the *quadratic residue* sequences and the *primitive root* sequences (see e.g. Schroeder (1999), Cox and Antonio (2004)). The sequence forming the base for making a quadratic residue diffuser (QRD) is given by

$$s_n = m^2 \text{MOD } N \quad \text{where} \quad m = 1,2,3,... \tag{4.65}$$

This means that s_n is the reminder when m^2 is divided by the prime number N. Taking $N=7$ as an example, we get the following sequence: 0, 1, 4, 2, 2, 4, 1. In a similar way as for the MLS diffuser the numbers are transformed into the corresponding depths d_n of the grooves or wells of the surface, but these are now not constant:

$$d_n = s_n \frac{d_{\text{max}}}{(s_n)_{\text{max}}}. \tag{4.66}$$

So how do we choose the maximum depth d_{max} and also the width of each well? Certainly, to make the diffuser work properly there should be plane wave propagation in each well and there must be a significant phase change for the waves reflected from the bottom. The design rule normally used for the latter, which determines the maximum workable wavelength or the equivalent minimum frequency, is expressed as:

$$d_n = \frac{s_n \lambda_{\text{max}}}{2N} \quad \text{or} \quad f_{\text{min}} = \frac{s_n c_0}{2N d_n}, \tag{4.67}$$

where c_0 is the speed of sound. This design rule implies that the mean depth of the wells at this frequency is of the order of a quarter of a wavelength. As for the width w of each well, we should ensure plane wave propagation, which implies being below the cut-off frequency giving

$$w = \frac{\lambda_{\text{min}}}{2} \quad \text{or} \quad f_{\text{max}} = \frac{c_0}{2w}. \tag{4.68}$$

The width w is normally chosen in the range of 5–10 cm. Making the wells too narrow may increase the surface area too much giving unwanted surface sound absorption, especially when the wells have separating walls (see Figure 4.17 a).

The other type of sequence having Fourier properties that makes them useful in the construction of broadband diffusers, giving little specular reflections, is the primitive root sequences. These are calculated in a slightly different way than the quadratic residue ones, given by:

$$s_n = p^n \text{MOD } N \quad \text{where } n = 1, 2, 3, \ldots \tag{4.69}$$

The number p is denoted a primitive root modulo N, also called a generating element because it generates a complete residue system in some permutation. As an example, choosing N equal to 7 there are two primitive roots, being 3 and 5. We shall use a higher number N in our example below, choosing N equal to 13 where the lowest primitive root is 2. Using (4.69) to calculate this sequence gives the values shown in Table 4.1.

Table 4.1 Primitive root sequence for N equal 13 and primitive root p equal 2. Well depths in mm for design frequency 1000 Hz.

n	1	2	3	4	5	6	7	8	9	10	11	12
s_n	2	4	8	3	6	12	11	9	5	10	7	1
d_n (mm)	28	57	113	43	85	170	156	128	71	142	99	14

In the last row the corresponding depths of the wells are given, calculated by equation (1.3) choosing a design frequency (f_{min}) of 1000 Hz. It should be noted that there is only $N - 1$ cells in the sequence. As is apparent from the table and also from Figure 4.17, where we have put three such periods on a row, diffusers based on a primitive root sequence (PRD) are unsymmetrical.

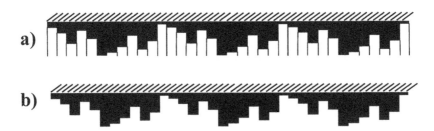

Figure 4.17 Sketch of a ceiling having three periods of a primitive root diffuser (PRD) with N equal 13. a) With dividing walls between the wells (grooves), b) Without dividing walls.

Prediction methods for the acoustic pressure field, i.e. the sum of the direct field from a source and the scattered field, is normally based on the Helmholtz-Kirchhoff integral equation (see e.g. Cox and Lam (1994)). This means using Equation (3.44) in Chapter 3 with an added term representing the direct field. If only the far field is of interest, a computational method based on the analogue Fraunhofer diffraction method in optics may be used. We shall not treat any of these methods here, but to illustrate the effect of these diffusers, especially to reduce the specular reflection, we shall present an example based on the FEM technique in two dimensions. The situation is depicted in Figure 4.18, showing the same three periods of the PRD depicted in Figure 4.17, where the wells (protrusions) are calculated in Table 4.1.

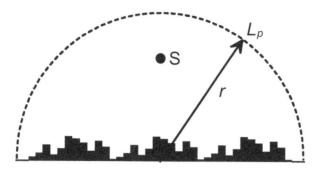

Figure 4.18 Sketch of situation for calculating the sound pressure level above a diffuser surface consisting of three periods of a primitive root sequence. Height of wells is given in Table 4.1 and height of point source (S) is 0.7 metres.

The resulting sound pressure level from a point source at height 0.7 metres is calculated on a circle with a radius of 1 metre above the diffuser. As the width of the wells is chosen equal to 5 cm, there will be a 10 cm flat (hard) surface added to each end of the diffuser. The calculations were performed using the Comsol Multiphysics™ software, modelling the field outside the semicircle to be a free field by adding a so-called perfectly matched layer (PML).

The results are shown in Figure 4.19, giving the total sound pressure level, at a design frequency of 1000 Hz, on the half-circle as a function of angle. The source acoustic power is arbitrarily set to 1 W, thus giving the rather high sound pressure levels. The FEM calculations are performed both for the situation described and also for a flat surface. The results are compared with a simple analytical calculation for an infinitely large flat surface. Apart from the discrepancies around the main lobe, the FEM calculations predict the flat surface situation quite well. The most important result, however, is the effect of the diffuser surface as compared by the flat one, giving a mean difference in the specular direction in the order of 6–8 dB.

4.7.2 Scattering by objects distributed in rooms

Big industrial halls, either production or assembly spaces, will always contain a large number of scattering objects. A realistic modelling of the sound propagation in such halls implies that one has to take scattering phenomena into account. Having objects covering a wide range of sizes, shapes and orientation in the room one certainly cannot take the influence of each object into account; one has to rely on rough characterizations and apply statistical concepts.

In presenting examples on calculating sound propagation in large rooms we shall use factory halls. It is therefore appropriate to give a short overview on the scattering theory used, which e.g. is outlined by Kuttruff (1981). Basically, two hypotheses are used:

- The sound scattering objects are assumed to be point like and the energy of the incident wave is scattered evenly in all directions.
- The scattering phenomenon follows a Poisson process. The energy emitted by the source is sent out in discrete quantities as "phonons" or sound packages having energy $W \cdot \Delta t$, where W is the sound power of the source.

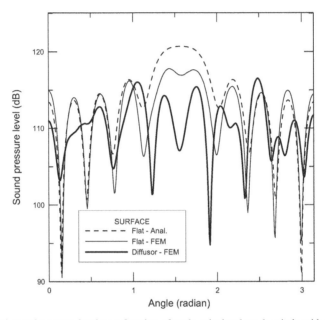

Figure 4.19 Total sound pressure level as a function of angle calculated on the circle with radius r equal one meter; situation as depicted in Figure 4.18. The source (S) power is equal to 1.0 W. Thick solid line – diffuser surface (FEM). Thin solid line – flat surface (FEM). Dashed line – flat, infinitely large surface (analytical).

The validity of the first hypothesis will depend on the ratio of the dimensions of the scattering object and the actual wavelength. Initially assuming that an object scatters sound, not only reflects sound in a specular way, we shall put up a limit on the relationship between a typical dimension D and the wavelength λ, demanding that $D/\lambda > 1/2\pi$.

From the second hypothesis follows that the probability density P_k of a phonon hitting a number k scattering objects within a time interval t_k is given by

$$P_k(c_0 t_k) = \frac{e^{-qc_0 t_k} \cdot \left(qc_0 t_k\right)^k}{k!}, \tag{4.70}$$

where c_0 is the usual wave speed and q is the average *scattering cross section* per unit room volume, a quantity also denoted the scattering frequency.

The determination of q is difficult for scattering objects having a complicated shape. A common practice is equalizing the scattering effect (at high frequencies) of an object having a total surface area S by the one offered by a sphere of equal surface area. The average scattering cross section may then be expressed as

$$q = \frac{1}{V} \sum_{i=1}^{N} \frac{S_i}{4}, \qquad (4.71)$$

when a total of N objects with surface areas S_i are present in a room of volume V.

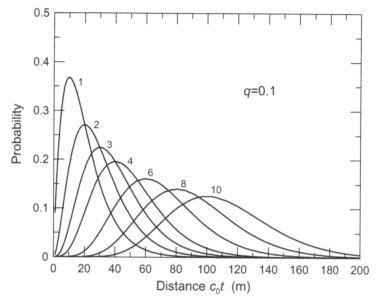

Figure 4.20 The probability of a wave (a phonon) hitting a given number of scattering objects, indicated by the number on the curves, having propagated a path of length $c_0 \cdot t$. The scattering cross section q is equal to 0.1m^{-1}.

Figure 4.20 shows the probability density P, according to Equation (4.70), of a phonon hitting a given number k of objects having propagated a path of length $c_0 \cdot t$. The number k is the parameter indicated on the curves calculated for a scattering cross section q equal to 0.1 m^{-1}. The Poisson distribution will typically give a high probability for hitting a single object; however, the corresponding width is small, whereas the probability for hitting many objects is small but the distribution is broad.

An important quantity relating to these aspects is the *mean free path* $\bar{R}$ of the sound. This quantity is generally used to characterize the path that the sound is expected to travel between two reflections. For an empty rectangular room having a volume V and a total surface area of S, we may show that $\bar{R}$ is equal to $4V/S$. Introducing scattering objects into the room (see Figure 4.21) we may, by using the probability function given by Equation (4.70), calculate the corresponding probability function of the free paths R and thereby the expected or mean value $\bar{R}$. The outcome is that $\bar{R}$ is equal to $1/q$.

4.8 CALCULATION MODELS. EXAMPLES

In the literature one will find reported a very large number of different models for predicting sound propagation in large rooms. A number of these are implemented in commercial computer software, e.g. CATT™, EASE™, EPIDAURE™ and ODEON™.

Most are developed for applications in performance rooms, i.e. for predicting the acoustics in rooms for speech and music. The trend is not only to give visual descriptions of the results but also to present the results by *auralization*. This implies that one may listen to music or speech "played" in a room at the design stage. This is accomplished by a process called *convolution*; the music or speech signal is convolved by the predicted impulse response belonging to a given source–receiver configuration.

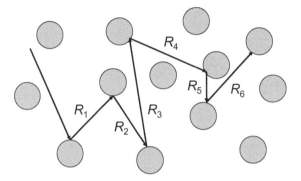

Figure 4.21 Sketch illustrating the concept of mean free path.

It is outside the scope of this book to give an overview or a closer description of this software based on the principles outlined in section 4.6. We shall, however, give examples on some special models primarily developed to predict sound propagation in large factory halls etc. The computer models mentioned above may certainly also be applied to such rooms but the ones we shall present cover the most important quantities to be predicted for such rooms: the attenuation of sound as a function of distance and the reverberation time. These are the analytical image-source models of Jovicic (1979) and Lindqvist (1982) together with the ray-tracing model of Ondet and Barbry (1989), the last including scattering in a very ingenious manner.

4.8.1 The model of Jovicic

The aim is to find an expression for the sound pressure level as a function of distance from a source of a given sound power level, which implies finding how the level decreases analogous to the results shown in Figure 4.9, however, without the constraint that the dimensions of the room should be fairly equal. Jovicic's models are confined to rooms of rectangular shapes, either "long" rooms, where one dimension is much larger than others (corridors etc.), or "flat" rooms, where two dimensions are much larger than the third one. We shall confine ourselves to the latter type, where the following assumptions are made:
- The influence of the sidewalls are neglected.
- The ceiling is treated as a plane surface like the floor. A serrated ceiling or ceilings with baffles etc. are treated as scattering objects.
- The absorption factor used is the mean value for the floor and ceiling.
- The sound source is placed midway between floor and ceiling.
- The scattering objects, which may also be assigned an absorption factor, are randomly distributed in the room.

The total energy density at a receiving point at a given distance r from the source is assumed to be given by

$$w_{\text{tot}} = w_{\text{d}} + w_{\text{s}}, \qquad (4.72)$$

where w_{d} is the contribution from the direct sound, i.e. the non-scattered part, and w_{s} is the contribution from the phonons arriving at the receiver position after one or more collisions with the scattering objects. Without these scattering objects, w_{d} will be given by Equation (4.63) but now we shall have to modify this expression by subtracting the part being scattered or attenuated in other ways than by specular reflections from the room boundaries. We shall start looking at the scattered sound.

4.8.1.1 Scattered sound energy

Starting from the probability density given in Equation (4.70), Kuttruff (1981) calculated the corresponding probability that a phonon after a time t should be at a distance r from the source. In an infinite space, this probability density will be given by

$$P(r,t) = \left(\frac{3q}{4\pi\, c_0 t} \right)^{\frac{3}{2}} \cdot e^{\frac{-3q}{4c_0 t} r^2}, \qquad (4.73)$$

assuming that $qc_0 t \gg 1$, which implies that the travelled distance $c_0 t$ must be much larger than the mean free path $\bar{R} = 1/q$. It may also be mentioned that P is a solution of the so-called diffusion equation used in fluid dynamics, which is

$$\nabla^2 Q = \frac{1}{D} \cdot \frac{\partial Q}{\partial t}, \qquad (4.74)$$

when setting the diffusion constant D equal to $c_0/(3q)$. The diffusion equation may e.g. describe how the concentration Q of a fluid, such as a dye, when injected into another fluid, changes with time. It should not be too difficult to envisage that this is a process quite analogous to how sound particles or phonons diffuse into a space containing scattering objects.

Jovicic assumes that the same probability $P(r,t)$ applies to the phonons from the image sources as all scattering objects are mirrored in the boundary surfaces (floor and ceiling) as well. The predicted total probability applicable to the phonons sent out from the original source and the image sources is then given by

$$P(r,t,h) = \frac{3q}{4\pi\, c_0 h t} e^{\frac{-3q}{4c_0 t} r^2}, \qquad (4.75)$$

where h is the height of the room. Inside a small volume element, containing the receiving position at a distance r from the source, we shall find phonons emitted from the source (and the image sources) at different points in time, thereby having different probability $P(r,t,h)$ of arriving at the chosen volume element. The shortest time of arrival will be r/c_0 and the longest one will be infinity.

On their way, the phonons are losing their energy, partly by hitting the scattering objects having absorption factor α_{s}, partly hitting the floor and ceiling having absorption factors α_{f} and α_{c}, respectively. In addition, we have the excess attenuation due to air absorption characterized by the power attenuation coefficient m. All these attenuation processes may be assembled in a factor $\exp(-bc_0t)$, where b is a total attenuation coefficient comprising all loss mechanisms.

Now, the idea is to assume that this attenuation takes place gradually along the whole path covered by a phonon. Thereby, we may assemble all the energy of phonons arriving by calculating the integral

$$w_s = W \int\limits_{r/c_0}^{\infty} P(r,t,h)\,e^{-bc_0t}\,dt. \tag{4.76}$$

An approximate solution to this integral, where e.g. the lowest limit is zero, is given by

$$w_s = \frac{3qW}{2\pi\,c_0h}\,K_0\left(r\sqrt{3qb}\right), \tag{4.77}$$

where K_0 is the modified Bessel function of zero order. The attenuation coefficient b may be expressed as

$$b = b'\left(\alpha',h,q\right) + \alpha_s q + m. \tag{4.78}$$

The quantity b', which expresses the attenuation due to the boundary surfaces is, as indicated, not only a function of the mean *absorption exponent* $\alpha' = -\ln(1-\alpha)$ for these surfaces but is also a function of the ceiling height and the scattering cross section.

4.8.1.2 "Direct" sound energy

The expression giving the direct energy density caused by the source and its infinite number of images (see Equation (4.63)) may approximately be solved by letting this row of sources be represented by a line source. The following solution is obtained:

$$w = \frac{WK}{2\pi\,rc_0h}\cdot F_0\left(\frac{\alpha'r}{h}\right), \tag{4.79}$$

where

$$K = \frac{\alpha'}{2}\cdot\frac{2-\alpha}{\alpha} \qquad \text{with} \qquad \alpha' = -\ln(1-\alpha)$$

and

$$F_0(x) = \sin(x)\cdot Ci(x) - \cos(x)\left[Si(x) - \frac{\pi}{2}\right].$$

The functions Ci and Si are the so-called cosine and sine integral function (see e.g. Abramowitz and Stegun (1970)). We have thereby arrived at a closed expression for the energy density in the direct field but without taking the scattered part into account. We shall have to correct it by the probability $\exp(-qc_0t)$ that a phonon has *not* been scattered during the time t. Also taking the excess attenuation due to air absorption into account, we finally may express the direct (or the non-scattered) energy density by

$$w_d = \frac{WK}{2\pi rc_0 h} \cdot F_0\left(\frac{\alpha' r}{h}\right) \cdot e^{-(q+m)r}. \tag{4.80}$$

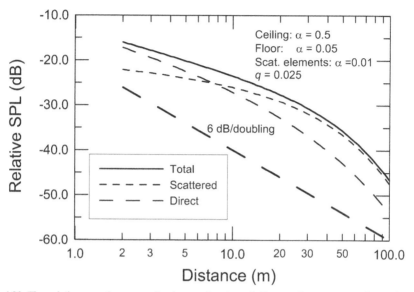

Figure 4.22 The relative sound pressure level as a function of distance from a source in a "flat" room. Contributions from scattered and non-scattered sound according to a model of Jovicic (1979). The room is 5 metres high.

4.8.1.3 Total energy density. Predicted results

The total energy density at a given distance from the source is then given by Equation (4.72) with w_s and w_d expressed by the Equations (4.77) and (4.80). We shall present some examples on using this equation where we, as in section 4.5.1.4, shall depict the relative sound pressure level, the difference between the sound pressure level L_p and the source sound power level L_w, as a function of the source–receiver distance. Assuming that the sound field is an assembly of plane waves having an intensity $w \cdot c_0$, we arrive at the ordinate for these curves by calculating the quantity $10 \cdot \lg(w \cdot c_0 / W)$.

The room height is chosen equal to 5 metres in all predictions shown. Furthermore, for simplicity the air absorption is put equal to zero. Figure 4.22 shows the total relative sound pressure level together with the separate contributions due to w_d and w_s for a room having a relatively small number of scattering objects; q is chosen equal to 0.025 m^{-1}. At large distances from the source, however, the level is still determined by the scattered field. For the sake of comparison, we have added a line representing the free field "distance law" for a monopole source, a 6 dB decrease per doubling of the distance. It should be obvious that one cannot apply any kind of "distance law", i.e. a constant number of decibels per distance doubling, in such rooms.

The next two figures show the total relative sound pressure level only but with different values for the absorption factor of the ceiling (see Figure 4.23) and in the mean scattering cross section q (see Figure 4.24). It should be noted that, even if the absorption exponents are entering into the equations above, the absorption factors α are used as input data when calculating the results.

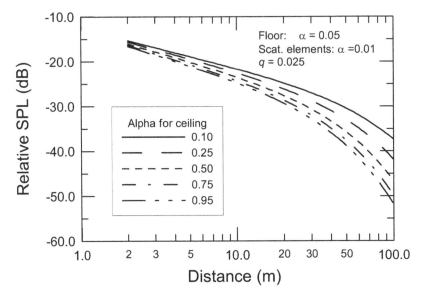

Figure 4.23 The relative sound pressure level as a function of distance from a source in a "flat" room. The room is 5 metres high. The parameter on the curves is the absorption factor α for the ceiling.

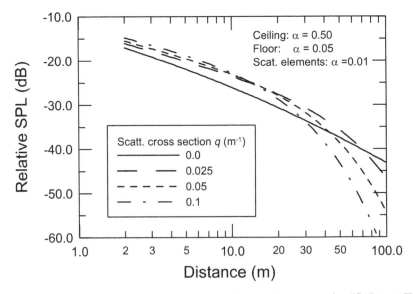

Figure 4.24 The relative sound pressure level as a function of distance from a source in a "flat" room. The room is 5 metres high. The parameter on the curves is the scattering cross section q (m^{-1}).

4.8.1.4 Reverberation time

Another effect to be observed in large rooms containing a large quantity of scattering objects is that the reverberation time is no longer a global quantity, but may vary systematically with the distance between source and receiver. This effect was observed by Jovicic (1971) by measurements in large industrial halls and confirmed theoretically by Vigran (1978) starting out from Jovicic's expressions given above.

The build-up of the scattered energy density in the room is given by Equation (4.76) As the build-up and the corresponding decay of sound energy are complimentary processes we may express the scattered energy density w_{rev} during decay as

$$w_{rev} = w_s - W \int_{r/c_0}^{t} P(r,t,h) \cdot e^{-bt} \, dt. \qquad (4.81)$$

Assuming that the mean scattering cross section q is relatively large, the scattered energy will dominate except when near to the source. In such a case we may use this equation directly to calculate the decay rate and thereby the reverberation time. A comparison between measured and predicted results is shown in Table 4.2. The reverberation time was measured by Jovicic (1971) in an industrial hall having a floor with dimensions 105 x 105 metres and a ceiling height of 11.5 metres. Measurements were performed in octave bands in the frequency range 125–4000 Hz at distances between source and receiver of 20 and 80 metre, respectively. The attenuation coefficient is given as a mean value, b equal to 1.22 m^{-1}, for this frequency range, and the mean scattering cross section q is stated to be 0.1 m^{-1}. The values in the table are average values for this frequency range and as seen, the fit between measured and predicted values are surprisingly good.

Table 4.2 Measured and predicted values for the mean reverberation time T at two different distances between source and receiver. Mean values for the frequency range 125–4000 Hz, in an industrial hall of volume 125 000 m^3.

Distance r (m)	Measured T (s)	Predicted T (s)
20	2.65	2.60
80	3.12	3.30

4.8.2 The model of Lindqvist

Lindqvist (1982) developed this analytically based image-source model further by also taking the reflections from the sidewalls into account, in addition, allowing for a random positioning of the source and receiver. The shape of the room is, however, still limited to rectangular, certainly a natural limitation for this kind of model. Based on the work of Kuttruff (see above), the scattering model applied by Lindqvist is more detailed than the one used by Jovicic but the scattering objects still have to be stochastically distributed in the room. The difference in predicted results using these two models will certainly depend on the actual situation. For relatively large rooms having not too much in the way of scattering object the differences is assumed to be relatively small, probably in the range of 1–2 dB.

In more recent time, the practical use of such analytical models is certainly reduced due to powerful computer simulations, either based on the ray-tracing or the image-source technique. The purpose of bringing forward the above works is primarily to illustrate some of the fundamental principles behind this type of modelling.

4.8.3 The model of Ondet and Barbry

An interesting solution to the problem of including scattering object was given by Ondet and Barbry (1989), which was implemented in the computer program RAYSCAT (RAYCUB in a later version). This is not, as the models discussed above, an image-source model but a ray-tracing one. Therefore, it does not impose any restrictions as to the shape of the room like the analytical models. The idea is to regard the areas of the room that contain scattering objects as zones having mean free paths depending on the density of these objects (see Figure 4.21). Each of these zones is allocated a certain mean free path $\bar{R} = R_k$, where the index k indicates the actual zone, whereas the areas without scattering object are allocated a mean free path $\bar{R} = \infty$. How is this idea compatible with a ray-tracing model where one certainly has to follow each ray around in the room?

Ondet and Barbry start by again using the Poisson distribution given by Equation (4.70), and they show that the paths lengths R_i covered between each hit have a probability density distribution given by

$$P(R) = q \cdot e^{-qR}, \tag{4.82}$$

which gives an expected value $E\{R\} = \bar{R} = 1/q$. Furthermore, one may generate these random distances R_i by using random numbers a_i between zero and one, thereafter inserting these numbers into the following expression:

$$R_i = -\bar{R} \cdot \ln(a_i). \tag{4.83}$$

The procedure is then as follows: One follows each ray in the normal manner until it crosses the border of a zone defined to contain scattering objects and thereby allocated a certain mean free path. A path length R_1 is then computed according to Equation (4.83) by drawing a random number a_1. This implies that it hits a scattering object after covering the distance R_1, thereafter directed in a random direction with a new random path length R_2. It may then hit another object within this zone or maybe escape from this zone.

A good fit between measured and predicted results is obtained by applying this procedure, both by Ondet and Barbry (1988) and others (see e.g. Vermeir (1992)). The computing time may, however, be quite long for rooms having complicated shapes, many zones with scattering objects of high density.

Later, other models have been developed (see e.g. Dance and Shield (1997)), limiting the room shape to rectangular where one may easily implement an image-source model, however, trying to keep the most important concepts from the Ondet–Barbry model; i.e. the subdivision of the room into zones containing scattering objects, the placement of absorbing element and barriers etc. The program CISM by Dance and Shield gives shorter computing times but at the expense of accuracy. It is not able to represent scattering in the same manner as the models treated above, which decreases the accuracy in areas far from the nearest source.

4.9 REFERENCES

ISO 9613–1: 1993, Acoustics – Attenuation of sound during propagation outdoors. Part 1: Calculation of the absorption of sound by the atmosphere.

ISO 3382: 1997, Acoustics – Measurement of the reverberation time of rooms with reference to other acoustical parameters. [Under revision to become ISO 3382 Acoustics – Measurement of room acoustic parameters. Part 1: Performance rooms; Part 2: Ordinary rooms.]

ISO 3741: 1999, Acoustics – Determination of sound power levels of noise sources using sound pressure – Precision methods for reverberation rooms.

ISO 354: 2003, Acoustics – Measurement of sound absorption in a reverberation room.

ISO 140–14: 2004, Acoustics – Measurement of sound insulation in buildings and of building elements. Part 14: Guidelines for special situations in the field.

ISO 16032: 2004, Acoustics – Measurement of sound pressure level from service equipment in buildings – Engineering method.

ISO/DIS 17497–1: 2002, Acoustics – Measurement of the sound scattering properties of surface. Part 1: Measurement of the random-incidence scattering coefficient in a reverberation room.

Abramowitz, M. and Stegun, I. A. (1970) *Handbook of mathematical functions.* Dover Publications Inc., New York.

Arau-Puchades, H. (1988) An improved reverberation formula. *Acustica*, 65, 163–180.

Cox, T. J., Avis, M. R. and Xiao, L. (2006) Maximum length sequences and Bessel diffusers using active technologies. *J. Sound and Vibration*, 289, 807–829.

Cox, T. J. and D'Antonio, P. (2004) *Acoustic absorbers and diffusers.* Spon Press, London and New York.

Cox, T. J. and Lam, Y. W. (1994) Prediction and evaluation of the scattering from quadratic residue diffusers. *J. Acoust. Soc. Am.*, 95, 297–305.

Dance, S. M. and Shield, B. M. (1997) The complete image-source method for the prediction of sound distribution in fitted non-diffuse spaces. *J. Sound and Vibration*, 201, 473–489.

Dance, S. M. and Shield, B. M. (2000) Modelling of sound fields in enclosed space with absorbent room surfaces. Part II. Absorptive panels. *Applied Acoustics*, 61, 373–384.

Davy, J. L., Dunn, I. P. and Dubout, P. (1979) The variance of decay rates in reverberation rooms. *Acustica*, 43, 12–25.

Ducourneau, J. and Planeau, V. (2003) The average absorption coefficient for enclosed spaces with non uniformly distributed absorption. *Applied Acoustics*, 64, 845–862.

Eyring, C. F. (1930) Reverberation time in "dead" rooms. *J. Acoust. Soc. Am.*, 1, 217–241.

Haas, H. (1951) Über den Einfluss eines Einfachechos auf die Hörsamkeit der Sprache. *Acustica*, 1, 49–58.

Heinz, R. (1993) Binaural room simulation based on an image source model with addition of statistical methods to include the diffuse sound scattering of walls and to predict the reverberant tail. *Applied Acoustics*, 38, 145–159.

Jovicic, S. (1971) Untersuchungen zur Vorausbestimmung des Schallpegels in Betriebgebäuden. Report No. 2151. Müller-BBN, Munich.

Jovicic, S. (1979) Anleitung zur Vorausbestimmung des Schallpegels in Betriebgebäuden. Report. Minister für Arbeit, Gesundheit und Soziales des Landes Nordrhein-Westfalen, Düsseldorf.

Krokstad, A., Strøm, S. and Sørsdal, S. (1968) Calculating the acoustical room response by the use of a ray tracing technique. *J. Sound and Vibration*, 8, 118–125.

Kuttruff, H. (1981) Sound decay in reverberation chambers with diffusing elements. *J. Acoust. Soc. Am.*, 69, 1716–1723.

Kuttruff, H. (1999) *Room acoustics*, 4th edn. Spon Press, London.

Lam, Y. W. (guest editor) (2000) Surface diffusion in room acoustics. *Applied Acoustics Special Issue*, 60, 2, 111–112.

Lewers, T. (1993) A combined beam tracing and radiant exchange computer model of room acoustics. *Applied Acoustics*, **38**, 161–178.

Lindqvist, E. (1982) Sound attenuation in factory spaces. *Acustica*, 50, 313–328.

Lubman, D. (1974) Precision of reverberant sound power measurements. *J. Acoust. Soc. Am.*, 56, 523–533.

Lundeby, A., Vigran, T. E., Bietz, H. and Vorländer, M. (1995) Uncertainties of measurements in room acoustics. *Acustica*, 81, 344–355.

Millington, G. (1932) A modified formula for reverberation. *J. Acoust. Soc. Am.*, 4, 69–82.

Naylor, G. (guest editor) (1993a) Computer modelling and auralisation of sound fields in rooms. *Applied Acoustics Special Issue*, 38, 2–4, 131–143.

Naylor, G. (1993b) ODEON – Another hybrid room acoustical model. *Applied Acoustics*, 38, 131–143.

Olesen, H. S. (1992) Measurements of the acoustical properties of buildings – Additional guidelines. *Nordtest Technical Report 203*.

Ondet, A. M. and Barbry, J. L. (1988) Sound propagation in fitted rooms – Comparison of different models. *J. Sound and Vibration*, 125, 137–149.

Ondet, A. M. and Barbry, J. L. (1989) Modelling of sound propagation in fitted workshop using ray tracing. *J. Acoust. Soc. Am.*, 85, 787–796.

Pierce, A. D. (1989) *Acoustics. An introduction to its physical principles and applications*. Published by the Acoustical Society of America through the American Institute of Physics, Melville, NY.

Schroeder, M. R. (1975) Diffuse sound reflection by maximum-length sequences. *J. Acoust. Soc. Am.*, 57, 149–150.

Schroeder, M. R. (1979) Binaural dissimilarity and optimum ceilings for concert halls: More lateral diffusion. *J. Acoust. Soc. Am.*, 65, 958–963.

Schroeder, M. R. (1999) *Number theory in science and communication*. Springer-Verlag, Berlin.

Sette, W. J. (1933) A new reverberation time formula. *J. Acoust. Soc. Am.*, 4, 193–210.

Simmons, C. (1997) Measurements of sound pressure levels at low frequencies in rooms. SP Report 1997: 27. SP Swedish National Testing and Research Institute, Borås, Sweden. (NORDTEST Project No. 1347-97.)

Stephenson, U. (1990) Comparison of the mirror image source method and the sound particle simulation method. *Applied Acoustics*, 29, 35–72.

Vermeir, G. (1992) Prediction of the sound field in industrial spaces. *Proceedings Internoise 92*, 727–730.

Vigran, T. E. (1978) Reverberation time in large industrial halls. *J. Sound and Vibration*, 56, 151–153.

Vigran, T. E. (1980) Corner microphone in laboratory rooms – Applicability and limitations (in Norwegian). *ELAB Report STF A80050*. NTH, Trondheim. (NORDTEST Project No. 129-78.)

Vigran, T. E., Lundeby, A. and Sørsdal, S. (1995) A versatile 2-channel MLS measuring system. Proceedings of the 15th ICA, 171–174, Trondheim, Norway.

Vorländer, M. (1995) Revised relation between the sound power and the average sound pressure level in rooms and consequences for acoustic measurements. *Acustica*, 81, 332–343.

Waterhouse, R. V. (1955) Interference patterns in reverberant sound fields. *J. Acoust. Soc. Am.*, 27, 247–258.

CHAPTER 5

Sound absorbers

5.1 INTRODUCTION

In the preceding chapter on room acoustics, we presupposed that an absorption factor and an accompanying equivalent absorption area could characterize the relevant sound absorbing surfaces. We did not, however, consider the kind of material parameters determining these quantities. Here we shall aim at giving a theoretical basis for the functioning of so-called *acoustic materials*; sound absorbing materials and constructions having their primary applications in controlling the acoustic conditions in rooms in buildings. This knowledge is in fact not only applicable in rooms but also for designing proper acoustic conditions in transport, e.g. for passengers and personnel in trains and buses and also for designing special devices for sound reduction, e.g. silencers for air-conditioning units. We shall in this chapter also deal with measurement and prediction methods for acoustic absorption, including how one measures the material parameters that determine the absorption.

The functioning of absorbing materials is linked to the behaviour of sound waves at the interface between two media (see Chapter 3). When a sound wave hits such a boundary it will normally be diffracted; a part of the energy will be deflected in a direction different from that of the incidence wave. If the boundary surface is large compared with the wavelength one characterizes the process as reflection. If the opposite is true, the word scattering is used. In many cases, we shall also be interested in what is happening on the other side of the boundary, i.e. we shall be concerned with the energy transmitted through the boundary surface. As in Chapter 3, we shall limit the treatment to simple cases of reflection and we shall, furthermore, assume plane wave incidence.

With reference to the preceding chapter on room acoustics, we shall remind the reader that the primary task of acoustic absorbers placed in a room is to ensure that only a controlled part of the sound energy is reflected back into the room. Seen from inside the room we want the rest of the energy, originating from whatever source, to be absorbed, which normally means that the energy is transformed into heat. It should be pointed out that according to internationally accepted conventions we class all non-reflected energy as absorbed energy. Seen from inside the room, an open window is a strongly absorbing surface even if no energy is dissipated in this surface, only transmitted out of the room.

The absorption factor of a given surface, defined by the ratio of the absorbed energy to the incident energy, may be determined by different measurement methods. For normal incidence the so-called Kundt's tube or standing wave tube may be used. This technique has as its background the determination of the absorption factor by scanning the maximum and minimum values of the standing wave set up in the tube. This classical method is considered to be a little outdated compared to methods based on modern signal analysis techniques. However, standing wave tube measurements are normally used for testing on small specimens, usually in development projects. Large-scale measurements for product data are normally performed in a laboratory reverberation room measuring the absorption factor by diffuse sound incidence, i.e. measuring the average value for all angles of incidence.

A number of other methods, most of them non-standard, are also in use both in laboratories and in the field (in situ methods). Development of in situ methods usable for low frequencies is particularly challenging, specifically from a few hundred Hertz and downwards.

For development purposes it should certainly be an advantage to make a direct calculation of the absorption factor based on material data and geometry for the absorber. Reasonable accurate analytical methods exist for homogeneous materials with simple geometry (plane absorbers), but certainly dependent on the accuracy of the material parameters going into the models.

5.2 MAIN CATEGORIES OF ABSORBER

Commonly used acoustic absorbers (absorbing surfaces) may be divided into two main groups:
   a)   Porous absorbers, e.g. mineral wool, plastic foams, fabric etc.
   b)   Resonator absorbers, either membrane or absorbers based on the Helmholtz resonator principle.

Basic forms of absorbers used in practice are depicted in Figure 5.1. Porous materials are often placed directly on to a hard surface or with a cavity behind to increase the absorption at low frequencies (see a) and b)). Membrane absorbers, depicted in c), may be a thin panel (or foil) of metal or hardboard, again placed at a certain distance from a hard surface. For a resonance absorber of the Helmholtz type, shown in d), the panel is perforated is various ways, normally by holes or slits. Combinations of the abovementioned types are also generally found.

When designing for proper acoustic conditions in a given room, one should be aware of other mechanisms present and capable of absorbing sound energy. In a room having lightweight wall constructions one may unintentionally induce plate vibrations by the sound field. This vibration energy may partly be dissipated in the plates themselves and partly radiated to a neighbouring room. With the latter mechanism the energy is also absorbed (non-reflected) as seen from the primary room. Thermal and viscous effects also add to the loss of acoustic energy in the room. These effects are contained in a "classical" part of the expression for the attenuation coefficient. More important, however, is the relaxation or hysteretic phenomena. Depending on the moisture content of the air in the room one may observe that these phenomena dominate the acoustic energy losses at the higher frequencies (in the kHz range). This type of air absorption has already been treated in the preceding Chapter 4 (see section 4.5.1.3).

5.2.1 Porous materials

Well-known porous materials are products of mineral fibres and plastic foams. Commonly used are blankets of mineral "wool", either glass or stone wool. These have fibres with a diameter in the range 2–20μ, commonly 4–10μ. Due to the manufacturing process the fibres will be distributed anisotropically. They will be randomly distributed in a plane parallel to the outer surface of the blanket but there will be few fibres oriented normally to this plane. One will find the mineral fibre products of the type "elastic" blankets as well as compressed into stiff boards, the latter normally used in suspended ceilings.

Plastic fibres products, e.g. polyester fibres materials, are also becoming popular for sound absorption. The diameter of the fibres in these products is normally larger than for the mineral wool products, being of the order 20–50μ. The aforementioned anisotropy also applies to these products. Other types of porous materials are also commercially available. Products comprise glass and metals in a sintered form. One may also find aluminium expanded as foam but to be effective as an absorber the pores must be interconnected. There also exist products using fibres of aluminium compressed into sheets of thickness 2–5 mm. These are intended for suspended ceilings.

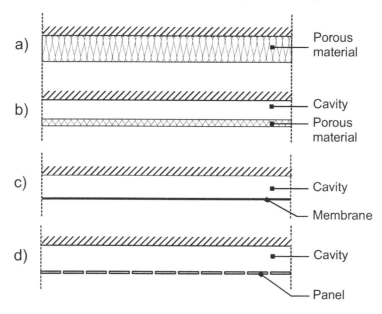

Figure 5.1 Basic types of acoustic absorber.

5.2.2 Membrane absorbers

By definition, a membrane shall have no stiffness but thin metal sheets are usually included when talking of membrane absorbers. The prerequisites for obtaining a reasonably high absorption factor are low surface weight and high internal losses in the membrane (plate) material. Aluminium or steel panels having a thickness of 0.5–1.0 mm mounted at a certain distance from a hard wall or ceiling normally absorbs in a limited frequency range only, having a statistical absorption factor α usually less than 0.5–0.6. By using thin sheets of plastic materials instead of metal much better performance may be achieved.

5.2.3 Helmholtz resonators using perforated plates

Helmholtz resonator absorbers are based on the principle that the air in the holes (or slots) of the plate represents a mass and that the air volume of the cavity behind the plate represents the spring stiffness in an equivalent oscillator, i.e. a simple mass-spring system. To absorb or dissipate acoustic energy one certainly must include a resistive component, which traditionally has been accomplished by filling the cavity behind the plate partly or wholly with a porous absorber as seen from Figure 5.2 a). It is not necessary, however, to fill the cavity wholly or partly to achieve high absorption. An adjustment of the resistance may be achieved by gluing a thin fabric to the plate (see b)). Such products are commercially available using panels of steel, aluminium, plaster and wood. A combination of fabric and a porous blanket in the cavity is sometimes used to achieve an even higher absorption.

Resonator absorbers with perforated plates or foils having holes of diameter less than 0.5 mm are commonly referred to as microperforated absorbers (MPA) (see c)). In this case the viscous losses in the holes give the necessary resistance without the use of any additional fabric. Instead of using perforations with holes, thin slits are an alternative. Commercial products are available made of steel, aluminium and plastics.

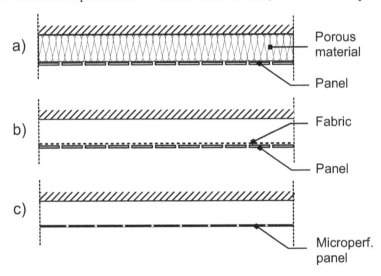

Figure 5.2 Resonator panels based on Helmholtz type.

5.3 MEASUREMENT METHODS FOR ABSORPTION AND IMPEDANCE

The absorption factor, the ratio of the absorbed and the incident acoustic energy, is determined by different methods according to the type of incident wave field. For measurements on small specimens, i.e. where the typical dimensions are smaller than the wavelength, the standing wave tube, also called the Kundt's tube, is used. The absorption factor for normal incidence is determined by measuring the maximum and minimum pressure amplitude in the standing wave set up in the tube by a loudspeaker. This basic technique is, as mentioned in the introduction, considered a little outdated in comparison with more modern methods based on transfer function measurements. It may then seem a little odd that this classical technique was implemented relatively late (1993) in an international standard, ISO 10534–1, after being used for at least 50 years. Commercial equipment has also been available for many decades. However, there exists a second part of the mentioned standard, ISO 10534–2, based on using broadband signals and measurement of the pressure transfer function between different positions in the tube.

The results from measurements of absorption factor and acoustic impedance, using the standing wave method, obviously are meaningful only when assuming these to be independent of the size of the specimen, which is normally quite small. The homogeneity of the material is therefore an important factor. If we wish to extrapolate the results to larger areas and other angles of incidence we also need to know whether we may assume that the material is locally reacting or not, which means whether or not we expect that the impedance is a function of the angle of incidence.

Traditionally, measurement of the absorption factor of larger specimens is performed in a reverberation room. One then determines the average value over all angles of incidence under diffuse field conditions. The product data normally supplied by producers of absorbers are determined according to the international standard ISO 354, which specifies measurement conditions for reverberation room testing. The area required for measurement is 10–12 square metres and there are requirements as to shape of the area. The reason for these requirements is that the absorption factor determined by this method always includes an additional amount due to the *edge effect*, which is a diffraction phenomenon along the edges of the specimen. This effect makes the specimen acoustically larger the geometric area, which may result in obtaining absorption factors larger than 1.0. Certainly, this does not imply that the energy absorbed is larger than the incident energy (!).

In the literature one may find a number of laboratory measurement methods for determining absorption factors as a function of incidence angle, applying relatively large specimens. None of these is yet standardized. There have recently been efforts put into the development of similar in situ techniques, i.e. methods for measurements of absorption and impedance both inside buildings and outside. These methods are mainly based on the same principle as used in ISO 10534–2, which implies the specified two-microphone method is extended to spherical wave fields. References to most of these methods may be found in Dutilleux et al. (2001). We shall not treat them here any further apart from one suggested by Mommertz (1995).

This method determines the reflection factor for a surface based on using a single microphone placed near to the surface of the specimen. The idea is to use MLS signals in a subtraction technique; the impulse response measured placing the microphone in a free field is subtracted from the impulse response measured near to the surface. Doing this, one is left with a signal representing the reflection. One obvious prerequisite is that the configuration (loudspeaker source and microphone) is identical in both impulse response measurements. With certain modifications this method is implemented in an ISO standard for determining absorption factors for road surfaces (see ISO 13472–1). Later efforts have been to make a reference measurement near to a very hard surface, i.e. a totally reflecting surface, instead of a measurement in the free field. The swept sine technique (SS) may of course be just as applicable as using MLS signals.

5.3.1 Classical standing wave tube method (ISO 10534–1)

Using this method the specimen is placed at one end of a tube (see Figure 5.3). A loudspeaker is used to create a standing wave field in the tube, a field that is detected by a probe microphone. To fulfil the requirement having only plane wave propagation in the tube, the linear dimension of the cross section must be less than the wavelength. More specifically, the frequency range of the measurements extends upwards to the frequency of the first cross mode of the tube, which for tubes of circular cross-section will approximately be given by $0.586 \cdot c/D$, where c is the speed of sound and D the diameter of the tube. For a 10 cm diameter tube this frequency will be approximately 2000 Hz.

As derived in Chapter 3 (section 3.5.1.1), the RMS-value of the sound pressure at a given frequency may be expressed as

$$\tilde{p}(x) = \frac{\hat{p}_i}{\sqrt{2}}\left[1 + \left|R_p\right|^2 + 2\left|R_p\right|\cos(2kx + \delta)\right]^{\frac{1}{2}},$$ (5.1)

where R_p is the pressure reflection factor having phase angle δ.

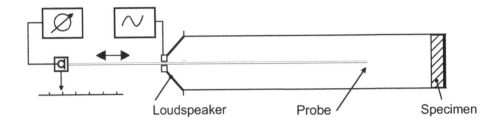

Loudspeaker   **Probe**   **Specimen**

Figure 5.3 Sketch of the set-up for the "classical" standing wave tube measurement method.

From the ratio of the maximum and the minimum sound pressure amplitudes, these amplitudes are given by

$$\left(\tilde{p}\right)_{\text{max}} = \frac{\hat{p}_i}{\sqrt{2}}\left[1 + \left|R_p\right|\right]$$

$$\text{and} \quad \left(\tilde{p}\right)_{\text{min}} = \frac{\hat{p}_i}{\sqrt{2}}\left[1 - \left|R_p\right|\right],$$ (5.2)

we may then determine the modulus of the pressure reflection factor R_p. The phase angle δ is determined by the position of the first minimum pressure close to the specimen. (Can you set up the expression for this phase angle using Equation (5.1)?). From these data both the input impedance Z_g and the absorption factor α are determined from the equations

$$Z_g = \rho_0 c_0 \frac{1 + R_p}{1 - R_p} = Z_0 \frac{1 + R_p}{1 - R_p}$$ (5.3)

and

$$\alpha = \frac{4\,\text{Re}\left\{\dfrac{Z_g}{Z_0}\right\}}{\left|\dfrac{Z_g}{Z_0}\right|^2 + 2\,\text{Re}\left\{\dfrac{Z_g}{Z_0}\right\} + 1}.$$ (5.4)

A sketch of the sound pressure level, given by the expression

$$L_p(x) = 20 \cdot \lg\left(\frac{\tilde{p}(x)}{\tilde{p}_{\text{ref}}}\right),$$

and exhibiting alternating maxima and minima, is shown in Figure 5.4. It should be noted that the equations above does not take into account any possible energy losses in the medium in front of the specimen. The figure, which is reproduced from the standard, gives however a much too exaggerated picture of how the ratio of the maximum and minimum varies along the tube. Obviously, the point here is to make one aware of this effect, which certainly must be observed when performing accurate measurements.

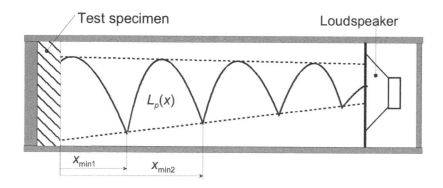

Figure 5.4 Standing wave pattern in the test tube.

5.3.2 Standing wave tube. Method using transfer function (ISO 10534–2)

The classical method is based on measurements on a standing wave having just one frequency component. We may obtain the same results by representing the sound pressure and particle velocity as simple functions of time. We can use an arbitrary time function to set up a sound field in the tube and then use the Fourier transform to revert to the frequency domain. As an example, we may represent the pressure reflection factor at an arbitrary position x in the tube as being the transfer function having the pressure p_i in the incident plane wave as the input variable and the pressure p_r in the reflected wave as the output variable:

$$R(x, f) = \frac{\text{F}\{p_r(x,t)\}}{\text{F}\{p_i(x,t)\}}, \qquad (5.5)$$

where F symbolizes the Fourier transform. The basic idea is to express the relationship between the wave components at two (or more) positions along the tube. Doing this it may be shown that we are able to make separate estimates of the intensity in the incident and the reflected wave. In fact, to determine the variables we shall be interested in, it is sufficient to measure *one* single transfer function, namely between the total pressure in two positions. Using Figure 5.5 as a starting point, we shall give a short description of the procedure. At two positions, having coordinates x_1 and x_2, we define

$$R(x_1, f) = \frac{\mathrm{F}\{p_\mathrm{r}(x_1,t)\}}{\mathrm{F}\{p_\mathrm{i}(x_1,t)\}} \quad , \quad R(x_2, f) = \frac{\mathrm{F}\{p_\mathrm{r}(x_2,t)\}}{\mathrm{F}\{p_\mathrm{i}(x_2,t)\}}, \qquad (5.6)$$

and a transfer function H_{12} for the total pressure in these two positions:

$$H_{12}(f) = \frac{\mathrm{F}\{p(x_2,t)\}}{\mathrm{F}\{p(x_1,t)\}} = \frac{\mathrm{F}\{p_\mathrm{i}(x_2,t) + p_\mathrm{r}(x_2,t)\}}{\mathrm{F}\{p_\mathrm{i}(x_1,t) + p_\mathrm{r}(x_1,t)\}}. \qquad (5.7)$$

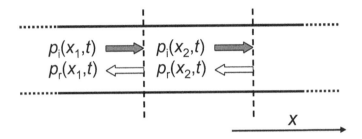

Figure 5.5 Wave components in a standing wave tube.

Correspondingly, for the pressure in the incident and reflected wave, respectively, we may define transfer functions

$$\left[H_{12}(f)\right]_\mathrm{i} = \frac{\mathrm{F}\{p_\mathrm{i}(x_2,t)\}}{\mathrm{F}\{p_\mathrm{i}(x_1,t)\}} \quad , \quad \left[H_{12}(f)\right]_\mathrm{r} = \frac{\mathrm{F}\{p_\mathrm{r}(x_2,t)\}}{\mathrm{F}\{p_\mathrm{r}(x_1,t)\}}. \qquad (5.8)$$

From Equations (5.6) to (5.8) we get by eliminating $R(x_2,f)$:

$$R(x_1, f) = \frac{H_{12} - \left[H_{12}\right]_\mathrm{i}}{\left[H_{12}\right]_\mathrm{r} - H_{12}}. \qquad (5.9)$$

We have assumed plane wave propagation only and we may then write

$$\left[H_{12}\right]_\mathrm{i} = e^{-jk_{12} \cdot (x_2 - x_1)} \quad \text{and} \quad \left[H_{12}\right]_\mathrm{r} = e^{jk_{21} \cdot (x_2 - x_1)}, \qquad (5.10)$$

where k_{12} and k_{21} are wave numbers for the incident and reflected wave, respectively. Furthermore, assuming no energy losses in the tube between these two positions, we may write $k_{12} = k_{21} = k_0$ and Equation (5.9) will become

$$R(x_1, f) = \frac{H_{12} - e^{-jk_0 d}}{e^{jk_0 d} - H_{12}}, \qquad (5.11)$$

where $d = x_2 - x_1$. This is the basic equation when using this two-microphone method to determine reflection factor, impedance and intensity.

Implementing this method to determine the reflection factor and impedance implies that the probe microphone in the classical method is substituted by two pressure microphones separated by a given distance d (see Figure 5.6). Here the distance from the microphone no. 1 (closest to the loudspeaker) to the surface of the specimen is denoted l. Feeding the loudspeaker with a broadband signal; the total pressure transfer function H_{12} between the positions 1 and 2 is determined. Again assuming that there are no energy losses in the tube we may transform the reflection factor, according to Equation (5.11), to the position of the surface of the specimen. We then get

$$R(0, f) = \frac{H_{12} - e^{-jk_0 d}}{e^{jk_0 d} - H_{12}} \cdot e^{j2k_0 l} .$$ (5.12)

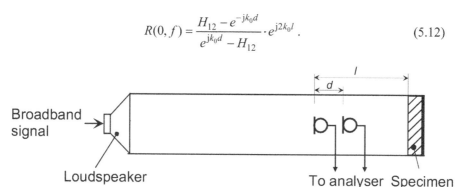

Figure 5.6 Impedance measurement set-up using a two-microphone method.

Knowing the refection factor we may easily calculate the input impedance of the specimen by inserting Equation (5.12) into Equation (5.3). After some algebra we get

$$Z_g(0, f) = j\rho_0 c_0 \frac{H_{12} \sin(k_0 l) - \sin[k_0(l - d)]}{\cos[k_0(l - d)] - H_{12} \cos(k_0 l)} .$$ (5.13)

We shall present some measurement results below using this method (see e.g. Figures 5.15 and 5.26).

Lastly, it should be mentioned that determining the basic transfer function H_{12} several types of broadband signals are in use; stochastic noise, MLS signals and swept sine (SS) signals. There is also the possibility of using just one microphone moved between positions assuming that the signal is reproducible. No calibration to account for the mismatch between microphones is then necessary.

5.3.3 Reverberation room method (ISO 354)

Product data specifying the absorption capability of materials are normally determined by measurements in a diffuse field using the so-called reverberation room method. The specifications applying to this method are found in the international standard ISO 354. We have already discussed the method in connection with the diffuse field theory for rooms (see section 4.5.1.2). We shall therefore just briefly repeat the basic principle. It is based on the Sabine formula for the reverberation time in a room:

$$T = \frac{55.3}{c_0} \cdot \frac{V}{A} = \frac{55.3}{c_0} \cdot \frac{V}{A_S + 4mV}, \tag{5.14}$$

where V is the volume of the room and A is the total equivalent absorption area. The total absorption area has, as is apparent from last expression, contributions A_S from the surfaces and objects in the room together with the air absorption, the latter specified by the power attenuation coefficient m.

The determination of the absorption factor is performed by measurements of the reverberation time before and after the specimen(s) is introduced into the room. Assuming the specimen to be a plane object having a total surface S (10–12 m^2), the absorption factor is expressed as

$$\alpha_{Sa} = \frac{55.3V}{c_0 S} \left(\frac{1}{T} - \frac{1}{T_0} \right), \tag{5.15}$$

where T and T_0 are the reverberation times in the room with and without the specimen, respectively. We have assumed that the environmental conditions are the same in both measurements and furthermore; we have neglected the absorption of the room surface being covered by the specimen, assuming this to be a hard surface of concrete having negligible total absorption. However, diffraction effects, denoted earlier on as edge effects, will often result in obtaining absorption factors in excess of 1.0. This phenomenon will be treated in more detail below (see section 5.5.3.2). For further details concerning this method the reader should consult the standard.

5.4 MODELLING SOUND ABSORBERS

Section 5.2 gave an overview of the main types of absorber being used in practice. These could roughly be divided into two groups: one based on the principle of viscous losses in a porous medium and the other utilizing a resonance principle. There is, however, no clear boundary between the two types. In practice, a given product may combine these two principles and the term used when specifying the absorber may follow the most dominant feature.

There are several design tools available, certainly of different complexity. Simple modelling based on lumped elements may often be sufficient, a modelling analogous to the one used when treating mechanical systems in Chapter 2 (section 2.5.1). An assembly of elements makes up the actual acoustical system, elements having their analogues in an equivalent mechanical or electrical system. For the latter, in particular, there are a substantial number of computer programs that may be applied for calculation on the analogous acoustical system. One should, however, be warned not to work beyond the range of validity. Acoustical systems imply *wave motion*, i.e. using lumped element models presupposes that the dimensions of the elements must always be less than the wavelength. Below, we shall give an example on such a modelling technique using a very simple acoustical system called a Helmholtz resonator.

A step further in modelling acoustical systems, allowing wave motion in *one* direction, is by using the transfer matrix method. In the analogous electrical system this is denoted four-pole theory. The method presupposes that we are able to set up a matrix that describes the relationship between the acoustical quantities on the input and output side of each element in the system. The matrices, representing all the elements in the actual acoustical system, may then be combined to calculate the sought-after quantities.

Examples will be given in section 5.7 where we shall include matrices representing porous materials based on models treated in section 5.5.

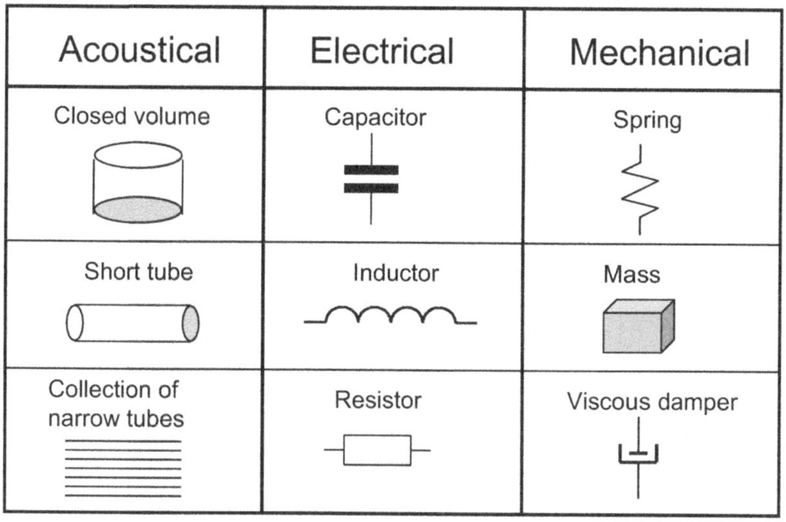

Acoustical	Electrical	Mechanical
Closed volume	Capacitor	Spring
Short tube	Inductor	Mass
Collection of narrow tubes	Resistor	Viscous damper

Figure 5.7 Analogue components in acoustic, electrical and mechanical systems.

5.4.1 Simple analogues

Figure 5.7 shows one of the simple analogies one may use for the relationship between acoustical, electrical and mechanical systems, the so-called impedance analogy. This implies that that the sound pressure in the acoustical system is equal to a voltage in the electrical system and to a force in the mechanical system. Correspondingly, an acoustical particle velocity (or volume velocity) will be a current and a vibration velocity. The relationship between acoustic impedance, specific acoustic impedance and mechanical impedance will then be:

$$Z_a = \frac{p}{v \cdot S} = \frac{Z_s}{S} = \frac{Z_{mec}}{S^2},$$ (5.16)

where S is an area (m^2).

As mentioned above, we shall start with a very simple acoustical system, the Helmholtz resonator, to illustrate the use of such analogies. In its simplest form, this may be considered as a harmonic oscillator; mechanically speaking it is a simple mass-spring system. The mass will be an oscillating air column in a tube (or in a slot) being driven by the sound pressure, this mass is coupled to a spring represented by a closed volume of air (see Figure 5.8 a)). We then have to find the spring stiffness, the mass and the damping coefficient, expressed by acoustical quantities, to calculate the resonance frequency and the energy dissipation.

5.4.1.1 The stiffness of a closed volume

We may find the mechanical stiffness of a closed volume by using the equation $P \cdot V^\gamma =$ constant, giving the relationship between pressure and volume under adiabatic conditions. However, it may be more appropriate from an acoustical viewpoint to use the acoustical equations derived previously. We shall use both approaches, starting with a resonator where the air volume are compressed by a small piston in the neck being driven by a alternating force F as sketched in Figure 5.8 b).

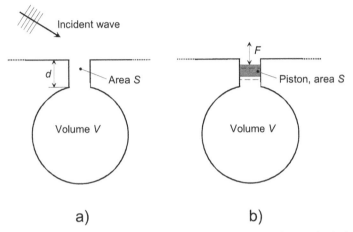

Figure 5.8 A simple Helmholtz resonator driven by a) a sound wave. b) a mechanical force.

By differentiating the adiabatic equation of state, we get

$$\frac{dP}{P} + \gamma \frac{dV}{V} = 0. \tag{5.17}$$

In this equation we may consider the differential dP as the equivalent sound pressure $p = F/S$ and the pressure P as the ambient pressure P_0. If the force is giving the piston a displacement Δx we get

$$\frac{F}{S \cdot P_0} = -\gamma \frac{\Delta x \cdot S}{V}, \tag{5.18}$$

and the mechanical stiffness k_{mec} will then be

$$k_{mec} = \frac{-F}{\Delta x} = \frac{\gamma P_0 S^2}{V} = \frac{\rho_0 c_0^2 S^2}{V} \qquad \text{where} \qquad c_0 = \sqrt{\frac{\gamma P_0}{\rho_0}}. \tag{5.19}$$

We have introduced the sound (phase) speed c_0 in the last equation. Considering the relationship in Equations (5.16), the corresponding acoustical stiffness will be equal to $\rho_0 c_0^2 / V$. We may now show by calculating the acoustical impedance in a tube closed at one end, as depicted in Figure 5.9, that we get the same result in the low frequency limit, i.e. when making the dimensions smaller than the wavelength. We shall use the equations in Chapter 3 (section 3.5.1), which gives us the acoustic impedance at a distance d from the reflecting closed end as

$$\left(Z_a = \frac{p}{v \cdot S} \right)_{x=d} = \frac{\rho_0 c_0}{S} \cdot \frac{\hat{p}_i e^{-jkd} + \hat{p}_r e^{jkd}}{\hat{p}_i e^{-jkd} - \hat{p}_r e^{jkd}}. \tag{5.20}$$

As we assume that the closed end is totally reflecting, the sound pressure amplitudes in the two partial waves must be equal. This gives us

$$Z_a = -j\frac{\rho_0 c_0}{S} \cdot \cot g(kd) \xrightarrow{\;kd \ll 1\;} -j\frac{\rho_0 c_0^2}{\omega S d} = -j\frac{\rho_0 c_0^2}{\omega V}. \tag{5.21}$$

Making the approximation $kd \ll 1$ we see that the result is the same as derived above. The equivalent acoustical stiffness is again $\rho_0 c_0^2/V$.

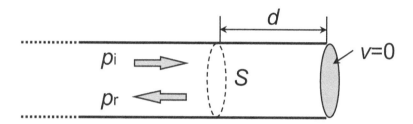

Figure 5.9 Plane waves in a tube. The tube is terminated by a totally reflecting surface.

5.4.1.2 The acoustic mass in a tube

Using a similar procedure, as when deducing the stiffness, we shall find an expression for the equivalent mass by calculating the acoustical impedance at a distance d from a *pressure release surface*, i.e. the pressure at the surface is zero as opposed to the above setting the particle velocity equal to zero. The same procedure that gave us Equation (5.21) will now give

$$Z_a = j\frac{\rho_0 c_0}{S} \cdot tg(kd) \xrightarrow{\;kd \ll 1\;} j\frac{\omega \rho_0 d}{S}. \tag{5.22}$$

The equivalent acoustical mass is therefore $\rho_0 d/S$. We are now in a position to calculate the resonance frequency of a resonator having a volume V and a "neck" of length d with a cross sectional area of S. We get:

$$f_0 = \frac{\omega_0}{2\pi} = \frac{1}{2\pi}\sqrt{\frac{k_a}{m_a}} = \frac{1}{2\pi}\sqrt{\frac{\rho_0 c_0^2 S}{V \rho_0 d}} = \frac{c_0}{2\pi}\sqrt{\frac{S}{Vd}}. \tag{5.23}$$

For a more accurate calculation we must take into account that the effective oscillating mass is larger than the one contained in the neck. We have to add the so-called *end correction*. Furthermore, making use of the resonator requires information on the energy losses of the system. We shall treat the latter item first.

5.4.1.3 Acoustical resistance

To calculate energy losses due to viscous forces, we have to modify the simple equation of force, the Euler equation, to include the effects of such forces. The linearized Navier-Stokes equation of force may be written as

$$\nabla p = -\rho_0 \frac{\partial \mathbf{v}}{\partial t} + \frac{4}{3}\mu\nabla(\nabla \cdot \mathbf{v}) - \mu\nabla \times \nabla \times \mathbf{v}, \qquad (5.24)$$

where we now have got two additional terms both containing the coefficient of viscosity μ. This coefficient is approximately equal to $2\cdot10^{-5}$ kg/(m·s) for air.

To carry out calculations on absorbers where perforated plates are an element, we shall have to predict the inherent viscous losses by sound propagation through perforations such as holes or slits. For the former and also for calculation on the single Helmholtz resonator in the next section, we shall start looking at thin tubes of circular cross section and calculate the sound propagation in the axis direction, again assuming harmonic time dependence. The equation of force may then be simplified to

$$\frac{\partial p}{\partial x} = -j\rho_0\omega v_x + \frac{\mu}{r}\frac{\partial}{\partial r}\left(r\cdot\frac{\partial v_x}{\partial r}\right). \qquad (5.25)$$

The variable v_x is the component of the particle velocity in the axial direction, and r the radius vector (cylindrical coordinates). Assuming that the velocity is zero at the tube walls, i.e. when r is equal to the radius a, the solution to the equation is (see Allard (1993))

$$v_x = -\frac{1}{j\omega\rho_0}\frac{\partial p}{\partial x}\left(1 - \frac{J_0(qr)}{J_0(qa)}\right), \qquad \text{where} \quad q = \sqrt{\frac{-j\omega\rho_0}{\mu}}. \qquad (5.26)$$

The symbol J_0 indicates a Bessel function of the first kind and zero order. From this equation we may now calculate the mean particle velocity in a cross section of the tube. We get

$$\langle v_x \rangle = \frac{\displaystyle\int_0^a v_x \cdot 2\pi r\, dr}{\pi a^2} = -\frac{1}{j\omega\rho_0}\frac{\partial p}{\partial x}\left(1 - \frac{2}{s\sqrt{-j}}\frac{J_1\left(s\sqrt{-j}\right)}{J_0\left(s\sqrt{-j}\right)}\right), \qquad (5.27)$$

where

$$s = a\sqrt{\frac{\omega\rho_0}{\mu}}. \qquad (5.28)$$

We may rewrite this expression as

$$j\omega\rho\langle v_x \rangle = -\frac{\partial p}{\partial x}, \qquad (5.29)$$

where ρ is the effective density of the air in the tube, given by

$$\rho = \rho_0 \left(1 - \frac{2}{s\sqrt{-j}} \frac{J_1\left(s\sqrt{-j}\right)}{J_0\left(s\sqrt{-j}\right)} \right)^{-1}. \tag{5.30}$$

Assuming that the length d of the tube in question is much shorter than the wavelength we may express the specific acoustic impedance as

$$Z_s = \frac{\Delta p}{\langle v_x \rangle} = j\omega\rho_0 d \left(1 - \frac{2}{s\sqrt{-j}} \frac{J_1\left(s\sqrt{-j}\right)}{J_0\left(s\sqrt{-j}\right)} \right)^{-1}. \tag{5.31}$$

At low frequencies, setting s in Equation (5.28) less than ≈ 2.0, a very good approximation for the term enclosed in the parenthesis is $4/3 - j \cdot 8/s^2$, which gives

$$Z_s \approx \frac{8\mu d}{a^2} + j\frac{4}{3}\omega\rho_0 d, \tag{5.32}$$

or expressed by the acoustic impedance

$$Z_a \approx \frac{8\mu d}{\pi a^4} + j\frac{4}{3}\frac{\omega\rho_0 d}{\pi a^2}. \tag{5.33}$$

The viscosity then gives us two effects. First, we get a resistive part being analogous to the mechanical damping coefficient, and we shall note the relationship with a quantity to be introduced later, the *airflow resistance*. This is a very important material parameter for all porous media. Using the definitions found in the international measurement standard, ISO 9053, which is treated further in section 5.6.1, the quantity $8\mu/a^2$ will be the flow resistivity of the tube, having symbol r and dimension Pa·s/m^2. Second, the viscosity also affects the mass term in the expression for the impedance. We get an increase of one-third compared with our earlier calculations (see Equation (5.22)).

For a panel, either a slatted one, i.e. an assemblage of parallel beams, or perforated by thin slits, we shall need an expression for the equivalent viscous losses in a single long slit (see Figure 5.10). We shall assume that the input pressure is the same along the whole length of the slit, the length being long in comparison with the wavelength. Furthermore, the pressure p in the slit varies only in the x-direction, and the velocity in this direction is only dependent on the z-coordinate. It may be shown (see e.g. Vigran and Pettersen (2005)), that the effective density of the air in the slit of width b may be written

$$\rho = \rho_0 \left(1 - \frac{\tan(\frac{k'b}{2})}{\frac{k'b}{2}} \right)^{-1}, \quad \text{where} \quad k' = \sqrt{\frac{\omega\rho_0}{j\mu}}. \tag{5.34}$$

For an (infinitely) long slit in a plate of thickness d the specific impedance will then be

$$Z_s = j\rho\omega d = j\rho_0\omega d \left(1 - \frac{\tan(\frac{k'b}{2})}{\frac{k'b}{2}}\right)^{-1}. \tag{5.35}$$

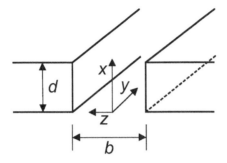

Figure 5.10 A single slit of width *b* in a plate of thickness *d*.

Using a series expansion in the angular frequency ω, the first three terms will be

$$Z_s \approx \frac{12\mu d}{b^2} + \frac{1}{700}\frac{d\,b^2\rho_0^2\omega^2}{\mu} + j\frac{6}{5}\omega\rho_0 d, \tag{5.36}$$

where the constant term in the real part corresponds to the value $8\mu/a^2$ found above for cylindrical tubes with radius *a*. We may also observe, looking at the imaginary part, that the viscosity again results in an added mass for the air in the slit.

5.4.1.4 The Helmholtz resonator. An example using analogies

A complete model for a Helmholtz resonator embedded in a hard wall, as depicted in Figure 5.8, must include all types of energy losses. The "natural" losses are represented by two components: the one caused by the viscous losses and the other due to the reradiated sound energy. Certainly, without the viscous losses the resonator will not act as an absorber! In designing for good room acoustics conditions it was noted that diffusers might be just as important as absorbers. In this connection, Helmholtz resonators are useful, contributing to the diffuse sound by partly reradiating the sound energy.

There will also be contributions to the viscous losses from the surfaces around the resonator neck but we shall neglect these for the moment. On the other hand, we shall have to find an expression for the acoustic impedance describing the sound radiation from the tube opening. Here we may make use of the formerly derived radiation impedance for a circular piston placed in an infinite baffle (see section 3.4.4). We then have to envisage that the air column in the neck moves like a rigid piston. Using a low frequency approximation for the radiation impedance expressed by its equivalent acoustic impedance, we get

$$\left(Z_a\right)_{\text{radiation}} \approx \frac{\rho_0 \omega^2}{2\pi c_0} + j \cdot \frac{1}{S}\left(\frac{8\omega\rho_0 a}{3\pi}\right), \quad\quad (5.37)$$

where S and a are the piston radius and area, respectively. The imaginary part of the impedance is cast into this form to make a comparison with Equation (5.22). It is easily seen that it gives an added mass equivalent to an increase $\Delta d = 8a/3\pi \approx 0.85 \cdot a$ in the length of the neck. It is commonly assumed that the same correction may be applied to both ends of the neck, thereby setting the effective length of the neck to be $d' = d + 1.7 \cdot a$. This so-called *end correction* will certainly depend on the actual cross sectional shape being different for noncircular openings, for slits etc. Data for these other shapes are listed in the literature. An important case in practice is long and narrow slits and we shall therefore include this case (see below).

How good is a single Helmholtz resonator when it comes to absorption? To arrive at the maximum absorption we have to adjust the system to make the two resistive terms, given by the Equations (5.33) and (5.37), equal at the resonance frequency. Doing this, the effective absorption area of the resonator opening will be given by

$$A_{\text{max}} = \frac{\lambda_0^2}{2\pi}, \quad\quad (5.38)$$

where λ_0 is the wavelength at resonance. The effective absorption area is therefore much larger than the physical size of the opening. However, the resonator will have a small bandwidth, e.g. the relative bandwidth $\Delta f / f_0$ could be as small as 0.01, which implies a Q factor as high as 100. In practice, one normally designs for a more broadband absorber by adding some resistance in the opening, which may be in the form of a porous material, a metal grid etc. To retain the absorption area one has to increase the volume of the resonator, which again means that the opening has to be adjusted to maintain the chosen resonance frequency.

5.4.1.5 Distributed Helmholtz resonators

Single Helmholtz resonators are used in many practical cases where the task is to remove single frequencies. More commonly used are the types that we may name distributed Helmholtz resonators, which we referred to in the introduction (see section 5.2.3). These are absorbers using perforated panels, perhaps in the form of slats, mounted at a certain distance ℓ from a hard wall or ceiling. To each opening (see Figure 5.11 a)), or to each slit, (see Figure 5.11 b)), we then allocate a part of the cavity volume that is used when calculating the resonance frequency by Equation (5.23). We then get

$$f_0 = \frac{c_0}{2\pi}\sqrt{\frac{\varepsilon}{\ell(d + \Delta d)}} \quad \text{with} \quad \varepsilon = \frac{S}{S_0}, \quad\quad (5.39)$$

where ε is the perforation or the "porosity" of the panel. The distance ℓ is here assumed to be much less than the wavelength. It should also be noted that for the assumption of a locally reacting absorber to be valid, the cavity volume has to be subdivided to minimize lateral wave propagation.

In the same way as for a single resonator one must introduce some resistance in addition to the natural viscous losses to obtain an absorber for practical use. One will certainly also have some additional viscous losses due to the air movements on the panel surfaces around the holes or slits but this is normally not enough. The exceptions to this situation, which was mentioned in section 5.2.3, are when using the so-called micro-perforated panels, panels where the perforations with holes or slits have a typical linear dimension less than 0.5 mm.

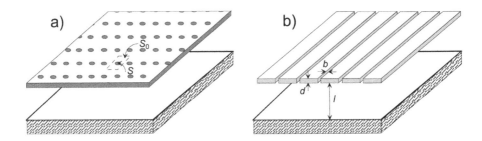

Figure 5.11 Distributed Helmholtz resonators. a) panel perforated by holes and backed by a cavity. b) slatted panel backed by a cavity.

It was also pointed out that one normally adjusted the resistance by filling, partly or wholly, the void behind the panel with a porous material, commonly mineral wool, or by gluing a thin fabric on to the backside of the panel. This was shown in Figure 5.2. There are several points to be noted concerning these solutions. The simple model used above is not valid when filling the void, wholly or partly, by a porous material. The model presupposes an empty air space and filling it modifies the resonance frequency calculated using Equation (5.39). The actual frequency will normally be lower. Furthermore, the particle velocity in the openings will be inversely proportional to the rate of perforation. We must take account of this when adjusting the resistance of the system. Using for example a porous material with a flow resistivity r and if we shall need a total flow resistance of R (Pa·s/m), this means that the thickness of the material required is $R \cdot \varepsilon / r$.

Figure 5.12 shows an example of calculated absorption factors at normal incidence for a distributed resonator of the type shown in Figure 5.11 a). The plate (panel) has a thickness of 1 mm and placed against a cavity having a depth of 100 mm. The plate is perforated with holes having a diameter of 3 mm and the area allotted to each hole is 100 mm^2, i.e. the perforation rate is approximately 7%. The lowest curve gives the result without any porous layer or fabric at the back of the plate, with the others a porous layer is added having the indicated total resistance. The calculations do not presuppose that the depth of the cavity is less than the wavelength. This means that, in addition to the broad peak around the resonance frequency according to Equation (5.39), we will see the effect of the standing waves in the cavity at higher frequencies.

A further example is given in Figure 5.13, showing measured results from a standard reverberation room test on a distributed resonator of the type discussed above. The panels, placed against a cavity of depth 50 mm, are measured having only a fabric of flow resistance 190 Pa·s/m glued to them as well as combined with the cavity filled with a high density porous material, rock wool 70 kg/m^3.

Furthermore, the rather thick panel of 14 mm allows the holes, normally cylindrical of diameter 8 mm, to have a conical shape (see insert to the figure), which may greatly enhance the absorption of this type of resonator absorbers (see Vigran (2004)). As seen from the results, filling the cavity completely with a porous absorber improves the low frequency absorption. However, by changing the shape of the perforations only, an amazing added improvement is attained.

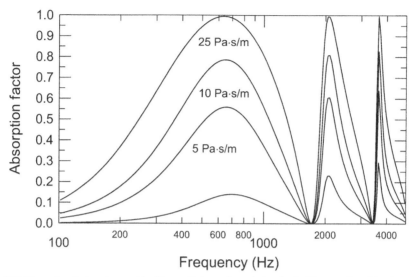

Figure 5.12 Absorption factor at normal incidence. Hole-perforated plate backed by a cavity of depth 100 mm. For further specifications, see text.

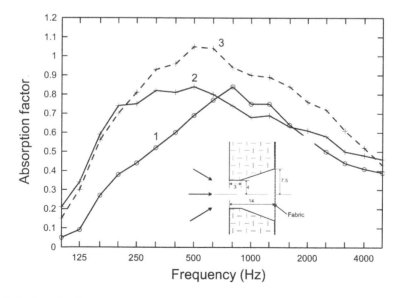

Figure 5.13 Absorption factor measured in a reverberation chamber. 14 mm perforated panel, perforation rate 12%, with a layer of fabric backed by a cavity of depth 50 mm. Curve no. 1 – cylindrical holes, empty cavity. Curve no. 2 – cylindrical holes, cavity filled with mineral wool. Curve no. 3 – conical holes, cavity filled with mineral wool.

As mentioned several times already, it is possible to obtain a sufficiently high resistance just by making the holes or slits small enough, which means a dimension of less than 0.5 mm. Traditionally, the hole-perforated models have been called micro-perforated absorbers or MPA due to the fact that these were the first on the market. However, as the same effect may be achieved by other shapes of perforations the term MPA comprises a larger range of products. We shall give two examples of data for such products, the first is 0.6 mm thick steel panels perforated by circular holes having diameter of 0.46 mm placed with a c-c distance of 5 mm. This corresponds to a rate of perforation of approximately 0.7%.

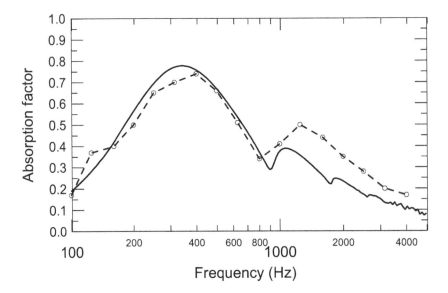

Figure 5.14 Absorption factor of microperforated plates (Gema Ultramicro® suspended 200 mm from a hard surface). Dotted curve – measured in a reverberation room. Solid curve – predicted.

Measured results, i.e. product data from the producer, are obtained in a standard reverberation room test, here with a cavity depth of 200 mm. These one-third-octave band data are shown in Figure 5.14 together with a prediction using the same type of calculation method as shown in Figure 5.12. Here, however, we have calculated a mean value over all incidence angles to compare with the diffuse field data from the reverberation room test. The prediction method for these absorption data is given in section 5.7.

Data allowing one to calculate the resonance frequency of resonator panels having perforations of other shapes than the circular holes may be found in the literature. We shall restrict ourselves to one important type, depicted in Figure 5.11 b). These absorbers are denoted "slatted panels", if they are constructed from parallel slats; which implies that the panel thickness normally is 9–10 mm or more. However, products of this type are often thin metal panels perforated by long slots. In that case the notion "slotted panels" may be more appropriate. As for the common slatted panels the width of the slots may be from some 5–10 mm upwards, which implies that a resistance fabric or porous material must be added. As for the calculation of resonance frequency, Equation (5.39) still applies but the end correction will be given by

$$\Delta d = -\frac{2b}{\pi} \cdot \ln\left[\sin\left(\frac{\pi \varepsilon}{2} \right) \right].$$ (5.40)

The quantity b is the width of the slot, and ε is the rate of perforation b/C, where C is the c-c distance between the slots. Similarly to the use of conically shaped holes instead of the normal cylindrical ones to enhance the absorption, one will obtain the same effect using a slatted panel where the slots are wedge-shaped instead of using the normal rectangular slats. However, this necessitates another prediction model (see Vigran (2004)).

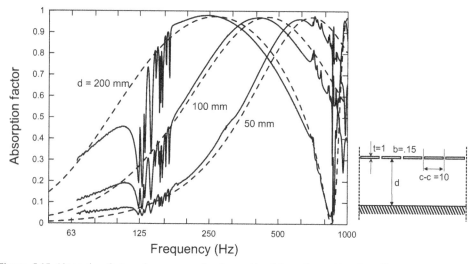

Figure 5.15 Absorption factor of resonance absorber. Aluminium plate with micro-slits. Measured and predicted results for normal incidence. The cavity depth d is indicated on the curves. Solid curves – measured. Dashed curves – predicted.

A recent development is using microperforations in the form of narrow slits of width of some tenth of a millimetre, again utilizing the "natural" viscous losses for obtaining the necessary resistance component (see section 5.4.1.3). We shall illustrate this by showing measured and predicted results where the measurements were performed in a standing wave tube. The tube had a square cross section with side length 200 mm, which limits the measurement range upwards to approximately 850 Hz. (Why is that ?)

A cross section of the plate used in shown in the insert to Figure 5.15. As indicated, the thickness of the aluminium plate used was 1.0 mm, and the slits made by laser were only 0.15 mm wide and 10 mm apart. Figure 5.15 shows measured and predicted results using a cavity depth of 50, 100 and 200 mm, respectively. As seen from the results, using a model based on Equations (5.36), (5.21) and (5.40), taking the perforation rate into account, predicts the general shape very well. However, this model presupposes that the plate itself does not move and cannot predict the excursions showing up in the frequency range 100–150 Hz, particularly pronounced at cavity depth 200 mm. These are due to mechanical resonances in the "bars" making up the plate.

5.4.1.6 Membrane absorbers

These absorbers are, as the name suggests, ideally, an impervious membrane stretched at a certain distance from a hard surface making up an airtight cavity. In common speech the notion has a wider use, also including cases where one is not using membranes but stiff materials such as metal or plastics. This means one is using materials giving bending forces and bending displacements, not only tensional ones. Modelling these absorbers, taking the bending stiffness into account, is relatively complicated. We shall therefore give an approximate model where we assume that the bending stiffness is negligible in comparison with the stiffness of the cavity. This implies that we return to the ideal membrane case. Assuming normal sound incidence, the input impedance may be written as

$$Z_g = R_m + j\omega m - j\rho_0 c_0 \cot g(k\ell)$$

$$\text{or} \quad Z_g \underset{k\ell \ll 1}{\approx} R_m + j\left(\omega m - \frac{\rho_0 c_0^2}{\omega \ell}\right), \tag{5.41}$$

where m is the mass per unit area of the membrane, and where R_m represents the resistance component in the system. By using the approximation shown in the last equation, where we assume that the depth ℓ of the cavity is small compared with the wavelength, we have returned to the simple mass-spring system having a resonance frequency f_0 given by

$$f_0 = \frac{\omega_0}{2\pi} = \frac{1}{2\pi}\sqrt{\frac{\rho_0 c_0^2}{m\ell}} \approx \frac{60}{\sqrt{m\ell}}. \tag{5.42}$$

In the approximation shown by the last term we have used $\rho_0 = 1.21$ kg/m^3 and $c_0 = 340$ m/s.

The absorption factor at resonance and the width of this resonance are certainly wholly dependent on the resistance. Estimating this quantity poses the real practical problem when designing membrane absorbers. Normally, thin metal panels are applied where the resistance is partly due to internal energy losses in the material itself, partly to frictional losses in the mechanical coupling between elements and partly to acoustic radiation. As pointed out above one also havs to take the possible influence of plate resonances into account, i.e. the eigenmodes of the single plates. This may give an absorption factor varying quite irregularly with frequency. This type of "membrane" absorber normally ends up having a badly adjusted resistance, which seldom gives a absorption factor in excess of 0.5. This also applies even when the cavity is filled, wholly or partly with a porous material. The simple model treated here does, however, not cover that case.

A quite different kind of membrane absorber is a collection of small, completely closed plastic "boxes" having wall thickness of some tenth of a millimetre (see e.g. Mechel and Kiesewetter (1981)). This gives a distributed resonant system where the effective resistance of each box is more optimal for the system than in the case of metal panels described above.

Another development, which deserves mention here, even if the membrane effect is not the primary concern, is absorbers using different types of flexible, microperforated sheets. These are, in fact, MPAs where the normal hole-perforated plates are replaced by flexible plastic sheets with holes made by heated spikes, a process much cheaper than making such holes in panels of hard materials as metal, glass etc. However, as these sheets are very light one cannot neglect the mass of the sheet itself and the calculation model has to be modified. Kang and Fuchs (1999) have presented measured and predicted results for such microperforated membranes and an example is given in Figure 5.16.

In this case the membrane has a thickness of 0.11 mm and a surface weight of 0.14 kg/m^2 and it is mounted at a distance of 100 mm in front of a rigid wall in a reverberation room. The diameter of the holes was 0.2 mm and the perforation rate was 0.79%. The measured results are presented together with predicted results using two slightly different models, the one by the authors and the other using a commercial software package WinFlag™. Both models represent the impedance of the perforated membrane as a parallel combination of the impedance of the membrane itself and the impedance represented by the perforations. For the latter, Kang and Fuchs use the approximations given by Maa (see e.g. Maa (1987)), whereas the other uses the Equation (5.31) directly.

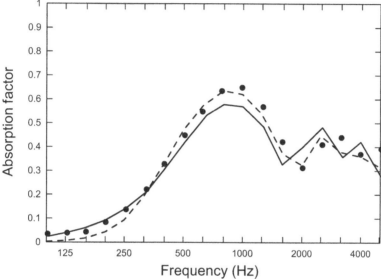

Figure 5.16 Absorption factor for a microperforated membrane mounted against a rigid wall at a distance of 100 mm. Measured (•) and predicted results (solid line) reproduced from Kang and Fuchs (1999). Dashed line – predicted results using the software WinFlag™.

5.5 POROUS MATERIALS

Modelling porous materials is still an active area of research due to the different fields of application. These are certainly not limited to the design of absorbers for use in room acoustics, but spans from materials for application in silencers over to the modelling of sound propagation in complex porous structures, which could be geological formations on the sea bottom as well as human tissue.

Traditionally, the porous materials used as sound absorbers have been of mineral wool fibre, either rock wool or glass wool. Later developments have been on cellular plastic foam materials, e.g. polyurethane, polyester etc. There are literally hundreds of different cellular foam materials on the market but only a few are actually suited as acoustic absorbers. The built-up structure of most porous materials is too complicated for a direct modelling of characteristic impedance and propagation coefficient based on the geometry of the frame or "skeleton" structure. This applies even if we assume that the frame is completely stiff resulting only in a movement of the air contained in the pores, this being subjected to viscous and thermal conditions. It should then not come as a surprise that most models describing sound propagation in porous materials may be characterized as being phenomenological. One will find models based on one or more macroscopic material properties, flow resistivity, porosity etc. We shall present some of these models, which in the literature are termed *equivalent fluid* models. The material behaves like a fluid in a macroscopic perspective, and the sound propagates in the form of a simple compressional wave.

When we cannot assume that the frame is completely stiff the modelling gets more difficult. The movement of the fame will be coupled to the movement of the air in the pores resulting in a significant influence on the properties in certain frequency ranges. There will be several types of wave propagating and, furthermore, it is generally not so that one type of wave propagates through the frame and another in the air particles in the pores. The models commonly used in this case are based on Biot theory (see e.g. Allard (1993)), which, in fact, was developed for quite another purpose, modelling sound propagation in porous, fluid-filled rock formations. We shall not go into details on this theory but give a short overview illustrated by examples.

5.5.1 The Rayleigh model

As a very simple model for a porous material we may envisage a bundle or matrix of very thin tubes. We assume that the tubes have a circular cross section and being sufficiently thin so as to make the air movements in them governed by viscous forces. We may then apply the calculations performed connected to the mode of operation of the Helmholtz resonator. A sketch of the cross section of the material is shown in Figure 5.17.

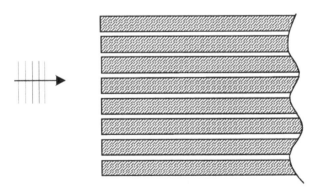

Figure 5.17 Simple model of a porous material, a bundle of thin tubes imbedded in a solid matrix.

Initially, we look at one of the tubes assuming that the particle velocity is represented by a mean value $\langle v_x \rangle$ taken over the cross section. We shall also use the approximation leading to Equation (5.32), assuming that the quantity s from Equation (5.28) is less than two. This implies that the diameter $(2a)$ of the tube should be less than 0.5 mm if the approximation is to be valid for frequencies up to 1000 Hz.

Denoting the mean value of the particle velocity as v_m will give the following equation of force

$$-\frac{\partial p}{\partial x} = j\omega\rho_0 v_m + \frac{8\mu}{a^2} v_m = j\omega\rho_0 v_m + r' v_m, \qquad (5.43)$$

where we have introduced the flow resistivity r' in the tube, having the dimension Pa·s/m^2. For simplicity, we have also replaced the constant 4/3 in the mass term by 1.

The corresponding continuity equation, i.e. the equation stating the conservation of mass, may in this one-dimensional case be written

$$-\rho_0 \frac{\partial v_m}{\partial x} = \frac{\partial \rho}{\partial t} = \frac{1}{c_0^2} \frac{\partial p}{\partial t}, \qquad (5.44)$$

when we take account of the relationship between the pressure and the density fluctuations. We shall look for solutions to these equations expressing the sound pressure and the particle velocity as

$$p = \hat{p} \cdot e^{j(\omega t - k' x)}$$

$$\text{and} \qquad v_m = \hat{v}_m \cdot e^{j(\omega t - k' x)}, \qquad (5.45)$$

where the wave number k' now is a complex quantity. Inserting these expressions into Equations (5.43) and (5.44) we get

$$k' p + (j r' - \rho_0 \omega) v_m = 0$$

$$\text{and} \qquad \omega p - \rho_0 c_0^2 k' v_m = 0, \qquad (5.46)$$

where a solution may only be found when the determinant is zero, i.e. when

$$\begin{vmatrix} k' & j r' - \rho_0 \omega \\ \omega & -\rho_0 c_0^2 k' \end{vmatrix} = 0.$$

This gives us the following equation for the wave number

$$k' = \frac{\omega}{c_0} \sqrt{1 - j\frac{r'}{\rho_0 \omega}} \qquad (5.47)$$

and thereby the impedance

$$Z' = \frac{p}{v_m} = \rho_0 c_0 \sqrt{1 - j\frac{r'}{\rho_0 \omega}}. \qquad (5.48)$$

We have then arrived at an expression giving the complex wave number and the characteristic impedance in one tube in the matrix of tubes sketched above. Our task now is to find the input impedance at the surface of the matrix. We have then to take account of the contraction of the "stream lines" when entering the matrix from the outside medium. This is accomplished by introducing the porosity σ of the matrix, the ratio of the pore (or tube) volume to the total volume. For this simple case the porosity will be the same as the quantity perforation rate ε used in section 5.4.1.5. Here we have to put $r = r'/\sigma$ and $Z = Z'/\sigma$, hence

$$k' = \frac{\omega}{c_0}\sqrt{1 - j\frac{r\sigma}{\rho_0\omega}}$$

$$\text{and}\qquad Z = \frac{\rho_0 c_0}{\sigma}\sqrt{1 - j\frac{r\sigma}{\rho_0\omega}}, \tag{5.49}$$

where Z will be the characteristic impedance for the equivalent fluid represented by this bundle of tubes. Z will also be the input impedance for a half-infinite thickness of such a porous medium, for which we may use Equation (5.4) to calculate the absorption factor for normal incidence. It should, however, be more realistic to calculate a situation where the porous medium has a finite thickness, also terminated by a hard reflecting surface. This will represent a first model simulating a porous sample of e.g. mineral wool placed against a hard wall in a room. We may modify Equation (5.21) by introducing the complex wave number k' and also exchanging the characteristic impedance $\rho_0 c_0$ for air by the impedance Z. The input impedance Z_g will then take the form

$$Z_g = -jZ\cdot\cot g(k'd). \tag{5.50}$$

Assuming that the thickness d of the material is much less than the wavelength ($k'd \ll 1$), we may use an approximation for the cotangent function, setting $\cot g(x) \approx 1/x - x/3$. Hence

$$Z_g \approx \frac{rd}{3} + j\left(\frac{\rho_0\omega d}{3\sigma} - \frac{\rho_0 c_0^2}{\omega\sigma d}\right). \tag{5.51}$$

This expression warrants several comments. First, the real part will only be one-third of the flow resistance. We will also get some sort of a resonance when the imaginary part is equal to zero. Normally, however, the stiffness part will dominate, which in practice gives a relatively high resonance frequency f_0. (Make a calculation of f_0 setting e.g. d equal 50 mm.)

5.5.2 Simple equivalent fluid models

A model suggested by Delany and Bazley (1970) is, due to its simplicity, widely used for describing the behaviour of porous materials, being applied to materials ranging from mineral wool products to porous soil. They developed their model, giving the complex wave number and the characteristic impedance, in a purely empirical way by measurements on a broad range of materials having a porosity of approximately one. This was done by fitting of data to a model having the flow resistivity and the frequency as parameters. Using the propagation coefficient $\Gamma = j\cdot k'$ instead of the complex wave number k', the expressions are

$$Z_c = \rho_0 c_0\left[1 + 0.0571\cdot E^{-0.754} - j\cdot 0.087\cdot E^{-0.732}\right]$$

$$\text{and}\qquad \Gamma = j\frac{\omega}{c_0}\left[1 + 0.0978\cdot E^{-0.700} - j\cdot 0.189\cdot E^{-0.595}\right] \quad\text{where } E = \frac{\rho_0 f}{r}. \tag{5.52}$$

It is assumed that the quantity E lies inside the range 0.01–1.0. As mentioned above, the model has been and still is widely used. One should, however, bear in mind that the materials used in the development were highly porous mineral wool products. A slightly different model is given by Mechel (1976) (see also Mechel (1988)), where the equations are partly theoretically based, i.e. when describing the behaviour at low frequencies, partly by a curve-fitting procedure on measured data at the higher frequency range. The model also includes the porosity as a parameter but, unfortunately, this parameter only affects the low frequency part. In effect, this is again a one-parameter model and to fit the various expressions together it is advisable to set the porosity parameter equal to 0.95.

When $E < E_x$ (see Table 5.1 for the transition between equations), we get

$$\Gamma = \Gamma_{\text{real}} + j\Gamma_{\text{imag}} = j\frac{\omega}{c_0}\sqrt{1 - j\frac{\gamma}{2\pi E}}$$

$$\text{and} \qquad Z_c = Z_{\text{real}} + jZ_{\text{imag}} = -j\frac{\rho_0 c_0^2}{\omega\gamma\sigma}\Gamma,$$

(5.53)

where γ is adiabatic constant for air (≈ 1.4). When $E > E_x$, Mechel gives the following expressions

$$\Gamma = \frac{\omega}{c_0}\left[0.2082 E^{-0.6193} + j\cdot\left(1 + 0.1087 E^{-0.6731}\right)\right]$$

$$\text{and} \qquad Z_c = \rho_0 c_0 \left[\left(1 + 0.06082 E^{-0.717}\right) - j\cdot 0.1323 E^{-0.6601}\right].$$

(5.54)

The limiting value for E, denoted E_x, that determines whether one shall use Equations (5.53) or (5.54) for the real and imaginary components of Γ and Z_c, respectively, is shown in Table 5.1.

Table 5.1 Limiting values for the components in the model by Mechel (1988).

Component	Limiting value E_x
Γ_{real}	0.04
Γ_{imag}	0.008
Z_{real}	0.006
Z_{imag}	0.02

Calculations on a 50 mm thick porous material having a flow resistivity of 10 kPa·s/m^2 and backed by a hard wall is shown in Figures 5.18 and 5.19, using the models of Delany-Bazley and Mechel. The porosity is put equal to 0.95 in Mechel's model. The first figure gives the real and imaginary part of the input impedance, the second one the corresponding absorption factor. It is observed that the differences are quite small in this example. However, Mechel's model does "repair" the anomaly at the lower frequencies.

Wilson (1997) has also developed a simple phenomenological model for sound propagation in porous materials, based on viewing the thermal and viscous diffusion in porous media as relaxational processes. Of particular interest compared to the models

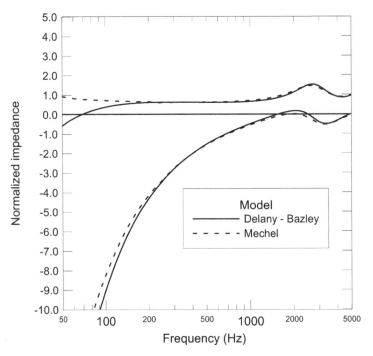

Figure 5.18 Real and imaginary part of the input impedance of a 50 mm thick material with a hard backing (normal incidence). Comparison of models by Delany-Bazley and Mechel.

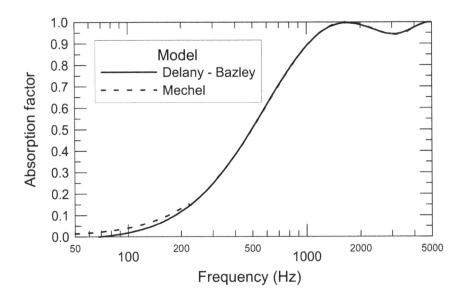

Figure 5.19 Absorption factor corresponding to the data shown in Figure 5.18.

above is that Wilson's model also repairs the anomaly of the Delany-Bazley model giving a realistic prediction over a far broader frequency range. In fact, predictions at low frequencies coincide very well with the ones by Mechel's model rendering it unnecessary to "splice" models in order to cover a broader frequency range. The latter comparison is not given by Wilson but is easily demonstrated.

5.5.3 Absorption as a function of material parameters and dimensions

It is interesting and of great practical importance as well, to know the way in which material parameters and dimensions do influence the absorption capabilities of a porous material. The influence of the parameters is easy to show in models involving just one or may be two parameters. For models having a great number of parameters, as the ones given below, it is rather difficult to give a complete overview. We shall therefore restrict the following illustrations, to the effect of varying the flow resistivity and thickness of the sample, to the model of Delany and Bazley. Furthermore, to make the illustrations simple, most data shown in this chapter apply to normal sound incidence on the actual absorbing surface. Referring back to the treatment in Chapter 3, on absorption by oblique and ultimately diffuse sound incidence, it will be appropriate also to illustrate this effect.

A further presumption to these calculations is that the actual surface is "infinitely" large. In practice, this implies that the surface is sufficiently large in comparison to wavelength thus enabling us to neglect any effects due to the outer free edges. On finite size samples there will, however, always be some diffraction effects along the edges, the so-called *edge effect*. The result is an increase in the effective absorption area, which means that the "acoustic area" is larger than the geometrical area. One may therefore end up with data for the absorption factor larger than one (1.0) when dividing the measured total absorption area by the area of the absorber.

As mentioned when presenting the measurement methods for absorption, this effect still shows up in results from reverberation room measurements in spite of the rather large area specified (10–12 m^2) and a ratio of width to length between 0.7 and 1.0 just to minimize this effect. Thomasson (1980), using the sound field distribution above a finite absorbing surface surrounded by a hard surface, which corresponds to a reverberation room situation, calculated the effective statistical absorption factor as a function of area. We shall illustrate the importance of absorber size by using Thomason's expression to compare with measured data from a reverberation room test.

5.5.3.1 Flow resistivity and thickness of sample

In the introduction it was mentioned that the normal mounting of a porous absorber is either directly on to a hard surface or at a certain distance from it, i.e. leaving a cavity behind it. For ceilings, the latter is the normal mounting, not only because there must be some space for the service equipment but one gains additional absorption at the lower frequencies. We will start giving some results where the absorber is directly attached to a hard and infinitely large surface.

A typical result when varying the thickness of porous absorber is shown in Figure 5.20. The absorption factor is calculated for normal incidence and for thickness in the range of 25 to 100 mm, the flow resistivity being 10 kPa·s/m^2. As is apparent from the figure, the thickness has to be large to obtain high absorption at the lower frequencies. The physical explanation is that waves having the larger wavelengths penetrate far into the material, also being less attenuated and thereby reflected from the back wall.

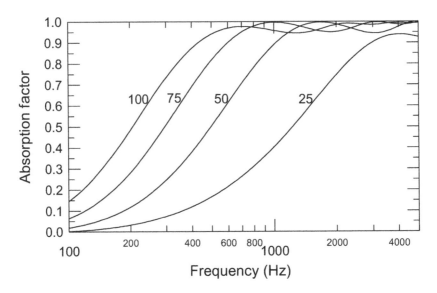

Figure 5.20 Absorption factor at normal incidence for a porous material attached directly on to a hard wall. The parameter in the graph is the thickness in mm. Dalany-Bazley model with r equal 10 kPa·s/m².

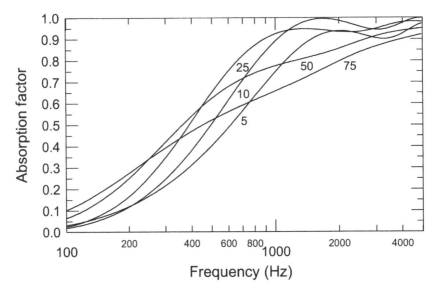

Figure 5.21 Absorption factor at normal incidence for a 50 mm thick porous material attached directly on to a hard wall. The parameter in the graph is the flow resistivity in kPa·s/m².

There is, however, a "trade-off" between thickness and flow resistance. We may increase the absorption in the lower frequency range, for a fixed thickness, by increasing the flow resistivity as the latter will increase the attenuation rate. But there will be an optimum value here due to increased reflection from the front surface caused by the increased flow resistivity. This is easily demonstrated by Figure 5.21 showing again the absorption factor for a 50 mm porous absorber at normal incidence, now where the flow resistivity is varied in the range 5 to 75 kPa·s/m^2. An increased absorption at the lower frequencies is obtained at the cost of a decreased absorption at the higher frequencies when making the flow resistivity too high. Some product data shown in Figure 5.33 may serve as an indication of the relationship between flow resistivity and the density.

We may, however, obtain a substantial increase in the absorption at lower frequencies by mounting the absorber at a certain distance from a wall or ceiling. This is, in fact, common practice when mounting absorbing ceiling panels. One will not obtain fully the same absorption as when applying the same total thickness of the material, but the effect is good. An example is shown in Figure 5.22, giving the absorption factor for three different combinations of a porous absorber and the cavity behind. The model used here is the one by Mechel using a flow resistivity of 10 kPa·s/m^2. As shown in section 5.5.2, this model gives a more correct result in the lower frequency range than the Delany-Bazley model but this is not important here. Making a comparison with the absorption offered by a 25 mm thick absorber placed directly against a hard wall (see Figure 5.20), the combination of a 25 mm absorber and a 75 mm cavity gives a substantial increase at low frequencies. The drawback of such a combination is that we get standing wave phenomena in the cavity, thereby a reduced effect in certain frequency ranges. We have seen these phenomena before when treating resonance absorbers (see e.g. Figure 5.12). The locations of these minima will, however, depend on the angle of incidence. For the absorption factor data measured by diffuse sound incidence this effect will hardly be noticeable, except when the ratio of cavity depth to the thickness of the porous sample is very large.

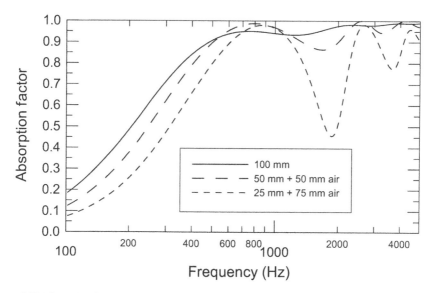

Figure 5.22 Absorption factor at normal incidence for combinations of a porous absorber with an airspace in front of a hard wall. Mechel's model for the porous absorber having $r = 10$ kPa·s/m^2.

5.5.3.2 Angle of incidence dependency. Diffuse field data

We have, to make it simple, used the condition of normal sound incidence in most illustrations. This is, however, not the angle of incidence giving the maximum absorption. Taking porous materials as an example, we find that an angle of incidence 50–60° will give a maximum absorption factor. The mean value obtained when averaging over all angles of incidence, i.e. the statistical absorption factor α_{stat}, is of even more practical interest. We may determine this factor by using our models to calculate the mean value for incidence angles in the range 0 to 90°. As shown earlier on, assuming local reaction such that the input impedance Z_g is independent of the angle of incidence φ, the statistical absorption factor is expressed as

$$\alpha_{stat} = 2 \int_0^{\pi/2} \alpha(\varphi) \sin\varphi \cos\varphi \, d\varphi = 2 \int_0^{\pi/2} \left[1 - \left| R_p \right|^2 \right] \sin\varphi \cos\varphi \, d\varphi, \qquad (5.55)$$

where R_p is the pressure reflection factor given by

$$R_p = \frac{Z_g \cos\varphi - Z_0}{Z_g \cos\varphi + Z_0} = \frac{Z_n \cos\varphi - 1}{Z_n \cos\varphi + 1}. \qquad (5.56)$$

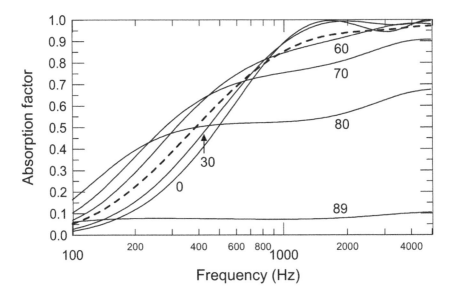

Figure 5.23 Absorption factor for a 50 mm porous sample with hard backing. The parameter on the curves is the angle of incidence in degrees. The dashed curve shows the statistical absorption factor. Delany-Bazley model with $r = 10$ kPa·s/m^2.

In the last expression we have normalized the input impedance by the characteristic impedance Z_0 for air. Inserting this last expression into Equation (5.55) we get

$$\alpha_{stat} = 8 \operatorname{Re}\{Z_n\} \int_0^{\frac{\pi}{2}} \frac{\sin\varphi \cos^2\varphi}{\left| Z_n \cos\varphi + 1 \right|^2} \, d\varphi, \qquad (5.57)$$

where Re again denote the real part of the actual quantity.

In Figure 5.23, we have again used a porous material with a thickness of 50 mm as an example. The angle of incidence is varied between zero and 89 degrees, i.e. between normal incidence and nearly grazing incidence. One can observe the larger absorption obtained for oblique incidence, however approaching zero by grazing incidence. The dotted curve shows the statistical absorption factor calculated from Equation (5.57).

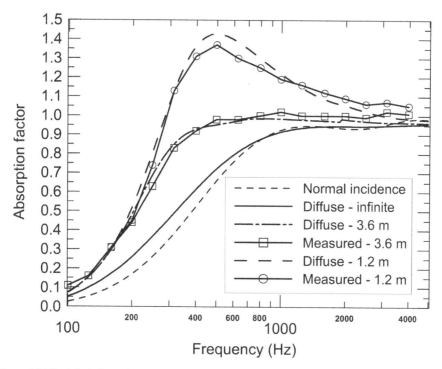

Figure 5.24 Statistical absorption factor of a porous absorber having different areas, 50 mm mineral wool (r = 30 kPa·s/m²) with hard backing. The test area is square with dimensions indicated. The corresponding predicted absorption factors for an infinitely large specimen, both for normal and diffuse sound incidence, are also shown. Prediction method for finite size specimen and corresponding measurement data (reverberation room) from Thomasson (1982).

It now remains to see how the last result would turn out if one cannot assume that the lateral dimensions of the sample are not very large compared to the wavelength. We have indeed up to now assumed that the area of the absorber was infinitely large. That we must take account of a finite size absorber does not only apply in a standard reverberation room measurement, but certainly in normal practical applications. Thomasson (1980) has shown, again assuming that the absorber is locally reacting, that we should substitute Equation (5.57) by

$$\alpha_{stat} = \frac{4\,\mathrm{Re}\{Z_n\}}{\pi} \int\limits_0^{\pi/2} \int\limits_0^{2\pi} \frac{\sin\varphi}{|Z_n + Z_f|^2}\,\mathrm{d}\varphi\,\mathrm{d}\theta, \tag{5.58}$$

where Z_f is denoted *field impedance*, a quantity that may be interpreted as a radiation impedance for a plane surface having the same shape as the absorber and the same velocity distribution. Thus, Z_f will be a function of the shape and dimensions of the specimen as well as a function of the frequency and the angle of incidence. It should be noted that the direction of incidence must be specified both by the angle φ with the surface normal as well as the azimuth angle θ.

One may interpret this in another way by using an electrical analogue, a circuit where $Z_f \cdot Z_0$ is the internal impedance of a generator twice the sound pressure in the incoming wave and where $Z_n \cdot Z_0$ is the outer load impedance. The expressions giving Z_f are quite complicated integrals that normally have to be solved numerically. As expected, Equation (5.58) is approaching Equation (5.57) when the linear dimensions of the absorber get large compared to the wavelength because then $Z_f \approx 1/\cos\varphi$.

Figure 5.24 shows predicted results compared with measurements data from a reverberation room test. According to Thomasson (1982), the test sample is mineral wool of 50 mm thickness having a flow resistivity of 30 kPa·s/m². His measurement data from the reverberation room tests are given in one-third-octave bands. Measurements and predictions are performed on three different sample areas of which we show the results for the two areas, 1.2 x 1.2 m² and 3.6 x 3.6 m².

For the calculated results in Figure 5.24 we have again used the model by Delany and Bazley to describe the mineral wool. For comparison, we have also given the result for the absorption factor at normal sound incidence as well as for diffuse sound incidence using Equation (5.57), i.e. data corresponding to the ones shown in Figure 5.23. As is evident from Figure 5.24, the increase in the statistical absorption factor for a finite sample size is quite dramatic, even for the sample having a side length 3.6 of metres; which in fact is a common size for reverberation room tests. As we also may observe the fit between measured and calculated results is generally very good. It should be mentioned that Thomasson's calculated values for one-third-octave bands are not shown in the figure. This is because the differences between his calculated results and the ones plotted are negligible.

We shall round off this discussion of the edge effect by presenting an illustration showing how it can be utilized in practice. As an alternative to attaching a certain amount of absorbers on to a wall or ceiling covering one single area one may, if this is not unsuitable, split the absorber into smaller patches separated by some distance. Figure 5.25 shows results from an experiment conducted in a reverberation room. Eight blankets of 25 mm thick mineral wool, each having a dimension of 0.6 times 1.2 metres, was first arranged as one single area and thereafter separated as shown in the sketch beside the figure. As the total area when placed adjacent to one another (5.8 m²) is less the 10 m² required for a normal test, the edge effect for the lowest curve will be larger than normal. However, when pushing the blankets away from each other the increase in the absorption is very large. (It should be noted that the absorption factor is calculated for an area of 5.8 m² in all cases).

Holmberg et al. (2003) have developed a prediction model to calculate the statistical absorption factor of such absorbing patches arranged in a periodic pattern. The measurement data shown above were, in fact, used to compare with their predictions. The predicted results, which compared favorably with the measured ones, are, however, not shown here.

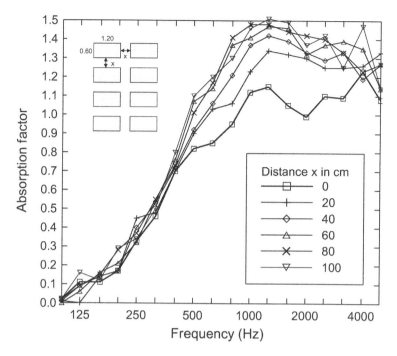

Figure 5.25 Reverberation room measurements of the absorption factor of eight mineral wool blankets, 25 mm thick and having a total area of 5.8 m². Dimensions and measured configuration is shown in the sketch. Student tutorial, NTNU.

5.5.4 Further models for materials with a stiff frame (skeleton)

In more recent years there has been development of several models using a more detailed description of the structure of the material, the aim being to use directly measurable material quantities. Early work on this (e.g. Zwikker and Kosten (1949)), introduced a *structure factor* in addition to the flow resistivity and porosity, a factor that is now termed *tortuosity* or *sinuosity*. In a popular way, we may say that this parameter gives us information on the directionality of the pores in the material. In a material having straight through pores of cylindrical shape making an angle φ with the outer surface, the tortuosity k_s is given by

$$k_s = \frac{1}{\cos^2 \varphi}, \tag{5.59}$$

which, as an example, gives k_s equals 2 for φ equal 45°. In many applications however, a model using these three parameters will be too simple even for materials where one assumes isotropy and a stiff frame.

5.5.4.1 The model of Attenborough

Attenborough (1983,1992), introduces an additional parameter s_f, denoted *pore shape factor*, as a phenomenological description of the geometrical form of the pores. The problem with this factor is that it cannot be measured separately but must be estimated by fitting the model to measured data. As shown in Chapter 3 (section 3.5.3), we may express the complex characteristic impedance Z_c and the propagation coefficient Γ by an equivalent or effective density ρ_{eff} and a corresponding bulk modulus K_{eff}, thus

$$Z_c = \sqrt{K_{eff}\,\rho_{eff}} \quad \text{and} \quad \Gamma = j\omega\sqrt{\frac{\rho_{eff}}{K_{eff}}}. \tag{5.60}$$

According to the model of Attenbourough we get

$$\rho_{eff} = \rho_0 k_s \cfrac{1}{\left[1 - \cfrac{2}{s_A\sqrt{-j}}\cfrac{J_1(s_A\sqrt{-j})}{J_0(s_A\sqrt{-j})}\right]} \quad \text{and}$$

$$K_{eff} = \cfrac{\gamma P_0}{\left[1 + (\gamma-1)\cfrac{2}{\sqrt{Pr}\cdot s_A\sqrt{-j}}\cfrac{J_1(\sqrt{Pr}\cdot s_A\sqrt{-j})}{J_0(\sqrt{Pr}\cdot s_A\sqrt{-j})}\right]}. \tag{5.61}$$

The symbol J denotes a Bessel function and the quantity s_A is given by

$$s_A = \frac{1}{s_f}\sqrt{\frac{8\omega\rho_0 k_s}{r\sigma}}. \tag{5.62}$$

The quantity Pr in Equation (5.61) is the so-called *Prandtl number* given by $\mu \cdot c_p/\kappa$. This number is a constant for a given fluid that describes the relationship between the coefficient of viscosity μ, the thermal conductivity κ and the specific heat capacity at constant pressure c_p. For air we get Pr $\approx$ 0.71.

Assuming $s_A \ll 1$, which implies low frequency and/or high flow resistivity r, Attenborough gives the following expressions for characteristic impedance and propagation coefficient:

$$Z_c \approx j\omega\rho_0\left[\frac{s_f^2 r}{\omega\rho_0} - j\frac{4k_s}{3\sigma}\right]/\Gamma \qquad \Gamma \approx j\frac{\omega}{c_0}\sqrt{\gamma\sigma}\left[\left(\frac{4}{3} - \frac{\gamma-1}{\gamma}Pr\right)\frac{k_s}{\sigma} - j\frac{s_f^2 r}{\omega\rho_0}\right]^{\frac{1}{2}}. \tag{5.63}$$

As an example on the use of these equations, Figure 5.26 shows measured and predicted absorption factors for discs of a porous wood (rattan palm) placed at given distances from a hard wall. In one set of curves the disc thickness d is 5 mm with an air cavity depth l of 85 mm and in the other set the disc thickness is 10 mm and the cavity depth is 40 mm. Measurements are performed in a standing wave tube, starting with measurements on a 50 mm thick sample placed directly on the hard backing surface. From the measured impedance data on this sample, the appropriate material parameters were extracted by fitting the Attenborough model to the data. We are then in the position to calculate Z_c and Γ according to Equations (5.60) and (5.61), which in turn enable us to calculate the results shown in Figure 5.26 using a general calculation routine based on transfer matrices (see section 5.7.1 and Equation (5.85)).

Certainly, in this case more simple models could probably have been used as the pores in the wooden material are straight tubes directed normally to the surface of the cut discs.

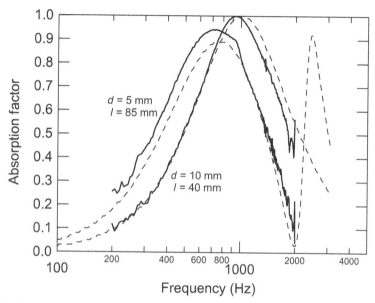

Figure 5.26 Normal incidence absorption factor of a disc of porous wood (rattan palm) mounted in front of a hard wall. Disc thickness (d) and cavity depth (l) is indicated. Solid curves – measured. Dashed curves – predicted. See also description in the text.

5.5.4.2 The model of Allard/Johnson

This model (see e.g. Allard (1993)) exchanges the non-measurable quantity s_f in the Attenborough formulation with two other parameters. These are the characteristic *viscous length* Λ and the characteristic *thermal length* Λ', which are defined in the following way

$$\frac{2}{\Lambda} = \frac{\oint_S v_i^2(r_w)\,\mathrm{d}S}{\int_V v_i^2(r)\,\mathrm{d}V} \quad \text{and} \quad \frac{2}{\Lambda'} = \frac{A}{V}. \tag{5.64}$$

In the expression for Λ the numerator is a surface integral where the velocity v_i of the fluid, as indicated by the index w, applies to the inner walls of the pores. The denominator is the corresponding volume integral that applies to the whole volume of pores. The thermal length Λ' is given by the ratio of the total inner surface area A to the total volume V of the pores.

The great advantage in using this description is that it is possible to determine both parameters separately by ultrasonic measurement technique (see below for measurements of material parameters). Concerning the size of these parameters, one will find values in the range of some tenths of a micrometer to some hundred micrometers for typical porous foam materials. The ratio between the two parameters will tell us something of what the pores look like. We have tried to illustrate this in Figure 5.27. Certainly, the two parameters will be equal in a material where the pores have simple tube like shape, whereas $\Lambda < \Lambda'$ when the connections between the pores are small and narrow. This is due to the fact that the viscous length Λ will mainly be determined by the contributions from areas having large velocity amplitudes, i.e. where the passages are narrow.

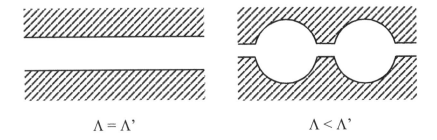

$$\Lambda = \Lambda' \qquad\qquad\qquad \Lambda < \Lambda'$$

Figure 5.27 Sketches of the form of the pores in a porous material.

The expressions for effective density and bulk modulus according to this model is

$$\rho_{\text{eff}} = \rho_0 k_s \left[1 + \frac{r\sigma}{j\omega\rho_0 k_s} G_J(\omega) \right] \qquad \text{and}$$

$$K_{\text{eff}} = \frac{\gamma P_0}{\gamma - (\gamma - 1)\left[1 + \frac{8\mu}{j\text{Pr}\cdot\omega\rho_0\Lambda'^2} G_J'(\text{Pr}\cdot\omega) \right]^{-1}}. \qquad (5.65)$$

The functions G_J and G_J' are given by

$$G_J(\omega) = \left[1 + j\frac{4k_s^2\mu\rho_0\omega}{r^2\Lambda^2\sigma^2} \right]^{\frac{1}{2}}$$

and $\qquad\qquad\qquad\qquad\qquad\qquad\qquad\qquad\qquad\qquad (5.66)$

$$G_J'(\text{Pr}\cdot\omega) = \left[1 + j\frac{\Lambda'^2\rho_0\omega\text{Pr}}{16\mu} \right]^{\frac{1}{2}}.$$

Figure 5.28 may serve as an example showing the importance of these characteristic lengths. Calculations are performed for a 50 mm thick sample directly on to a hard backing, keeping the characteristic thermal length constant while varying the corresponding viscous length between 10 and 100 μm. The other parameters used in this example are given in the figure caption.

When both lengths are relatively large and of equal size the absorption characteristic is similar to the ones found for ordinary mineral wool products. By decreasing values of Λ, however, the characteristic resembles the ones found for certain types of plastic foam materials (see next section).

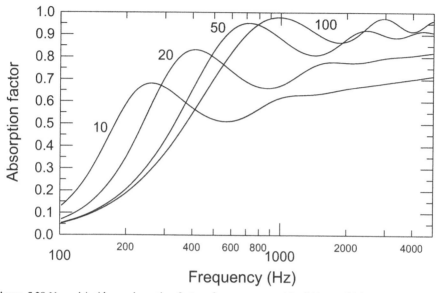

Figure 5.28 Normal incidence absorption factor of a porous material of 50 mm thickness with hard backing predicted by the Allard/Johnson model. The curve parameter is the characteristic viscous length Λ (μm). Other data are: $r - 20000$ Pa·s/m², $\sigma - 0.95$, $k_s - 2.0$, $\Lambda' - 100$μm.

5.5.5 Models for materials having an elastic frame (skeleton)

The models are getting even more complicated where we cannot assume that the frame stays motionless under the sound field impact. There will be a coupled motion of the frame and the air in the pores, which may give pronounced effects in certain frequency ranges. Up to now we have tacitly assumed that that there will be just one wave type propagating in the medium, i.e. a compressional wave. Now we will have three types of wave: two compressional waves denoted a fast and a slow wave, respectively, together with a shear wave. These waves will have quite different properties depending on the coupling between the fluid and the frame.

In common porous materials, where the fluid is air, there is a weak coupling between the frame and the fluid. In the audio- and ultrasound frequency range one of the compressional waves will mainly propagate in the air contained in the pores leaving the frame motionless. The second one will propagate in both media, having a velocity approximately equal to the velocity of a compressional wave in the frame situated in vacuum. One then gets a situation where the motion or vibration of the frame results in a corresponding motion of the air in the pores.

A movement in the air will, on the other hand, not affect the frame in the same way, as the latter is much heavier. The air will move in the frame without affecting it. One should then envisage that we could talk about an airborne and a frame-borne wave, but the former, for which the attenuation is large due to the viscous coupling between frame and air, is traditionally denoted the slow wave.

The effect of an elastic frame may be quite pronounced for certain types of absorber, especially for plastic foam materials. A simulated result is shown in Figure 5.29, where the shear modulus G of the frame is reduced from a maximum value (G1) of $2.0 \cdot 10^7(1+j \cdot 0.1)$ Pa in steps of 10. The other parameters for the material are identical to the ones used in Figure 5.28 having Λ equal to 20μ. The calculations are performed using a full Biot-model following the procedure and formulae given by Brouard et al. (1995). To compare, we have also plotted the data, shown by the circular points, from the latter figure that assumes an infinitely stiff frame according to the Allard/Johnson model. Apparently, the results using the maximum value of the shear modulus gives nearly identical result as a calculation assuming an infinitely stiff frame.

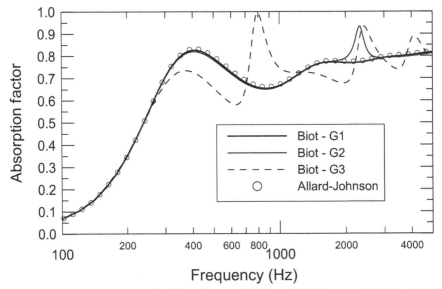

Figure 5.29 Normal incidence absorption factor of a material with an elastic frame modelled using Biot theory. The shear modulus is reduced in steps with a factor of 10 from G1 = $2.0 \cdot 10^7(1+j \cdot 0.1)$ Pa. Other data are the same as used in Figure 5.28 ($\Lambda = 20\mu$). Symbols indicate data calculated by the Allard/Johnson model given by Equation (5.65).

Another example is given in Figure 5.30, which shows a comparison between measured and calculated results, again using a full Biot model. The measurements are conducted in a free field environment using a two-microphone technique. The calculations are performed using two different methods, an analytical one and one using a finite element method (FEM). The analytical method is the same as used in Figure 5.29. Apparently, the two calculation methods are consistent with each other and the general appearance is validated by the measurement results. It should be noted that a linear frequency axis is used in this case.

The absorption has a maximum value in the frequency range 700–800 Hz, which is caused by a rather strong movement of the frame at these frequencies. Calculating the displacement at the surface of the foam material, a calculation possible by using FEM technique, is shown in Figure 5.31. The displacement is calculated for a normal incidence plane wave having a sound pressure of 1 Pa. We observe that the displacement has a pronounced maximum in the same frequency range as found for the absorption curve.

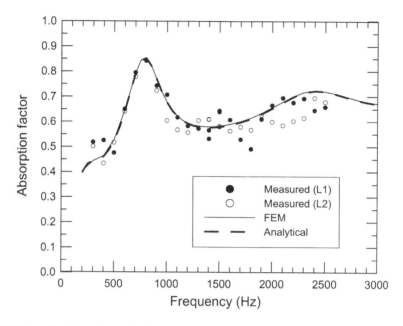

Figure 5.30 Normal incidence absorption factor of 50 mm polyurethane with hard backing. Measurement data, using a two-microphone free field technique and two loudspeaker positions (L1, L2), are compared both with analytical results and results using a finite element method (FEM). From Vigran et al. (1997).

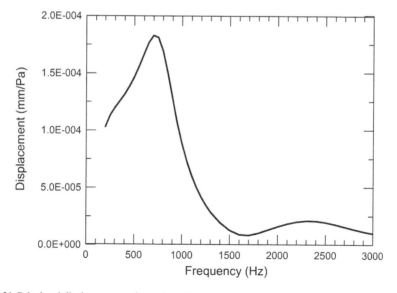

Figure 5.31 Calculated displacement at the surface of 50 mm thick polyurethane foam. Incident sound pressure equal 1 Pa.

5.6 MEASUREMENTS OF MATERIAL PARAMETERS

In general, there exist several methods for determining the material parameters needed for the models described here. Most of them are laboratory methods and it is certainly an advantage having a method which allows direct determination of each individual parameter. There is an international standard for determination of the important parameter airflow resistance.

5.6.1 Airflow resistance and resistivity

The airflow resistivity is one of the most important parameters to characterize porous materials. Porous materials, as applicable in building acoustics, could be mineral wool products, plastic foam materials as well as porous fabric, curtains etc. The terms and symbols used for characterising the resistance of materials may sometimes be confusing. We shall here use the international standard, ISO 9053, as a directive, represented in Table 5.2.

Table 5.2 Terms and symbols for airflow resistance.

	Quantity	Symbol	Unit
ISO 9053	1) Airflow resistance	R	Pa·s/m^3 (N·s/m^5)
	2) Specific airflow resistance	R_S	Pa·s/m (N·s/m^3)
	3) Airflow resistivity	r	Pa·s/m^2 (N·s/m^4)

The ISO standard specifies two different methods for determining these quantities. Both methods are based on measuring the pressure difference across a disk, cut out from the test material, when a known volume of air is passing through. The difference between the methods is that one method (method a) uses a constant airflow (DC flow) while the other uses alternating low frequency airflow (AC flow) of frequency 2–4 Hz. In the former case one measures the pressure difference by a manometer while the latter uses a microphone. Sketches showing these two principal methods are shown in Figure 5.32.

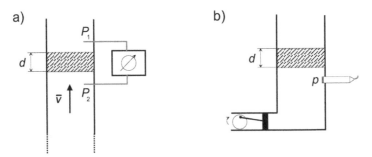

Figure 5.32 Principal set-up for measurement of airflow resistance according to ISO 9053. a) Direct airflow method; b) Alternating airflow method.

The quantities R, R_s and r are defined by the following

$$R = \frac{\Delta P}{q_v}, \qquad R_s = R \cdot S \quad \text{and} \quad r = \frac{R_s}{d}. \qquad (5.67)$$

The quantity q_v is volume velocity (m³/s) of the airflow through the specimen having an area S, which implies that the mean flow velocity is equal to q_v/S (m/s). The quantity d is the thickness of the sample in the direction of the flow. The *specific airflow resistance* R_s is the linear airflow resistance of the sample, having the same dimension as the specific impedance. In the old centimetre-gram-second (CGS) system of units, this quantity had its own unit, the Rayl in honour of the physicist, Lord Rayleigh. One may find it still in use but then as mks Rayls reflecting the SI system of units.

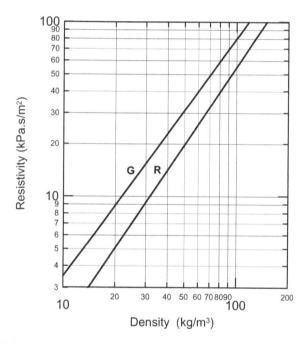

Figure 5.33 Typical data for airflow resistivity of mineral wool, glass wool (G) and rock wool (R) as a function of density.

For a homogeneous material we may find the flow resistance per unit length, the airflow resistivity having the symbol r. This is the quantity normally found as product data for porous materials. Some typical data (Norwegian products) for mineral wool types of porous material are given in Figure 5.33.

Finally, it should be mentioned that in other circumstances an inverse quantity is used, this in order to characterize the "openness" of a porous material for airflows. This is the quantity *permeability B* defined by

$$B = \frac{\mu}{r} \qquad \text{(m}^2\text{)}, \qquad (5.68)$$

where μ is the coefficient of viscosity.

5.6.2 Porosity

By the porosity of a porous material, it is understood as the relative part of the volume of the open pores in the material, which in our case is the relative volume part of air. This applies to fibrous materials as well as granular ones. As is apparent from the models described above, the parameter is important when it comes to sound propagation in porous materials. In connection with geophysical characterization, it is common practice to determine this parameter by filling up the pores with water or some other fluid. It goes without saying that this will not be a practical procedure in the case of the porous materials used as acoustic absorbers.

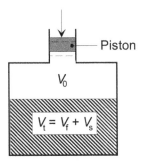

Figure 5.34 A principal set-up for determining porosity.

Champoux et al. (1991) have developed an accurate method based on using air. The principle is not new but by using modern equipment they arrive at accuracy better than 1%. As shown in Figure 5.34, the material having a given total volume $V_t = V_f + V_s$ is placed in a closed compartment. Here V_f and V_s denote the volume of the pores and the volume of the solid frame, respectively. By definition, the porosity σ is given by

$$\sigma = \frac{V_f}{V_t}. \tag{5.69}$$

When we talk about the volume of the pores we shall infer that the pores in this volume are interconnected, i.e. excluding the volume of closed pores. The rest of the air volume in the chamber is denoted V_0. The procedure is now to give the piston a controlled displacement resulting in a precise change ΔV in the volume and a resulting pressure change ΔP. Assuming that the pressure in the chamber initially is equal to the barometric pressure P_0 and, furthermore, that the change of state takes place isothermally we get

$$P_0 V = (P_0 + \Delta P)(V + \Delta V)$$

or   (5.70)

$$V = -(P_0 + \Delta P)\frac{\Delta V}{\Delta P}.$$

We have then determined the unknown total volume V of air in the chamber, $V = V_0 + V_f$. Since the volume V_0 is easily measured, we have determined the sought after volume V_f of the pores.

This is an outline of the basic principle of the measurement. The practical implementation certainly involves problems such as ensuring an isothermal change of state, the determination of small pressure changes etc. In the set-up mentioned, one is using a piston of diameter 4 mm, positioned to an accuracy of 1 micrometer. A differential pressure transducer detects pressure changes within 10^{-6} mm of mercury. The material samples have a maximum volume of 1.5 litres. The general accuracy is, as mentioned above, better than 1%.

5.6.3 Tortuosity, characteristic viscous and thermal lengths

There are several methods for the determination of tortuosity. For materials having a non-conducting frame one may compare the conductivity when saturating the material with an electrical conducting fluid with the conductivity of the fluid itself (see e.g. Allard (1993:73)).

Later developments apply more efficient methods based on high frequency measuring techniques. The principle is based on measuring the sound transmission through the material, utilizing the high frequency asymptotic behaviour to determine the tortuosity and well as the characteristic viscous and thermal lengths. At sufficiently high frequencies, a practical frequency range for these measurements is 100–800 kHz, one may assume that the frame is motionless. The complex wave number may then be approximated to read

$$ k' \approx \frac{\omega}{c_0} \sqrt{k_s} \left[1 + (1-j) \frac{\delta}{2} \left(\frac{1}{\Lambda} + \frac{\gamma - 1}{\sqrt{\text{Pr}} \, \Lambda'} \right) \right] \quad \text{where} \quad \delta = \sqrt{\frac{2\mu}{\rho_0 \omega}}. \quad (5.71) $$

The quantity δ is denoted the *viscous skin depth*. Allard et al. (1994) used this expression to determine the tortuosity k_s, utilizing the fact that the viscous skin depth approaches zero at sufficiently high frequencies, i.e. that the following apply:

$$ \left(\frac{c_0}{c_{\text{eff}}} \right)^2 \xrightarrow[\frac{1}{\sqrt{f}} \to 0]{} k_s \quad \text{where} \quad c_{\text{eff}} = \frac{\omega}{k'}. \quad (5.72) $$

The quantity c_{eff} is thereby the effective speed of sound through the material. The measurements are relatively easy to perform by placing a disc of the material between an ultrasound source and receiver. One then compares the transit time of a broadband pulse of ultrasound with and without the disc between source and receiver. A Fourier transformation into the frequency domain then gives the sound speed as a function of frequency, which enables one to use the extrapolation given in Equation (5.72). A practical problem is to find suitable ultrasound transducers for air, both having sufficient power and bandwidth. The range of ultrasound propagation in air is generally short in addition to a normally large attenuation through the material sample. This implies that the sample may have to be just a few millimetres in thickness, thus not being representative of the material as such.

Leclaire et al. (1996) took the use of Equation (5.71) a step further by utilizing the differences in the physical properties of air and helium to obtain an independent determination of the characteristic viscous and thermal lengths. A procedure to do this is to plot the quantity

$$y = \mathrm{Re}\left\{\left(\frac{c_0}{c_{\mathrm{eff}}}\right)^2\right\} \quad \text{as a function of } \left(\frac{1}{\sqrt{f}}\right), \text{ a linear function given by}$$

$$y \approx k_s\left[1 + \frac{1}{\sqrt{f}}\sqrt{\frac{\mu}{\rho_0 \pi}}\left(\frac{1}{\Lambda} + \frac{\gamma - 1}{\sqrt{\mathrm{Pr}}\Lambda'}\right)\right]. \tag{5.73}$$

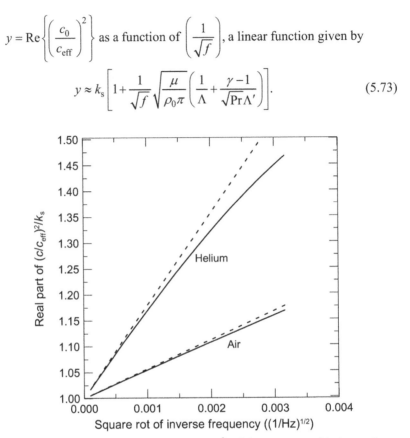

Figure 5.35 The relationship between the real part of $(c_0/c_{\mathrm{eff}})^2$ and the square root of the inverse frequency for a porous material having air-filled and helium-filled pores, respectively. The ordinate is normalized by the tortuosity k_S. Material parameters: r – 30000Pa·s/m², σ – 0.95, Λ – 50μm, Λ' – 100μm. Solid lines – complete model. Dashed lines – approximate model.

The slope of this function gives the possibility of determining the length L in the expression

$$L = \left[\frac{1}{\Lambda} + \frac{\gamma - 1}{\sqrt{\mathrm{Pr}}\Lambda'}\right]^{-1},$$

whereas k_S again is determined by the intercept with the ordinate. Performing this measurement twice, once by having the pores filled with air and, second, by helium, we can determine Λ and Λ' by finding the slopes b_{air} and b_{He} given by

$$b_{\text{air}} = \frac{k_s}{L_{\text{air}}} \sqrt{\frac{\upsilon_{\text{air}}}{\pi}} \quad \text{and}$$

$$b_{\text{He}} = \frac{k_s}{L_{\text{He}}} \sqrt{\frac{\upsilon_{\text{He}}}{\pi}} \quad \text{where} \quad \upsilon_i = \left(\frac{\mu}{\rho_0}\right)_i. \tag{5.74}$$

How this linear approximation based on Equation (5.71) looks like compared with calculated results using the complete model is shown in Figure 5.35. By the notion complete model we understand the one where the complex wave number k' is calculated from Equations (5.65) and (5.66). The suitability of the linear approximation is certainly dependent on the ratio of Λ to Λ', which in the example is 1:2. By using a frequency range upwards from 250 kHz, the method seems to be accurate within 10–15 %. More information may be found in the referenced paper.

5.7 PREDICTION METHODS FOR IMPEDANCE AND ABSORPTION

Both commercial and specially made sound absorbers are rarely a simple and homogeneous structure. As an example, a fabric of some kind, a plastic membrane, a perforated panel etc., may cover a porous material. The whole structure may then be mounted at a certain distance from a wall or ceiling. We have given several examples of data for such absorbers in the sections above. Here we shall give a short review of the prediction method used.

A number of the elements or layers making up the structure of a given absorber may not be characterized as being locally reacting. This implies, not only that the input impedance will depend on the angle of sound incidence, but we will also get a lateral wave movement, i.e. along the actual surface. We then have to take the dimensions and the boundary conditions into account. We may accomplish this by applying models using finite element methods (FEM). There is software, also commercially available, to perform acoustic calculations on absorbers, especially on porous materials. We gave an example earlier (see also Figure 5.30).

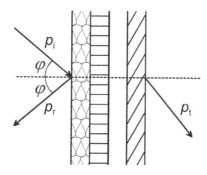

Figure 5.36 A construction composed of several layers, each of infinite extent.

Another procedure is by way of analytical modelling using transfer matrices. Basically, each layer in the combination, assumed to be of infinite extent, is represented by a matrix giving the relationship between a set of physical variables on the input and output side of the layer. These matrices may then be combined to give the relationship between the relevant physical variables for the whole combination. Characteristic data as absorption factor, input impedance and sound reduction index (transmission loss) may then be calculated assuming plane wave incidence. The size and complexity of these matrices, however, are totally dependent on the specific material in the actual layer, i.e. how many physical variables one has to use describing the wave motion in the material and then how many material parameters that are necessary to specify the material. In many cases, two physical variables are sufficient i.e. the sound pressure and the particle velocity. A simple 2 by 2 matrix then describes the relationship between these variables on the input and the output side. We shall use this description below to illustrate the method.

5.7.1 Modelling by transfer matrices

For the description of layers e.g. thin plates (panels), either perforated or non-perforated, two physical variables are always sufficient. The word "thin" here signify that we do not need to worry about the wave motion inside the plate itself; the wavelength being much longer than the thickness of the plate. With thicker elastic materials this simple model is no longer feasible (see the discussion below).

Porous materials may also be included in a simple 2 by 2 matrix description if they are modelled as an equivalent fluid. Such a model is applicable to many porous materials, e.g. mineral wool type absorbers. The basic assumptions are that the material is homogeneous and isotropic, having pores filled with air embedded in an infinitely stiff matrix or skeleton. Again, if the elastic properties of this skeleton have to be taken into account a description using two physical variables only is not feasible.

In our outline of the transfer matrix method we shall only use the 2 by 2 matrices, which in the analogue electrical case is denoted a two port or four pole. Using the sound pressure p together with particle velocity v as the variables, we can express the relationship between these variables on each side of the layer numbered n:

$$\begin{bmatrix} p_{n-1} \\ v_{n-1} \end{bmatrix} = \begin{bmatrix} a_{11} & a_{12} \\ a_{21} & a_{22} \end{bmatrix} \begin{bmatrix} p_n \\ v_n \end{bmatrix}. \tag{5.75}$$

The matrix for the total system, i.e. the one describing the relation between variables on the input and output side of the system, is arrived at by multiplying together the matrices representing each of the contributing layers. Denoting the elements in this matrix as A_{11}, A_{12}, A_{21} and A_{22}, the input impedance Z_g is given by

$$Z_g = \frac{p_0}{v_0} = \frac{A_{11}Z_L + A_{12}}{A_{21}Z_L + A_{22}}, \tag{5.76}$$

where Z_L is the impedance on the output side, the load impedance. As shown earlier, when Z_g is known we are able to calculate the absorption factor by using the expressions

$$\alpha = 1 - \left| R_p \right|^2 \quad \text{and} \quad R_p = \frac{Z_g \cos \varphi_0 - \rho_0 c_0}{Z_g \cos \varphi_0 + \rho_0 c_0}. \tag{5.77}$$

The angle φ_0 denotes the direction of plane wave incidence on the first layer. Furthermore, we have assumed that the medium on the input side is air having characteristic impedance $\rho_0 c_0$.

As mentioned above, our model will turn out much more complicated if we want to include elastic materials either solid or porous. In the former case, we need at least four physical variables for description, e.g. the particle velocity and the stress in two directions. The corresponding matrix to the one given in Equation (5.75) will therefore be a 4 by 4 matrix instead of a 2 by 2 matrix. It may be shown, however, that if there are fluid layers on both sides of the elastic layer, this 4 by 4 matrix may be reduced to a simple 2 by 2 matrix.

Having a porous elastic material, on which we want to use the Biot theory, we end up with a 6 by 6 matrix. If we want to combine layers described in such a different way we cannot just multiply the matrices to make a model for the complete system; we shall have to construct coupling matrices expressing the boundary conditions between the layers. We have presented results above that have been calculated using such a technique (see section 5.5.5). Here we shall just illustrate the technique by showing how to find the components of the 2 by 2 matrix for a porous material described as an equivalent fluid. For a corresponding description using the Biot theory we shall refer to the literature (see e.g. Brouard et al. (1995)).

5.7.1.1 Porous materials and panels

We shall characterize a porous material by using the wave number k and the characteristic impedance Z_c, both normally complex quantities. Alternatively, we could have used the effective density and the bulk modulus. However, the conversions between these variables are simple if one wishes to use the alternative description. For simplicity, we shall consider plane wave incidence normal to a layer of thickness d (see Figure 5.37). Our task is, first, to set up the relationships between the sound pressure and the particle velocity on the two sides of the layer, second, to cast these into the form given in Equation (5.75).

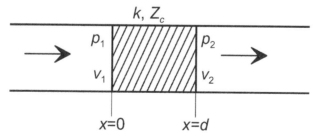

Figure 5.37 Sound transmission through a porous layer of thickness d.

Generally, we may express these variables by the following equations, when assuming harmonic time dependence

$$p(x) = A \cdot e^{-jkx} + B \cdot e^{jkx}$$

$$\text{and} \quad v(x) = \frac{1}{Z_c}\left(A \cdot e^{-jkx} - B \cdot e^{jkx}\right). \tag{5.78}$$

The quantities A and B will be determined by the boundary conditions on each side of the layer. On the left-hand side, where x is equal zero, we get

$$p_1 = A + B$$

$$\text{and} \quad v_1 = \frac{1}{Z_c}\left(A - B\right). \tag{5.79}$$

On the right-hand side, where x is equal to d, the pressure is given by

$$p_2 = A \cdot e^{-jkd} + B \cdot e^{jkd} = (A + B) \cdot \cos kd - j \cdot (A - B) \cdot \sin kd. \tag{5.80}$$

Hence, using the Equations (5.79)

$$p_2 = p_1 \cdot \cos kd - j \cdot Z_c v_1 \cos kd. \tag{5.81}$$

Correspondingly, the particle velocity on the output side will be

$$v_2 = v_1 \cdot \cos kd - j \cdot \frac{p_1}{Z_c} \sin kd. \tag{5.82}$$

We may now cast the Equations (5.81) and (5.82) into the form sought after

$$\begin{bmatrix} p_1 \\ v_1 \end{bmatrix} = \begin{bmatrix} \cos kd & j \cdot Z_c \sin kd \\ j \cdot \dfrac{\sin kd}{Z_c} & \cos kd \end{bmatrix} \begin{bmatrix} p_2 \\ v_2 \end{bmatrix}. \tag{5.83}$$

As an example, we shall assume that the layer is placed on to an infinitely hard wall, which implies that v_2 is equal to zero or that the load impedance Z_L in Equation (5.76) is infinite. The input impedance will then be

$$Z_1 = \frac{p_1}{v_1} = -j \cdot Z_c \cot g(kd), \tag{5.84}$$

which was expected from our formerly derived result (see Equation (5.50)).

Equation (5.83) may, however, be put into a general form. First, it is common to substitute the wave number by the propagation coefficient Γ, which is given by $j \cdot k$. Second, we shall not assume normal incidence but introduce oblique incidence giving the wave vector propagating through the material an angle φ with the normal. The result will be

$$
\begin{bmatrix} p_1 \\ v_1 \end{bmatrix} = \begin{bmatrix} \cosh\left(\Gamma d \cdot \cos\varphi\right) & \dfrac{Z_c \sinh\left(\Gamma d \cdot \cos\varphi\right)}{\cos\varphi} \\ \dfrac{\sinh\left(\Gamma d \cdot \cos\varphi\right)}{Z_c} \cdot \cos\varphi & \cosh\left(\Gamma d \cdot \cos\varphi\right) \end{bmatrix} \begin{bmatrix} p_2 \\ v_2 \end{bmatrix}. \tag{5.85}
$$

It should be noted that $\cos\varphi$ could be a complex quantity; see discussion on Snell's law in Chapter 3 (section 3.5.3).

A concentrated layer, such as a membrane, thin plate etc. will result in a much simpler transfer matrix. In this case we get

$$
\begin{bmatrix} p_1 \\ v_1 \end{bmatrix} = \begin{bmatrix} 1 & Z_w \\ 0 & 1 \end{bmatrix} \begin{bmatrix} p_2 \\ v_2 \end{bmatrix}, \tag{5.86}
$$

where Z_w is the so-called *wall impedance*, which is given by the ratio of the sound pressure difference across the layer to the velocity of the layer. We thereby assume that the velocity is equal on both sides of the layer. Equations (5.85) and (5.86) are the ones used in a number of the illustrations in this chapter (see e.g. Kristiansen and Vigran (1994)).

5.8 REFERENCES

ISO 9053: 1991, Acoustics – Materials for acoustical applications – Determination of airflow resistance.

ISO 10534: 1996, Acoustics – Determination of sound absorption coefficient and impedance in impedance tubes. Part 1: Method using standing wave ratio. Part 2: Transfer-function method.

ISO 13472–1: 2002, Acoustics – Measurement of sound absorption properties of road surfaces in situ. Part 1: Extended surface method.

ISO 354: 2003, Acoustics – Measurement of sound absorption in a reverberation room.

Allard, J. F. (1993) *Propagation of sound in porous media. Modelling sound absorbing materials.* Elsevier Applied Science, London and New York.

Allard, J. F., Castanede, B., Henry, M. and Lauriks, W. (1994) Evaluation of the tortuosity in acoustic materials saturated by air. *Review of Scientific Instruments,* 65, 754–755.

Attenborough, K. (1983) Acoustical characteristics of rigid fibrous absorbents and granular materials. *J. Acoust. Soc. Am.,* 73, 785–799.

Attenborough, K. (1992) Ground parameter information for propagation modelling. *J. Acoust. Soc. Am.,* 92, 418.

Brouard, B., Lafarge, D. and Allard, J. F. (1995) A general method for modelling sound propagation in layered media. *J. Sound Vib.,* 183, 129–142.

Champoux, Y., Stinson, M. R. and Daigle, G. A. (1991) Air-based system for the measurement of porosity. *J. Acoust. Soc. Am.,* 89, 910–916.

Delany, M. E. and Bazley, E. N. (1969) Acoustical characterisation of fibrous absorbent materials. NPL Aero Report AC 37.

Delany, M. E. and Bazley, E. N. (1970) Acoustical properties of fibrous materials. *Applied Acoustics,* 3, 105.

Dutilleux, G., Vigran, T. E. and Kristiansen, U. R. (2001) An in situ transfer function technique for the assessment of the acoustic absorption of materials in buildings. *Appl. Acoustics*, 62, 555–572.

Holmberg, D., Hammer, P. and Nilsson, E. (2003) Absorption and radiation impedance of finite absorbing patches. *Acta Acustica/Acustica*, 89, 406–415.

Kang, J. and Fuchs, H. V. (1999) Predicting the absorption of open weave textile and micro-perforated membranes backed by air. *J. Sound and Vibration*, 220, 905–920.

Kristiansen, U. R. and Vigran, T. E. (1994) On the design of resonant absorbers using a slotted plate. *Applied Acoustics*, 43, 39–48.

Leclaire, P., Kelders, L., Lauriks, W., Melon, M., Brown, N. and Castanede, B. (1996) Determination of the viscous and thermal characteristic length of plastic foams by ultrasonic measurements in helium and air. *J. Appl. Physics*, 80, 2009–2012.

Maa, D. Y. (1987) Microperforated-panel wideband absorbers. *Noise Control Engineering Journal*, 29, 77–84.

Mechel, F. P. (1976) Ausweitung der Absorberformel von Delany und Bazley zu tiefen Frequenzen. *Acustica*, 35, 210–213.

Mechel, F. P. (1988) Design charts for sound absorber layers. *J. Acoust. Soc. Am.*, 83, 1002–1013.

Mechel, F. P. and Kiesewetter, N. (1981) Schallabsorber aus Kunstoff-Folie. *Acustica*, 47, 83–88.

Mommertz, E. (1995) Angle dependent in situ measurement of the complex reflection coefficient using a subtraction technique. *Applied Acoustics*, 46, 251–263.

Thomasson, S.-I. (1980) On the absorption coefficient. *Acustica*, 44, 265–273.

Thomasson, S.-I. (1982) Theory and experiments on the sound absorption as function of the area. Report TRITA-TAK-8201, Dept. Tech. Acoustics, Royal Institute of Technology, Stockholm.

Vigran, T. E. (2004) Conical apertures in panels; sound transmission and enhanced absorption in resonator systems. *Acta Acustica/Acustica*, 90, 1170–1177.

Vigran, T. E., Kelders, L., Lauriks, W. and Leclaire, P. (1997) Prediction and measurements of the influence of boundary conditions in a standing wave tube. *Acustica/ Acta Acustica*, 83, 419–423.

Vigran, T. E. and Pettersen, O. K. Ø. (2005) The absorption of slotted panels revisited. Proceedings of the Forum Acusticum 2005, Budapest.

Wilson, D. K. (1997) Simple, relaxational models for the acoustic properties of porous media. *Applied Acoustics*, 50, 171–188.

Zwikker, C. and Kosten, C. W. (1949) *Sound absorbing materials*. Elsevier, New York.

CHAPTER 6

Sound transmission. Characterization and properties of single walls and floors

6.1 INTRODUCTION

In the preceding chapter, our concern was directed at the task of removing acoustic energy from a medium (air) by transport into an absorber that effectively could convert the energy into heat. Absorbing materials have a wide range of applications: as sound absorbers in rooms, included in noise barriers along roads, in silencers for air-conditioning systems etc. One could then envisage that e.g. a porous material could be effective in reducing the transport of sound energy from an air-filled space to another, i.e. act as an isolator for sound energy in the same way as the material works for heat energy. Unfortunately, this is not the case. We normally demand that a dividing wall of reasonably good quality between rooms has a sound reduction index in the range 40–50 dB. This corresponds to a transmission factor of 10^{-4}–10^{-5} (see definitions below). To achieve this, our porous material must have an absorption factor of 0.9999 in the actual frequency range, indicating that this is not a workable solution. Effective partition walls are based on a major jump in the impedance, i.e. a very high reflection factor. Reducing sound transmission between rooms is therefore based on reflecting the energy back as opposed to trying to dissipate the energy in the partition.

These considerations apply to the transmission of airborne sound; we shall use the concept *airborne sound insulation* when talking of the ability of a construction to isolate against airborne sound. With the notion *structure-borne sound* is meant vibrations in solid structures, which in turn may radiate sound in the audible frequency range. Human walking or jumping is the major source of these vibrations in buildings, and one will find in the literature the terms *impact sound* and *footfall noise* used for describing the radiated sound. We shall use the first term, *impact sound insulation* as the corollary to airborne sound insulation. It should be noted that here we strictly are concerned with the transmission phenomena, i.e. not with the sound generated in the source room itself.

This chapter will be devoted to methods and techniques for prediction and measurement of the sound insulation properties of partitions, i.e. the sound transmission properties both by excitation of airborne sound and impacts. These properties may be quite different due to the different type of excitation by these two sources. In the case of airborne sound, a distributed pressure field will drive the construction, whereas the other excitation will be by "point" forces. This, however, does not prevent us, subject to some conditions, of calculating the airborne sound insulation when the impact sound insulation is known and vice versa.

As an introduction to the subject we shall use a practical example involving both types of excitation, in this case also a more general type of impact. A piece of machinery, mounted on a floor as shown in Figure 6.1, may induce vibrations (structure-borne sound) in the floor due to unbalanced forces. Furthermore, the machine will radiate sound energy setting up a sound field in the room, which may in turn excite the floor. Which one of these processes will dominate the sound energy being radiated into the room below will not only depend on the dynamic properties of the floor, but also on its radiating properties, i.e. the efficiency of the floor to radiate sound. One may consider this process as analogous to a two-stage rocket. The vibration pattern set up in the floor is determined by the excitation and the dynamic properties and the radiated sound energy will be determined by how well this vibration pattern couples to the room below. This does not exclude possible feedback in the system; the floor vibration may influence the sources at the same time as the surrounding medium (air) may influence the movement of the floor. The latter type of feedback is normally neglected when dealing with building constructions; the inertia of the constructions is normally of quite another order of magnitude compared with the corresponding ones for air. The situation would have been quite different if the medium were water!

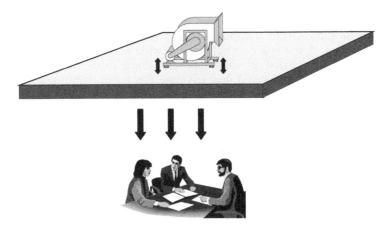

Figure 6.1 Machinery on the floor in an equipment room. Sound transmission to neighbouring room.

6.2 CHARACTERIZING AIRBORNE AND IMPACT SOUND INSULATION

We shall introduce the quantities used to characterize sound insulation, quantities that will be found in common building regulations and requirements for the sound insulating properties of building elements and constructions. We shall derive the expressions used when measuring these properties both in laboratories and in the field and we shall show how these results may be converted to give a single number rating.

6.2.1 Transmission factor and sound reduction index

The transmission factor τ of a given surface is defined by the sound power, the ratio of the transmitted power W_t and the power W_i incident on the surface:

$$\tau = \frac{W_t}{W_i}.$$

$$(6.1)$$

The sound reduction index R is the corresponding logarithmic quantity defined as

$$R = 10 \cdot \lg\left(\frac{1}{\tau}\right) = 10 \cdot \lg\left(\frac{W_i}{W_t}\right) \qquad \text{(dB)}. \qquad (6.2)$$

In the literature one often finds the term *transmission loss* used for R.

In a traditional laboratory measurement procedure to determine the sound reduction index of a building element, it is presupposed that all sound energy transmission from the sending room to the receiver room takes place by way of the actual element (see Figure 6.2). In practice, there are always limitations in spite of the massive flanking constructions surrounding the element under test. When the sound insulation gets sufficiently high there is bound to be additional sound transmission by way of these flanking constructions in much of the same way as normally encountered in buildings. This is discussed below.

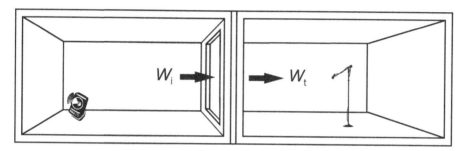

Figure 6.2 Laboratory set-up to determine the sound reduction index.

We shall assume that the sound field in the sending room, as well as in the receiving room, is diffuse. The sound intensity at the wall in the sending room will then be given by

$$I_i = \frac{\tilde{p}_S^2}{4\rho_0 c_0}, \qquad (6.3)$$

where p_S is the sound pressure in the sending room. The power transmitted through the building element having the surface area S will be

$$W_t = I_t \cdot S = \frac{\tilde{p}_R^2}{4\rho_0 c_0} \cdot A_R, \qquad (6.4)$$

where p_R and A_R is the sound pressure and the total absorption area, respectively, in the receiving room. Hence, the transmission factor will be given by

$$\tau = \frac{I_t \cdot S}{I_i \cdot S} = \frac{\tilde{p}_R^2}{\tilde{p}_S^2} \cdot \frac{A_R}{S}. \qquad (6.5)$$

The sound reduction index then becomes

$$R = 10 \cdot \lg\left(\frac{1}{\tau}\right) = 20 \cdot \lg\left(\frac{\tilde{p}_S}{\tilde{p}_R}\right) + 10 \cdot \lg\frac{S}{A_R} = L_S - L_R + 10 \cdot \lg\frac{S}{A_R}, \qquad (6.6)$$

where $D = L_S - L_R$ is the difference in the mean sound pressure level in the sending and receiving room. This is the expression used in a standard laboratory procedure based on measurements of the sound pressure levels (see ISO Standard 140 Part 3).

An alternative procedure is based on determination of the transmitted power to the receiving room by way of measuring the intensity. An important reason for applying such a method is when the traditional method breaks down due to substantial flanking transmission. Using this method one determines the mean transmitted intensity I_R over a surface S_R that completely encloses the actual element having an area S. Instead of Equations (6.5) and (6.6) we get

$$\tau = \frac{I_R \cdot S_R}{\dfrac{\tilde{p}_S^2}{4\rho_0 c_0} \cdot S} \qquad (6.7)$$

and

$$R_I = 10 \cdot \lg\left(\frac{\tilde{p}_S^2}{4\rho_0 c_0 \cdot I_R}\right) + 10 \cdot \lg\left(\frac{S}{S_R}\right) \approx L_{pS} - L_{IR} + 10 \cdot \lg\left(\frac{S}{S_R}\right) - 6\,\mathrm{dB}. \quad (6.8)$$

The latter expression corresponds to the one to be found in ISO 15186 Part 1, having introduced the mean sound pressure level L_{pS} in the sending room and the mean intensity level L_{IR} taken over the measuring surface S_R. In addition, the characteristic impedance $\rho_0 c_0$ for air is set equal to 400 Pa·s/m.

Do we get the same results when applying these two methods? In theory, there will be a difference due to a certain underestimating of the transmitted power by the traditional method, resulting in a higher sound reduction index. When determining the mean sound pressure level in the receiving room certain limits are imposed on the distance to the room boundaries, i.e. the power is determined by measurements in the "inner part" of the room. A modified sound reduction index $R_{I,M}$ will therefore be more in accordance with R determined using Equation (6.6). This is defined as

$$R_{I,M} = R_I + K_c = R_I + 10 \cdot \lg\left(1 + \frac{S_b c_0}{8Vf}\right), \qquad (6.9)$$

where the last term is the so-called Waterhouse correction (see section 4.5.1.1). The quantities S_b and V are the total surface and volume of the receiving room, respectively. For the frequency f one uses the centre frequency in the actual frequency band.

6.2.1.1 Apparent sound reduction index

As opposed to the measuring situation in a laboratory, the normal situation in a building is the existence of a large number of transmission paths for the sound energy (see Figure 6.3). As indicated in the figure, sound energy may, in addition to being directly transmitted through the wall partition, be transmitted via flanking constructions, via crack formations, out and in through windows, via a common ventilation duct, via cable ducts etc.

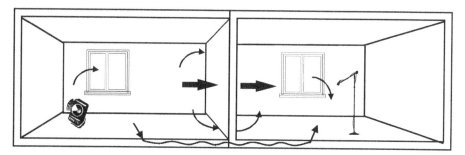

Figure 6.3 Measurement of sound insulation in a building. Examples of transmission paths between rooms.

It should be noted that the notion *flanking transmission* only implies that the transmission takes place in the manner of setting the flanking constructions into vibration, a part of which is transferred to constructions on the receiving side that are capable of radiating sound energy. We shall use the common term *transmission path* when referring to all the other mechanisms for transmission of sound between rooms.

The requirements for airborne sound insulation in buildings are in most countries given by certain limiting values of the sound reduction index. The same measurement procedure and calculation method as for the laboratory is applied. However, as one does not quantify the various contributions to the sound pressure level in the receiving room, we will write instead of Equation (6.6):

$$R' = D + 10 \cdot \lg \frac{S}{A_\mathrm{R}},\tag{6.10}$$

where D again is the difference in sound pressure level between rooms, but now the sound reduction index R' is called the *apparent sound reduction index* of the partition. A popular way of expressing this is to say that the partition takes the blame when other transmission paths contribute significantly to the level in the receiving room. There are, however, other measures that may be used to specify the sound insulation. As stated in ISO 140 Part 4, which concerns airborne sound insulation measurements in the field, we may also use the sound pressure level difference referred to a given reverberation time T_0. Here we shall write

$$D_{\mathrm{n}T} = D + 10 \cdot \lg \left(\frac{T}{T_0} \right),\tag{6.11}$$

where the quantity $D_{\mathrm{n}T}$ is denoted *standardized level difference* with T_0 set equal to 0.5 seconds for dwellings. Some countries are using this quantity in their requirements concerning airborne sound insulation, and it could be argued that this quantity is more in line with the actual sound insulation experienced by the users than the R'-value.

6.2.1.2 Single number ratings and weighted sound reduction index

When specifying the sound insulation capability of constructions, in particular when connected to acoustical requirements in building codes, it is sensible to use a single number instead of the whole frequency curve. Normally, the latter is composed of data in one-third-frequency bands from 100 to 3150 Hz or in an extended range going from 50 to 5000 Hz. The procedure to substitute these data by a single number is based on the use of a *reference curve*, which means that we have agreed on a suitable sound reduction curve to be compared with our measurement data. We shall not delve into the history of how these reference curves were developed, it is sufficient to state that there exists such reference curves both for airborne and impact sound, these being internationally accepted (see ISO 717 Part 1 and Part 2). The reference curve for the sound reduction index is shown in Figure 6.4 together with a measured result, a laboratory measurement of a double wall of 13 mm plasterboards on separate studs. The distance between the boards is 150 mm and the void between the boards is filled with mineral wool.

To calculate the single number value R_W, the reference curve is shifted in 1 dB steps towards the measured curve until the sum of unfavourable deviations is as large as possible but not more then 32.0 dB when using 16 measurement frequencies. An unfavourable deviation at a given frequency occurs when the measurement result is less than the reference value. Only the unfavourable deviations are taken into account, i.e. high sound insulation data in the higher frequency range does not compensate for bad insulation at low frequencies.

As is apparent from Figure 6.4, a maximum shift of 7 dB is possible in this case, giving a sum of unfavourable deviations of 28 dB and hence an R_W of 59 dB. In former standards there existed the so-called "8 dB rule", which was based on the non-acceptance of very low sound insulation at one or more frequency bands. Unfavourable deviations of maximum 8 dB was accepted, a rule that was often the determining factor when calculating R_W.

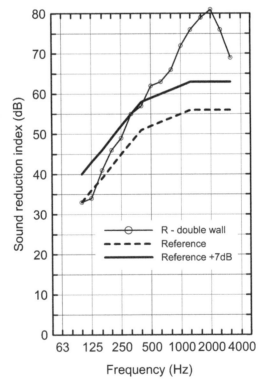

Figure 6.4 Calculation of the weighted sound reduction index R_W. In this example R_W is equal to 59 dB, which is the reference value at 500 Hz. The sum of unfavourable deviations is 28 dB; see text.

To a certain extent, the "8 dB rule" is compensated for by the introduction of the so-called *spectral adaptation terms*, which is added to the weighted sound reduction index. These terms, specified by the symbol C, are in this case defined as:

$$C = X_{A,1} - X_w \quad \text{and}$$
$$C_{tr} = X_{A,2} - X_w, \tag{6.12}$$

where $X_{A,1}$ is the normalized difference in the A-weighted sound pressure level between the sending and receiving room, the source spectrum being pink noise. Correspondingly, $X_{A,2}$ is the normalized difference in the A-weighted sound pressure level between the sending room (or in the free field in front of a façade) and receiving room, the source spectrum being road traffic noise. The symbol X_w stands for the single number calculated using the reference curve (e.g. R_w or R'_w).

For the specification of the sound insulation of façades, in particular against road traffic noise, a traffic noise sound insulation index R_A has been in use in the Nordic countries. This index is adapted by ISO stating

$$R_A = R_w + C_{tr}. \tag{6.13}$$

6.2.1.3 Procedure for calculating the adaptation terms

We shall illustrate the calculation procedure using the adaptation term C_{tr} as an example. We shall calculate the sound reduction index R_A according to Equation (6.13) using the sketches in Figure 6.5 as a basis. We will assume that the source, being road traffic noise, sets up a diffuse sound field in the room with sound pressure level L_{in}, the corresponding driving sound pressure level at the façade is L_{out}. We shall define a sound reduction index for the actual part of the façade, having an area S, by the equation

$$R = L_{out} - L_{in} + 10 \cdot \lg\left(\frac{S}{A}\right), \tag{6.14}$$

where A is the total absorption area in the room. Normalizing the sound reduction index by setting $S/A \equiv 1$, we may express the A-weighted sound pressure levels outside and inside, respectively, as

$$(L_{pA})_{out} = 10 \cdot \lg\left[\sum_j 10^{\frac{(L_{out})_j - \Delta A_j}{10}}\right] \tag{6.15}$$

and

$$(L_{pA})_{in} = 10 \cdot \lg\left[\sum_j 10^{\frac{(L_{out})_j - R_j - \Delta A_j}{10}}\right]. \tag{6.16}$$

The quantity ΔA_j in the last equation is the A-weighting factor[1] for the frequency band having centre frequency j. The frequency bandwidth may be octave or one-third-octave, whichever is appropriate.

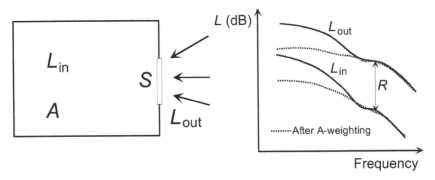

Figure 6.5 Sound transmission through a window. Sketch showing sound pressure levels outside and inside before and after A-weighting.

The quantity of interest, the difference in the A-weighted sound pressure levels, is given by

$$\Delta L_{pA} = (L_{pA})_{\text{out}} - (L_{pA})_{\text{in}}, \qquad (6.17)$$

a quantity that we will be able to calculate knowing the spectrum of the outside noise. Here one uses a standard spectrum for road traffic noise. When expressing this spectrum in frequency band values in such a way that a summation after A-weighting gives zero dB, i.e.

$$\left[\sum_{j} 10^{\frac{(L_{\text{out}})_j - \Delta A_j}{10}} \right] \equiv 1 \quad \text{or} \quad (L_{pA})_{\text{out}} \equiv 0,$$

we arrive at the sound reduction index for traffic noise

$$R_{\text{A}} = (\Delta L_{pA}) \xrightarrow{(L_{pA})_{\text{out}} = 0} -(L_{pA})_{\text{in}} = -10 \cdot \lg \left(\sum_{j} 10^{\frac{L_j - R_j}{10}} \right). \qquad (6.18)$$

The spectrum level values L_j is given by

$$L_j = (L_{\text{out}})_j - \Delta A_j. \qquad (6.19)$$

These values are tabled in ISO 717 Part 1 for one-third-octave bands as well as for octave bands. We may then readily calculate R_{A} if we have laboratory measurement data

---

[1] The A-weighting curve is specified in the IEC standard for sound level metres; see references to Chapter 1.

for the sound reduction index of the actual façade element. Tabled data for L_j covers not only the usual frequency range of 100–3150 Hz, but also the extended range 50–5000 Hz.

A corresponding procedure is used to calculate the adaptation term C, which is used for the internal sound insulation in buildings. The only difference is that here one uses a pink noise spectrum instead of a traffic noise spectrum. It is recommended in ISO 717 to state the performance of building elements by adding the adaptation terms to the calculated weighted sound reduction index in the manner shown in the following example:

$$R_w (C; C_{tr}) = 41(0; -5) \text{ dB}.$$

Several countries are now using the sum of the relevant single number quantity and the appropriate adaptation term in their requirements for sound insulation, e.g. between dwellings

$$R'_w (\text{limit}) = R'_w + C_{50-5000}. \tag{6.20}$$

6.2.2 Impact sound pressure level

Sound insulation against impact sound is normally concerned with isolation of sound energy generated by footfalls. A *hammer apparatus,* normally called a *tapping machine,* for testing of floors was as far back as 1960 standardized by ISO; see ISO 140 Part 6 for specifications. In spite of many criticisms over the years, complaints that measured data do not rank the floors in accordance with the subjective impression etc., no international agreement has been reached concerning another test source. We shall treat this tapping machine in greater detail later on. At this stage we shall just point out that it has five hammers, each weighing 0.5 kg, arranged in such a way that they fall freely against the test object (the floor) twice per second, i.e. the tapping frequency is 10 Hz. The quantity actually measured is the sound power radiated into the room in question. This is commonly the room below the floor but there may as well be situations as with rooms sharing a common floor, transmission to a room from footfalls on a staircase etc. The measure in question, reflecting the sound power, is the *normalized impact sound pressure level* L_n. Under laboratory conditions, it is defined by

$$L_n = L_i + 10 \cdot \lg \left(\frac{A}{A_0} \right), \tag{6.21}$$

where L_i is the sound pressure level in the receiving room due to the tapping machine. The room has a total equivalent absorption area A, and the reference area A_0 is 10 m². In a building we will define in a similar way an *apparent normalized impact sound pressure level*

$$L'_n = L'_i + 10 \cdot \lg \left(\frac{A}{A_0} \right), \tag{6.22}$$

where L'_i is the sound pressure level in the receiving room. In this level, there may now be contributions from other surfaces than the primary floor. These two situations, measurements in a laboratory and in the field are illustrated in Figure 6.6.

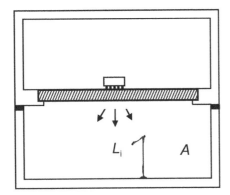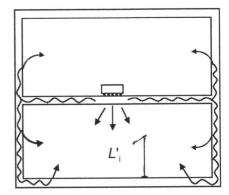

Figure 6.6 Measurement of impact sound pressure level caused by the standard tapping machine. The laboratory situation is illustrated in the left figure; the figure to the right shows how flanking transmission may contribute in a field situation.

In the laboratory situation, where the task is to determine the radiated power from the actual test specimen, the receiving room is equipped with elastic layers in such a way that other surfaces are structurally separated from the primary one. In a building, on the other hand, flanking constructions may contribute but normally this is a lesser problem than in the case of airborne sound.

Finally, it should be mentioned that the impact sound pressure level measured in the field might, in an analogous way to the airborne sound pressure level difference in Equation (6.11), be referred to a standard reverberation time T_0. The quantity in question is then called *standardized impact sound pressure level*

$$L'_{nT} = L'_i + 10 \cdot \lg\left(\frac{T}{T_0}\right),\qquad (6.23)$$

where T_0 for dwellings is equal to 0.5 seconds.

6.2.2.1 Single number rating and adaptation terms for impact sound

As for airborne sound insulation, we may characterize the impact sound insulation by a *weighted normalized impact sound pressure level* $L_{n,w}$ for a building element and by $L'_{n,w}$ when measured in a building. As the quantities measured represent the radiated sound power, a high impact sound isolation implies that the values of L_n are low. When comparing measured data with the reference curve unfavourable deviations are characterized by the measured data being greater than the corresponding reference data. Apart from this difference, the procedure is the same as for airborne sound insulation. The reference curve is shifted in steps of 1.0 dB against the measured one until the sum of unfavourable derivations is as large as possible but no greater than 32 dB.

An example showing laboratory data for a wood joist floor is given in Figure 6.7 together with the reference curve. When shifting the reference curve by 5 dB as shown we obtain a normalized impact sound pressure level $L_{n,w}$ of 65 dB as the sum of unfavourable deviations then becomes 27 dB. Shifting the reference by 4 dB only, one gets a deviation sum of 33 dB, which is larger than the limit. It should be mentioned that the standard demands that the measured data should be given to one decimal place but that is not used in this example.

The "8 dB rule" mentioned above was formerly also used for impact sound insulation. This has been superseded by an adaptation term having the symbol C_I where the index signifies "impact". The reason for introducing this term is that the impact sound pressure level $L_{n,w}$ does not take sufficient account of level peaks at low frequencies, especially for wood joist floors. There is clear evidence that the unweighted impact level of the tapping machine is more representative of the A-weighted impact levels caused by walking for all types of floors. The adaptation term C_I is therefore given as the difference between the unweighted sum $L_{n,sum}$ of the normalized impact levels and the weighted impact sound pressure level $L_{n,w}$, such that

$$C_I = L_{n,sum} - L_{n,w} - 15 \text{ dB} \qquad \text{where} \qquad L_{n,sum} = 10 \cdot \lg \left(\sum_j 10^{L_j/10} \right). \quad (6.24)$$

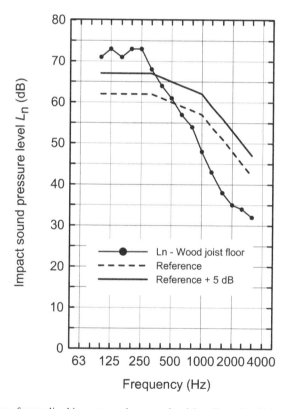

Figure 6.7 Calculation of normalized impact sound pressure level $L_{n,w}$. Example of laboratory measurement of a wood joist floor. $L_{n,w}$ is equal to 65 dB and the sum of unfavourable deviations is 27 dB; see text.

The sum is either taken in one-third-octave bands in the frequency range 100 Hz to 2500 Hz or in octave bands in the frequency range 125 to 2000 Hz. An extended range including the one-third-octave bands 50, 63 and 80 Hz is also used. The term is then denoted $C_{I, 50-2500}$.

6.3 SOUND RADIATION FROM BUILDING ELEMENTS

The sound insulation offered by a building element or a complex construction, either for airborne sound or impacts will depend on two factors: 1) the dynamic response to the actual excitation, being an acoustic field or a direct mechanical force or moment and 2) the efficiency as a sound radiator given the actual response pattern. In this section, we shall deal with the last item, in particular the sound radiation from plane elements when given a bending wave velocity distribution. We shall give a definition of a quantity that is used to characterize the efficiency of a surface as a sound radiator, the *radiation factor*. In this connection we shall return to the simple and idealized sound sources, the monopole and dipole, to illustrate the idea. Following this presentation we shall treat the problems connected to the generation of the bending wave field and further on the transmission properties.

6.3.1 The radiation factor

A commonly used quantity to characterize the efficiency of a given vibrating surface, as a sound radiator is the *radiation factor* σ, also called *radiation efficiency* or *radiation ratio*. By definition

$$\sigma = \frac{W_{\text{rad}}}{\rho_0 c_0 S \langle \tilde{u}^2 \rangle}, \tag{6.25}$$

where W_{rad} is the radiated power from the actual vibrating surface, having the area S, to the surrounding medium with characteristic impedance $\rho_0 c_0$. The quantity $\langle \tilde{u}^2 \rangle$ is the mean square velocity amplitude taken over the surface. The denominator in the expression is the power radiated from a partial area S of an infinitely large plane surface, all parts vibrating in phase with a velocity equal to this mean value, i.e. a plane wave radiation condition. We shall here refer back to the calculation of the radiated power from a plane circular piston set in a baffle (see section 3.4.4). Here we found the same expression when the piston dimensions become much larger than the wavelength.

The brackets in the expression signify that we are taking the mean value in the spatial domain, i.e. of the square RMS-value taken over all points on the surface. The condition for doing this, in a practical sense, is that the velocity does not vary too much from point to point, making it sensible to represent the velocity as a mean value. How large variations should be allowed will obviously depend on the application. In addition to taking the mean value in the time and spatial domain, a third type of averaging must be performed in practice; averaging inside frequency bands of width one-third-octave or octave. We then assume that the applied bandwidth is large enough to contain several natural frequency modes of the actual structure.

Determination of the radiation factor is often performed by way of measurements, as a direct prediction is difficult except for idealized cases. However, there are a number of analytical expressions available, both for plane surfaces and for shell constructions. We shall limit our discussions to plane surfaces.

6.3.1.1 Examples using idealized sources

We shall start using the idealized type of sources, monopoles and dipoles, to illustrate the concept of radiation factor. For a monopole we found in section 3.4.1 that the radiated power could be expressed as

$$W = \rho_0 c_0 \tilde{u}_a^2 \frac{k^2 a^2}{1 + k^2 a^2} S, \qquad (6.26)$$

where k is the wave number and a the radius of the sphere with area $S = 4\pi a^2$. Inserting this expression into Equation (6.25) giving the radiation factor, we get

$$\sigma_{\text{monopole}} = \frac{k^2 a^2}{1 + k^2 a^2}. \qquad (6.27)$$

Examples on the radiation factor for a monopole source having radii of 5 and 25 cm, respectively, are shown in Figure 6.8. The radiation factor is given on a logarithmic scale as $10 \cdot \lg \sigma$, a quantity commonly denoted *radiation index*.

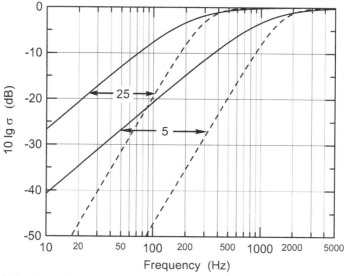

Figure 6.8 Radiation factor of a monopole (pulsating sphere) and a dipole (oscillating sphere) with radius 5 and 25 cm, respectively. Solid curves – monopole. Dashed curves – dipole.

The technical report ISO/TR 7849 (1987)[2] is using Equation (6.27) as an upper limit when calculating the radiated noise from machinery based on measured vibration levels. In this report it is expressed as

---

[2] Currently (2007) under revision.

$$10 \cdot \lg \sigma = -10 \cdot \lg \left[1 + 0.1 \cdot \frac{c_0^2}{(fd)^2} \right], \tag{6.28}$$

where f is the frequency and d is a typical dimension, which for a monopole is the diameter of the sphere. This implies that $d = \sqrt{S/\pi}$ or $d = \sqrt[3]{2 \cdot V}$, where S and V are equal to the source radiating area and volume, respectively. It is easy to see that the two expressions are identical.

Figure 6.8 also shows the corresponding radiation factor for a dipole source exampled by an oscillating sphere having the same radius. As pointed out earlier on, a dipole is a much less effective source than a monopole at low frequencies. A practical example is the radiation from a loudspeaker mounted in a large baffle or in a closed box as compared to being freely suspended in the air. For illustration see section 3.4.1.

Calculating the radiation factor for an oscillating sphere is however a little more involved than for a pulsating one. We shall therefore just give the result, which is

$$\sigma_{\text{dipole}} = \frac{1}{\left| 1 - \dfrac{2}{k^2 a^2} - j \cdot \dfrac{2}{ka} \right|^2} \tag{6.29}$$

and where $| \; |$ indicates the modulus of the expression. The expression is furthermore based on setting the mean squared velocity of the oscillating sphere equal to one-third of the same for the pulsating sphere. Hence

$$\left\langle \tilde{u}_{\text{osc.sphere}}^2 \right\rangle = \tilde{u}_{\text{puls.sphere}}^2 \Big/ 3 \; .$$

This may also be formulated by saying that the mean particle velocity on the surface of the oscillating sphere is $(1/3)^{1/2}$ of the maximum velocity.

6.3.2 Sound radiation from an infinite large plate

We shall use an idealized example to show which parameters are important in sound radiation from plates, namely radiation from an infinitely large plate where a simple plane bending wave is propagating (see Figure 6.9). We shall calculate the sound pressure p in a point with coordinates (x,y) above the plate and further on, the radiation factor when the velocity is given by

$$u_{\text{B}} = \hat{u} \cdot e^{j(\omega t - k_{\text{B}} x)}, \tag{6.30}$$

where k_{B} is the wave number for the bending wave that is propagating in the x-direction. We now assume that the sound pressure above the plate can be expressed as

$$p(x, y) = \hat{p} \cdot e^{j(\omega t - k_x x - k_y y)}, \tag{6.31}$$

where k_x and k_y are the components of the wave number in the medium around the plate (air). This expression has then to be a solution of the ordinary wave equation:

$$\nabla^2 p - \frac{1}{c_0^2}\frac{\partial^2 p}{\partial t^2} = 0. \tag{6.32}$$

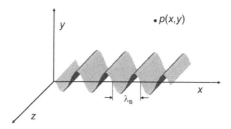

Figure 6.9 Sketch showing a plane bending wave on an infinitely large plate. The plate lies in the x–z plane and the pressure is calculated in points (x,y).

Inserting Equation (6.31) into (6.32) we immediately see that the wave number k for the sound field above the plate must be expressed by

$$k = \frac{\omega}{c_0} = \sqrt{k_x^2 + k_y^2}. \tag{6.33}$$

A further condition is that the component v_y of the particle velocity, i.e. the component normal to the plate, must be equal to u_B at the surface of the plate ($y = 0$). Since v_y is given by

$$v_y = -\frac{1}{j\omega\rho_0}\cdot\frac{\partial p}{\partial y} = \frac{\hat{p}\,k_y}{\rho_0\omega}\cdot e^{j(\omega t - k_x x - k_y y)}, \tag{6.34}$$

we get when setting $y = 0$,

$$\hat{u}\cdot e^{-jk_B x} = \frac{\hat{p}\,k_y}{\rho_0\omega}\cdot e^{-jk_x x}. \tag{6.35}$$

Hence,

$$\hat{p} = \frac{\rho_0\omega}{k_y}\cdot\hat{u} \quad and \quad k_x = k_B.$$

The sound pressure may thereby be expressed as

$$p(x, y) = \frac{\rho_0 c_0 \hat{u}}{\sqrt{1 - \dfrac{k_B^2}{k^2}}}\cdot e^{j(\omega t - k_B x)}\cdot e^{j\sqrt{k^2 - k_B^2}\,y}. \tag{6.36}$$

This result shows that the important factor for the sound radiation is the ratio of the wave numbers in the plate and the surrounding medium. When $k_B > k$, i.e. the wavelength λ_B in plate is smaller than the wavelength λ in the air, the sound pressure will decrease exponentially with the distance y. We only get an exponentially diminishing near field, as the exponent containing y becomes a real number.

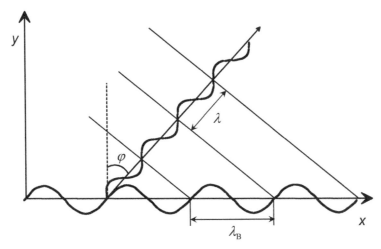

Figure 6.10 Sound radiation from a plate. The wavelength of the bending wave in the plate is larger than the wavelength in the surrounding medium.

If, on the other hand, $k_B < k$ (or $\lambda_B > \lambda$) we have an ordinary propagating plane wave where the sound pressure increases with increasing ratio k/k_B. This may be expressed by the angle φ of the radiated wave (see Figure 6.10). We get

$$\frac{k}{k_B} = \frac{1}{\sin\varphi} \qquad \text{or} \qquad \lambda_B = \frac{\lambda}{\sin\varphi}. \qquad (6.37)$$

The condition having $\lambda_B > \lambda$ is sometimes called *trace matching*; the wavelength of the radiated wave is equal to the plate bending wavelength projected in the direction of the wave. In this case we may calculate the radiation factor by finding the radiated power from a partial surface S. This power may be expressed as

$$W = \frac{1}{2}\mathrm{Re}\{p \cdot v^*\} \cdot S = \frac{\rho_0 c_0 |\hat{u}|^2 S}{2\sqrt{1 - \frac{k_B^2}{k^2}}} = \frac{\rho_0 c_0 |\tilde{u}|^2 S}{\sqrt{1 - \frac{k_B^2}{k^2}}}. \qquad (6.38)$$

Hence, the radiation factor is given by

$$\sigma = \frac{1}{\sqrt{1 - \frac{k_B^2}{k^2}}}, \qquad (6.39)$$

where we assume that $k_B < k$. For increasing k, i.e. when the wavelength λ_B in the plate increases in relation to the wavelength λ in air, the radiation factor approaches 1 (one). This applies, as already demonstrated, for the idealized source types but also for plates of finite dimensions.

6.3.3 Critical frequency (coincidence frequency)

In the above we introduced the notion of *trace matching*. The term was first introduced in German literature (see e.g. Cremer et al. (1988)), describing the condition of the trace wavelength in an incident wave equal to the wavelength of the plate, i.e. a reversed situation of the one being described in the last section. In either case, there will be a limiting or *critical frequency* where this coincidence phenomenon may occur, also called the *coincidence frequency*. We shall use the former notion. At this frequency f_c the wavelength λ_B is equal to the wavelength λ in the surrounding medium. In other words: the phase speed c_B in the solid medium is equal to the phase speed c_0 in the surrounding medium (air).

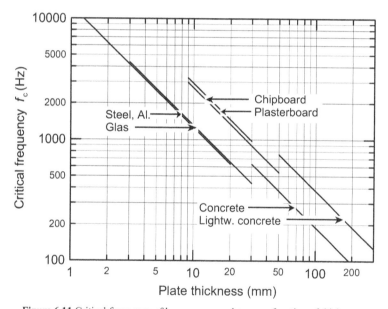

Figure 6.11 Critical frequency of homogeneous plates as a function of thickness.

For thin plates, i.e. when the wavelength is larger than approximately six times the plate thickness, we have shown that the phase speed is expressed as

$$c_B = \sqrt{\omega} \cdot \sqrt[4]{\frac{B}{m}}, \tag{6.40}$$

where B is the bending stiffness per unit length and m is the mass per unit area. By putting c_B equal to c_0 and solving with respect to frequency, we get

$$f_c = \frac{c_0^2}{2\pi}\sqrt{\frac{m}{B}}. \qquad (6.41)$$

For homogeneous plates we may write

$$f_c \approx \frac{c_0^2}{1.8 \cdot c_L \cdot h}, \qquad (6.42)$$

where c_L is the phase speed for longitudinal waves and h is the plate thickness. Figure 6.11 shows the critical frequency for plates of some typical building materials.

6.3.4 Sound radiation from a finite size plate

We showed in the previous section that for frequencies lower than the critical frequency no radiation could occur from a plate of infinite size. This is certainly not the case for real plate structures of finite size but how shall this be calculated, e.g. for a rectangular plate having edges of length a and b? This will be rather more complicated than the idealized example with the infinite plate. Assuming that the vibration of the plate is determined by its natural modes, the radiation will depend on the actual modal pattern, which in turn is determined by the modes taking part and their individual vibration amplitudes. This implies that the mean surface velocity of the plate does not uniquely determine the radiated power. In principle therefore, one cannot calculate the radiation factor solely from the dimensions and material properties. The vibration generating mechanism or the form of excitation must also be known. With the latter we have knowledge of the actual source, what kind of source and how it actually is driving the plate.

In most practical cases, having a stationary mechanical excitation, the structure will be forced into vibration by a more or less broad banded source. This means that the vibration pattern is a combination of the natural modes having eigenfrequencies inside the actual frequency band being excited into resonance. The contribution from each of these modes will depend on how the structure is driven by the source. In our case, concerning the rectangular plate, we shall be quite pragmatic assuming that all modes having their natural frequency within the actual frequency band have the same velocity amplitude. Data given in standards, e.g. EN 12354–1 is calculated using this assumption and we shall give some examples below.

However, it will be quite useful to calculate the radiation factor for a single mode to see how critical the vibration pattern is concerning the radiated power. We shall therefore calculate the radiation factor for a simply supported plate set in an infinite baffle. We assume that the plate is vibrating in a simple harmonic way with a velocity given by

$$u_y(x,z) = \hat{u}\sin\left(\frac{n_x\pi x}{a}\right)\sin\left(\frac{n_z\pi z}{b}\right) \qquad 0 \le x \le a,\ 0 \le z \le b, \qquad (6.43)$$

where n_x and n_z are the modal numbers in the x- and z-direction, respectively. This is illustrated in Figure 6.12, where the plate vibrates in a (5, 4) mode. The corresponding wave number are, as shown in Chapter 3, given by

$$k_{n_x n_z} = \left[\left(\frac{n_x \pi}{a} \right)^2 + \left(\frac{n_z \pi}{b} \right)^2 \right]^{\frac{1}{2}} = \left[k_{n_x}^2 + k_{n_z}^2 \right]^{\frac{1}{2}}. \tag{6.44}$$

Hence, the corresponding eigenfrequencies are given by

$$f(n_x, n_y) = \frac{\pi}{2} \sqrt{\frac{B}{m}} \left[\left(\frac{n_x}{a} \right)^2 + \left(\frac{n_z}{b} \right)^2 \right], \tag{6.45}$$

where B and m again is the bending stiffness per unit length and mass per unit area, respectively.

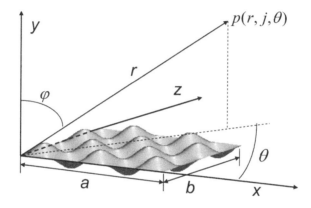

Figure 6.12 Sketch for calculating sound radiation from a finite size plate.

Before we perform the calculation of the radiated power, we shall give some qualitative comments on the situation. In the same manner as for the infinite plate, the relationship between the wave number in the plate and the wave number in the surrounding medium must be the determining factor for the radiation. In this case, however, we have two partial wave numbers (or partial wavelengths) to consider. Commonly, these are divided into three groups:

a) Surface modes $k_{n_x}, k_{n_z} < k$ ($\lambda_{n_x}, \lambda_{n_z} > \lambda$)

b) Edge modes $k_{n_x} > k > k_{n_z}$ or $k_{n_z} > k > k_{n_x}$ ($\lambda_{n_x} < \lambda < \lambda_{n_z}$ or $\lambda_{n_z} < \lambda < \lambda_{n_x}$)

c) Corner modes $k_{n_x}, k_{n_z} > k$ ($\lambda_{n_x}, \lambda_{n_z} < \lambda$)

The reason for these terms should be evident from Figure 6.13, with sketches showing a simply supported plate vibrating in a given corner mode and an edge mode, respectively. The modal pattern is for simplicity indicated by alternating signs; the left-hand sketch is equivalent to the one shown in Figure 6.12 having modal numbers (5, 4).

In this case, we envisage that the wavelength in the surrounding medium is larger than both partial wave numbers of the bending wave in the plate. The various parts of the plate are vibrating in opposite phase separated by small distances, i.e. small when compared with the wavelength in the medium around. The plate becomes a multipole source; the movements of the various parts are not correlated or coordinated to be an effective sound source. The only effective radiating areas are the areas situated near to the corners, these are sufficiently far apart so as not to be mutually destructive. For the edge mode, *one* of the partial wavelengths is larger then the wavelength in the medium around, which results in a larger effective radiation area.

Both types of mode are also called slow acoustic modes because the phase speed of the wave is smaller than the same in the surrounding media. As for the last mentioned type of mode, the surface mode, the partial wavelengths and the speed is larger than in the media around. We get fast acoustic modes where the whole surface is an efficient radiator bringing the radiation factor towards the value of one.

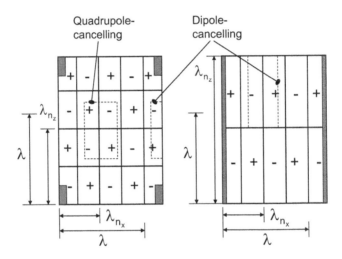

Figure 6.13 Examples of modal pattern for a rectangular plate. Left figure – "corner mode"; both partial wavelengths are smaller than the wavelength in the surrounding medium. Right figure – "edge mode"; one of the partial wavelengths is larger than the wavelength in the surrounding medium.

6.3.4.1 Radiation factor for a plate vibrating in a given mode

Wallace (1972), based on the Rayleigh integral introduced in Chapter 3, calculated the sound pressure and thereby the intensity in the far field from a plate where the velocity is given by Equation (6.43). When integrating the intensity over a hemisphere over the plate we get the radiated power and thereby the radiation factor by using Equation (6.25). We shall not give the details here but for completeness we shall give the end result, which is

$$\sigma(n_x, n_z) = \frac{64\,k^2 ab}{\pi^6 n_x^2 n_z^2} \int\limits_0^{\pi/2} \int\limits_0^{\pi/2} \left\{ \frac{\cos\left[\dfrac{\alpha}{2}\right] \cdot \cos\left[\dfrac{\beta}{2}\right]}{\sin\left[\dfrac{\alpha}{2}\right] \sin\left[\dfrac{\beta}{2}\right]} \right\}^2 \cdot \sin\varphi\, d\varphi\, d\theta. \quad (6.46)$$

The quantities α and β are given by

$$\alpha = ka \sin\varphi \cos\theta, \; \beta = kb \sin\varphi \cos\theta, \qquad \text{and} \qquad \begin{matrix} \cos \\ \cdot \\ \sin \end{matrix} \qquad (6.47)$$

is to be understood in the following way: cosine should be used when n_x or n_z is an uneven number, sine when they are even. The radiation factor, given by the radiation index $10 \cdot \log \sigma$, is shown in Figure 6.14 as a function of relative frequency, i.e. relative to the critical frequency. The plate is square ($a = b$), and the index is calculated for a number of the lower modes, the mode numbers (n_x, n_z) are indicated on the curves.

A Gaussian numerical integration is used to evaluate the integral in Equation (6.46). The accuracy is relatively low for $f > f_c$ and high mode numbers (>8–10). The important point is, however, to show the behaviour of the radiation factor at low frequencies and, at the same time, to link the results to the observations above and to compare with calculated results using the idealized source types. A plate vibrating in the fundamental mode (1, 1) will represent a monopole, whereas the vibration pattern in the (1, 2) mode or (2, 1) mode will represent a dipole. (Do compare Figure 6.8 and Figure 6.14).

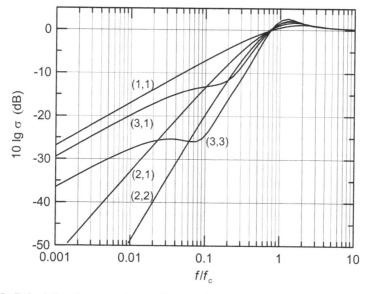

Figure 6.14 Radiation index of a square plate as a function of frequency relative to the critical frequency f_c. The mode number (n_x, n_z) is indicated on the curves.

6.3.4.2 Frequency averaged radiation factor

It is not possible to give a simple formula to calculate the radiation factor for a plate driven at a single frequency. The response will normally contain contributions from several modes each with different amplitude, depending on the location of these modes relative to the driving frequency and on the damping of the plate. The best one can do for practical use is to find a frequency averaged radiation factor. One will then assume that the excitation is relatively broadband as compared with the distance between the natural frequencies, furthermore, that all modes inside the frequency band are equally excited. Several expressions exist in the literature; see e.g. EN 12354–1, but it is not obvious that one is better than another. We will show data given by Leppington et al. (1982), who give the following expressions covering three frequency ranges

$$\sigma = \frac{Uc_0}{2\pi^2 \sqrt{f \cdot f_g} \cdot S \sqrt{\chi^2 - 1}} \cdot \left[\ln \frac{\chi + 1}{\chi - 1} + \frac{2\chi}{\chi^2 - 1} \right] \qquad \text{for} \qquad f < f_g,$$

$$\sigma = \sqrt{\frac{2\pi f}{c_0}} \cdot \sqrt{a} \left(0.5 - 0.15 \frac{a}{b} \right) \qquad \text{for} \qquad f \approx f_g, \quad (6.48)$$

$$\text{and} \qquad \sigma = \frac{1}{\sqrt{1 - \frac{f_g}{f}}} \qquad \text{for} \qquad f > f_g.$$

The quantities a and b again give the dimensions of the rectangular plate, where it is assumed that $a < b$. Further, the quantities S and U are the plate area and circumference, respectively, i.e. $S = a \cdot b$ and $U = 2(a + b)$. The parameter χ is the square root of the ratio f_c/f. It should be noted that the critical frequency should be much higher than the first eigenfrequency of the plate, due to the assumption that there should be resonant radiation by an assembly of modes.

An example using these equations is shown in Figure 6.15. The radiation index is calculated for plates of aluminium or steel where the length of one edge is 2 metres whereas the other is varied between 0.5 and 2 metres.

Comparing the lowermost three curves in the diagram, representing plates having identical thickness h, we observe that a long and narrow plate is a more efficient source than a square one, of course provided the velocity is the same. Similarly, when comparing the two uppermost curves we find that an increased thickness increases the radiation. Increasing the thickness implies a reduced critical frequency, here by a factor of two. This is the reason for the choice showing the radiation factor as a function of the ratio of frequency to the critical frequency. Additionally, the curves are more general than indicated by the examples. They may be applied to plates where the relationship between circumference U, area S and thickness h, i.e. the quantity $U \cdot h/S$ is equal to the ones given in the diagram. To calculate the critical frequency based on material properties and thickness one should apply Equation (6.41) or Figure 6.11.

6.3.4.3 Radiation factor by acoustic excitation

The results given in the preceding section only apply to *resonant multimode* vibration of a plate. It is presupposed that the plate is mechanically excited by a vibration source having a given bandwidth, either directly excited or by vibrations transmitted from a connected structure. A typical example of the latter type in buildings is the so-called *flanking transmission.*

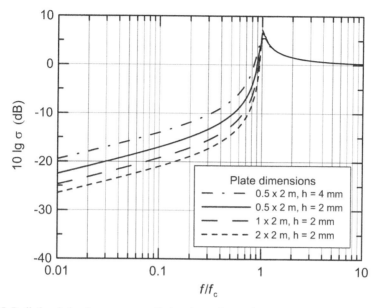

Figure 6.15 Radiation index by resonant radiation from plates of steel or aluminium. Calculated from expressions by Leppington et al. (1982).

Hence, we cannot use these data when a sound field is driving the plate in a forced vibration pattern which is not "natural". This is illustrated by the data in Figure 6.16, which are collected from a series of measurements by Venzke et al. (1973) on panels of 4 mm thick aluminium. The radiation factor is measured using two different types of excitation: directly by an electrodynamic exciter and by a diffuse sound field, respectively, the latter is used in a standard sound insulation measurement. For the former we have compared the results by calculations according to the Equations (6.48), which shows that the fit between these data is quite good for frequencies above some 400–500 Hz. Similar results are also reported by others (see e.g. Macadam (1976)).

As shown, the radiation factor will be larger for the case of sound field excitation than for a mechanical excitation in the frequency range below the critical frequency f_c. The wave field in the plate will partly be determined by the sound pressure distribution imposed by the sound field, a forced vibration field, partly by the free waves originating from the edges of the finite plate. Of these partial wave types, the non-resonant (forced) and the resonant one, the former will be dominant when it comes to sound radiation. This implies, when we shall be able to predict the sound transmission through a panel or wall, which we will treat later, one must take both the resonant and the non-resonant radiation into account.

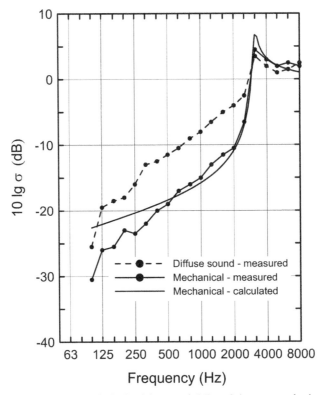

Figure 6.16 Radiation index of a 4 mm thick aluminium panel, 2.7m x 3.4m, measured using different kinds of excitation. Data from Venzke et al. (1973). Calculated data from Equation (6.48).

The radiation factor for forced vibrations by a sound field will necessarily be dependent on panel dimensions and the actual wavelength, but also on the angle of sound incidence. In building acoustics we shall primarily be interested in the radiation factor for an incident diffuse field. Several alternative expressions exist in the literature, e.g. Sewell (1970), Ljunggren (1991) and Novak (1995). We shall quote the first mentioned, which may be written

$$\sigma_f = \frac{1}{2}\left(\ln(k\sqrt{S}) + 0.16 - F(\Lambda) + \frac{1}{4\pi k^2 S} \right), \quad \text{where} \quad \Lambda = \frac{b}{a} \quad (\Lambda > 1) \tag{6.49}$$

and where k is the wave number and $F(\Lambda) = F(1/\Lambda)$ is denoted a *shape function*. Data for this shape function may be taken from a table but a more practical solution will be to use a polynomial approximation or similar.

Based on this equation, EN 12354–1 gives an approximate formula, where an upper limit of 2 is applied to the value of σ_f, i.e. $10 \cdot \lg(\sigma_f)$ has a maximum value of 3 dB. Without making unduly large errors one may also leave out the shape function, because F will vary between zero and 0.5 when b/a varies between one and 10, and also the last term in the expression. One easily observes that the variability of σ_f is much smaller than the corresponding one for resonant radiation. The radiation index $10 \cdot \lg(\sigma_f)$ seldom goes below −5 dB.

6.3.4.4 Radiation factor for stiffened and/or perforated panels

Finally, we shall name some additional factors which are important concerning sound radiation from plates. The first deals with the effect of studs or stiffeners. One might argue that the stiffeners will divide the plate into a number of smaller plates, with the effect that the total circumference comprising all partial plates becomes larger. This will then increase the radiated power due to the resonant modes when $f < f_c$; see Equation (6.48). This effect is experimentally confirmed as apparent from Figure 6.17. The radiation index is shown for the same aluminium panel used for Figure 6.16 but now the panel is stiffened by aluminium studs attached to the panel in a centre-to-centre distance of 400 mm. In one case, the stiffeners are running in one direction, in the other there were crosswise stiffeners as well.

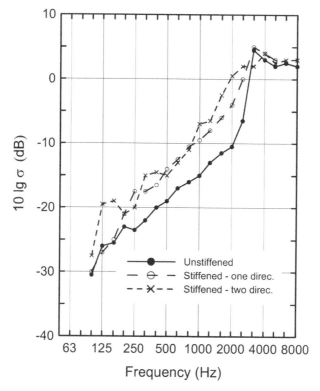

Figure 6.17 Radiation index of a 4 mm thick aluminium panel, 2.7 x 3.4 metres, measured using mechanical point excitation. The panel is stiffened using Al-profiles in one and two (crosswise) directions. Measurements according to Venzke et al. (1973).

The results shown are measured using mechanical point excitation. However, Venzke et al. also present similar data using diffuse field excitation. These exhibit considerably less differences between stiffened and non-stiffened panel. (How would you explain this?)

Furthermore, it should be mentioned that a clamped plate would normally radiate sound better than a simply supported one. On the other hand, a freely suspended plate could create an acoustical "short-circuiting" between back and front, i.e. behaving like a dipole and thereby reducing the radiation. A pronounced example on such acoustical short-circuiting is to be found in perforated plates, which may exhibit very small radiation factors. In noise control of machinery this effect is well known and perforated panel are used in enclosures for e.g. rotating parts. This will reduce the radiated noise from the enclosure if being mechanically excited by the machine. Certainly, it will not act like an acoustical enclosure or barrier, if this should be the purpose of the shielding.

6.4 BENDING WAVE GENERATION. IMPACT SOUND TRANSMISSION

In section 6.3, the subject was sound radiation from a plate assuming that, in one way or another, it had been set into vibrations and thus obtained a velocity or velocity distribution. The pertinent question to be asked is therefore: How shall we find the velocity or velocity distribution induced by a given excitation of a structure having a certain shape, dimensions and material properties? Again, we shall only be concerned with plate structures and the types of excitation will either be a sound pressure distribution over the surface or mechanical point forces. We shall start with the latter being relevant for the problem of impact sound.

6.4.1 Power input by point forces. Velocity amplitude of plate

In Chapter 3 (section 3.7.3.4), we presented an example of the response of a plate excited by a point force. We calculated the input mobility M in a given point defined by

$$M = \frac{u_0}{F} = \frac{1}{Z} \qquad \left(\frac{\mathrm{m}}{\mathrm{N} \cdot \mathrm{s}}\right). \tag{6.50}$$

The quantities u_0 and F represent the velocity and force at this point, respectively, and Z the corresponding point impedance or input impedance. Corresponding quantities are defined for moment and angular velocity but we shall limit our treatment to point forces.

For an infinitely large plate, excited into bending vibrations, Cremer et al. (1988) have shown that the input mobility is a real quantity given by

$$M_\infty = \frac{1}{8\sqrt{m \cdot B}} \approx \frac{1}{2.3 \cdot \sqrt{E \cdot \rho} \; h^2}, \tag{6.51}$$

where the last term is an approximation applicable for homogeneous plates. The important point to be made concerning Equation (6.51) is that it also represents the mean value of the mobility of a *finite* plate, i.e. the mean value taken over all input points and over frequency. This fact has already been referred to in Chapter 3, where we showed in Figure 3.23 that the natural modes resulted in a mobility and impedance strongly space and frequency dependent. The expected or mean value, however, is equal to the one found for an infinite plate.

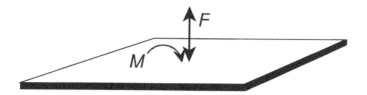

Figure 6.18 Dynamic point force input F to a plate having point mobility M.

The mechanical power imparted to a structure by a point force may in general be written

$$W_{\text{mec}} = \tilde{F}^2 \cdot \text{Re}\{M\} = \tilde{F}^2 \cdot \text{Re}\left\{\frac{1}{Z}\right\}, \qquad (6.52)$$

where $\text{Re}\{\ldots\}$ denotes the real part. Setting out to calculate the part of this power being radiated as sound energy we shall need information, not only on the velocity amplitude of the driving point but on the global velocity distribution as well (see Equation (6.25)). In practice, we shall assume that the plate is driven by a dynamic force having a certain frequency bandwidth $\Delta\omega$, and we shall further assume that a number of modes have their natural frequencies inside this band. (Normally, having 5–6 eigenfrequencies inside $\Delta\omega$ will give reasonable estimates). It may then be shown (Cremer et al. (1988)) that the mean square velocity amplitude may be expressed as

$$\left\langle \tilde{u}^2 \right\rangle_{\Delta\omega} = \frac{1}{\Delta\omega} \int_{\Delta\omega} \left\langle \tilde{u}^2 \right\rangle d\omega = \frac{\tilde{F}_{\Delta\omega}^2}{S^2 m^2} \cdot \frac{\pi}{2\eta\omega} \cdot n(\omega), \qquad (6.53)$$

where S, m and η are, respectively; the area, the mass per unit area and the loss factor of the plate. The angular frequency ω will be the centre frequency in the band of width $\Delta\omega$. The function $n(\omega)$ is the modal density of the plate, formerly derived in Chapter 3, section 3.7.3.5 and given by

$$n(\omega) = \frac{S}{\pi} \cdot \sqrt{\frac{m}{B}}. \qquad (6.54)$$

Inserting this expression into Equation (6.53) we get

$$\left\langle \tilde{u}^2 \right\rangle_{\Delta\omega} = \frac{\tilde{F}_{\Delta\omega}^2}{S^2 m^2} \cdot \frac{\pi}{2\eta\omega} \cdot \frac{S}{4\pi} \cdot \sqrt{\frac{m}{B}} = \frac{\tilde{F}_{\Delta\omega}^2 k_B^2}{8 S \eta \omega^2 m^2}, \qquad (6.55)$$

where we have introduced the wave number k_B in the last expression. We arrive at an alternative expression by introducing the mobility from Equation (6.51). Hence

$$\left\langle \tilde{u}^2 \right\rangle_{\Delta\omega} = \frac{\tilde{F}_{\Delta\omega}^2 M_\infty}{S m \eta \omega}. \qquad (6.56)$$

The numerator in this expression represents the power imparted by the force, thereby giving us the following important relationship between that power and the resulting mean square velocity

$$W_{\text{mec}} = Sm\,\eta\,\omega \left\langle \tilde{u}^2 \right\rangle_{\Delta\omega} = E_{\text{mec}}\eta\,\omega. \tag{6.57}$$

The quantity E_{mec} denotes the mechanical (modal) energy of the plate. It should be noted the validity of this expression is not limited to point excitation but it gives the general relationship between mechanical power and the mean square velocity of a structure. It is, however, presupposed that the structure is neither so large nor so heavily damped that we cannot reasonably determine a representative mean velocity.

We can at this point find an interesting analogue in the relationship between the sound power W_{ac} injected by a source into a room and the resulting mean square sound pressure in the diffuse field. See the derivation in section 4.5.1, where we found that

$$W_{\text{ac}} = \frac{\left\langle \tilde{p}^2 \right\rangle_{\Delta\omega}}{4\rho_0 c_0} \cdot A = \frac{\left\langle \tilde{p}^2 \right\rangle_{\Delta\omega}}{4\rho_0 c_0} \cdot \frac{4(\ln 10^6)}{c_0} \cdot \frac{V}{T}, \tag{6.58}$$

where A, V and T are the total absorbing area, volume and reverberation time of the room, respectively. Substituting the latter quantity with the equivalent loss factor of the room using the relationship

$$T = \frac{\ln 10^6}{2\pi f \eta} \approx \frac{2.2}{f \eta}, \tag{6.59}$$

we get

$$W_{\text{ac}} = \frac{V}{\rho_0 c^2} \cdot \eta\,\omega \left\langle \tilde{p}^2 \right\rangle_{\Delta\omega} = w \cdot V \eta\,\omega = E_{\text{ac}}\,\eta\,\omega, \tag{6.60}$$

where w is the acoustical energy density in the room and E_{ac} is the corresponding total acoustical energy. The last expressions in Equations (6.57) and (6.60) we shall meet again in Chapter 7 when dealing with statistical energy distribution and energy flow in multimode systems; i.e. when dealing with methods and prediction models labelled as *statistical energy analysis* (SEA).

We shall return to our point excited plate, and we shall address the problem of estimating the part of the mechanical energy transformed into sound power. Knowing both the mean velocity and the radiation factor we should be able to calculate the radiated power directly. However, as shown in the next section, the conditions in the neighbourhood of the driving point do complicate matters.

6.4.2 Sound radiation by point force excitation

The mechanical power imparted to a structure will be dissipated partly by internal losses, partly transmitted to connected structures and partly radiated as sound. When the surrounding medium is air, which is the case we shall be concerned with, the latter part will be small; normally a maximum of 1–2%. As indicated in the preceding section, radiation will partly be due to the reverberant wave field set up due to reflections from the plate boundaries. This part will be determined by the resonant modes giving a wave field having a mean square velocity given by Equation (6.55) or (6.56). In addition, there will be a discontinuity at the driving point, which is called a *bending wave near field*. This field will determine the minimum amount of sound power radiated from a plate, the plate is of finite or infinite size, when it is driven by single point force or a collection of such forces. This fact is of great importance in the design of wall linings attached to heavier walls to minimize radiation (see Chapter 8). Figure 6.19 illustrates a possible situation at frequencies $f < f_c$, where there is sound radiation caused by edge modes and bending near field around the excitation point.

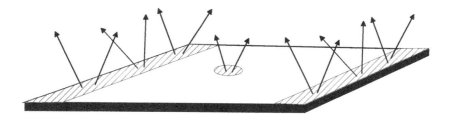

Figure 6.19 Sound radiation from a plate excited by a point force. Radiation due to bending wave near field and edges modes.

The total sound power radiated from a plate having finite dimensions may then be expressed as

$$W_{ac} = W_{near\ field} + W_{reverberant} = W_{near\ field} + \rho_0 c_0\ S\langle\tilde{u}^2\rangle\sigma. \qquad (6.61)$$

Before treating this expression in more detail and applying it to the problem of impact sound we shall have a look at the first term, radiation due to the bending near field.

6.4.2.1 Bending wave near field

The radiated power caused by the bending near field may be calculated from an expression of the bending wave field set up on a thin, infinitely large plate by a point force. The derivation is given in Cremer et al. (1988) and we shall cite only the end result which is

$$W_{near\ field} = W_{point} = \frac{\rho_0 c_0\ \tilde{F}^2 k^2}{2\pi\ \omega^2 m^2} = \frac{\rho_0\ \tilde{F}^2}{2\pi\ c_0 m^2} \quad \text{assuming} \quad k \ll k_B, \qquad (6.62)$$

and where k is the acoustic wave number, the wave number in the surrounding air. We have presented the first expression to show the analogy with the reverberant part of the radiated power (see below). However, looking at the second expression we observe that W_{point} is dependent neither on frequency nor on the bending stiffness. The latter fact may seem odd, as we know that an increased stiffness will result in a longer wavelength and thereby an increase in the radiated power. This effect is however offset by an increased "resistance" against movement; the input impedance is increasing and the mobility will be less as seen from Equation (6.51). It should be noted that Equation (6.62) applies only for frequencies below the critical frequency.

It could be useful to give an example illustrative of the sound power due to this bending near field. We shall then make a comparison with the radiated sound power from a given area of an infinite plate, in conformity with defining the radiation factor. We envisage this area as a piston having radius a. Setting the velocity amplitude equal to u_0 the radiated sound power will be given by

$$W_{\text{piston}} = \rho_0 c_0 \pi\, a^2 \tilde{u}_0^2. \qquad (6.63)$$

The task is now to find the radiated sound power due to the near field according to Equation (6.62) having velocity u_0 in the driving point. Second, we shall calculate the radius a of the piston when W_{point} is equal to W_{piston}. From Equations (6.50) and (6.51) together with Equation (6.41), we obtain

$$W_{\text{point}} = \frac{8\rho_0 c_0^3}{\pi^3} \cdot \frac{\tilde{u}_0^2}{f_c^2}. \qquad (6.64)$$

Equating this sound power with W_{piston} we obtain

$$a = \frac{2\sqrt{2}}{\pi^2} \cdot \frac{c_0}{f_c} \approx 0.29 \frac{c_0}{f_c}. \qquad (6.65)$$

Example The critical frequency f_c of a concrete plate, having a thickness of 50 mm, will be approximately 380 Hz (see Figure 6.11). The radius a of the equivalent piston source will thereby be $\approx$ 26 cm. Using e.g. an applied force of 10 N (RMS-value) we get from Equation (6.62) a radiated sound power of approximately $4.2 \cdot 10^{-6}$ watts or a sound power level L_W re 10^{-12} watts of 66 dB. Assuming a semicircular radiation centred on the driving point the sound pressure level L_p will be 58 dB at a distance of 1 metre.

In several practical cases, it is important to know the radiated power from the bending near field, not only when a point force drives the plate, but also equally well when driven along a line. The latter applies to cases where vibrations are transmitted to a panel or wall by studs or stiffeners. Corresponding expressions to the ones given in Equations (6.62) and (6.64) are

$$W_{\text{line}} = \frac{\rho_0 \tilde{F}_\ell^2}{m^2 \omega} \cdot \ell = \frac{2\rho_0 c_0^2 \tilde{u}_0^2}{\pi f_c} \cdot \ell \qquad \text{when} \quad f \ll f_c. \qquad (6.66)$$

The quantities F_ℓ and ℓ are the force per unit length and the length of the line, respectively.

6.4.2.2 Total sound power emitted from a plate

We have already presented an expression giving the total acoustical power emitted from *one* side of a point-excited plate (see Equation (6.61)). By inserting the expression for the power radiated from the near field, Equation (6.62) and using Equation (6.55), giving the mean square velocity, we obtain

$$W_{ac} = \frac{\rho_0 \tilde{F}^2}{2\pi c_0 m^2} + \frac{\rho_0 c_0 \tilde{F}^2 k_B^2}{8\eta\omega^2 m^2}\sigma = \frac{\rho_0 \tilde{F}^2}{2\pi c_0 m^2}\left[1 + \frac{\pi f_c}{4\eta f}\cdot\sigma\right]. \qquad (6.67)$$

Concerning the force F it is still presupposed that it has a given bandwidth although we allowed ourselves to leave out this index. Also, as pointed out above, the first term is only applicable at frequencies below the critical frequency f_c. Another presumption is that the total energy loss of the plate is dominated by the inner material losses and boundary losses; i.e. the radiated acoustical power makes up a minor part of the total power dissipated. The latter condition is normally fulfilled when the surrounding medium is air (refer to the introduction to section 6.4.2).

From Equation (6.67) we may arrive at a couple of conclusions of great practical interest in general noise abatement. The radiation factor will, as seen from Figure 6.15, strongly increase when the frequency approaches the critical frequency, which implies that above a given frequency the reverberation field will dominate the radiated power. This again implies that the loss factor η will be of great importance. By artificially increasing this factor, adding e.g. viscoelastic layers to thin plates is therefore favourable.

Conversely, in the lower frequency range, the contribution from the near field could be dominant given that the loss factor is not too small. We may prove this by using Equation (6.48) to find a low frequency approximation for the radiation factor, this being

$$\sigma \approx \frac{2c_0 U}{\pi^2 S}\cdot\sqrt{\frac{f}{f_c^3}} \quad \text{when} \quad f \ll f_c. \qquad (6.68)$$

Thus, the second term in Equation (6.67) will be proportional to $1/(\eta \cdot f^{1/2})$ in the low frequency range, whereas the first term, representing the radiation from the near field, will be constant and frequency independent; thereby determining the radiated power above a certain frequency. Increasing the loss factor will in this case have no effect on the radiated power. Denoting the crossover frequency by f_k, i.e. the frequency where W_{point} is equal to $W_{\text{reverberant}}$, we get

$$f_k = \left(\frac{c_0 U}{2\pi S\eta\sqrt{f_c}}\right)^2, \qquad \text{where} \quad f_k \ll f_c. \qquad (6.69)$$

If the task is to reduce the radiated power for excitation frequencies f in the range given by $f_k < f \ll f_c$, no further increase in η will accomplish this.

Example We shall assume that a steel panel, of 1 mm thickness, has a loss factor of 0.05, partly due to material losses and partly to boundary losses. For simplicity, we will further assume that the panel is square with dimensions 2.5 meters. The crossover frequency f_k will then be approximately 250 Hz. A loss factor $\eta > 0.05$ will therefore not reduce the radiated power in the frequency range from 250 Hz and upwards to several kHz (the critical frequency for 1 mm steel is 12.5 kHz). It must, however, be noted that by increasing the loss factor with a viscoelastic layer, glued or sprayed on to the panel, there will be a reduction in the radiated power due to the added mass.

6.4.2.3 Impact sound. Standardized tapping machine

As mentioned in section 6.2.2, a standard tapping machine is used to quantify the impact sound insulation in buildings. It was also pointed out that there have been many objections against this machine, criticism on the practical use as regards to calibration etc., claims that its ranking of the test objects is not coinciding with a subjective ranking of the impact sound insulation. Concerning the ranking, problematic cases arise when used on e.g. wood joist floors. A great deal of research has be directed towards finding alternative methods for testing impact sound insulation, on methods using other types of sources that in a better way simulates the impact of human footfalls. Some alternative sources are suggested and explored; some are also included in national standards (e.g. Japan). The topic will, however, be too extensive to treat in this book. We shall therefore only give specifications on the standard tapping machine.

The tapping machine has five hammers, each having a mass m_h of 0.5 kg. The hammers fall freely from a height H of 4.0 cm; each of them falls twice per second making the tapping frequency f_s of the machine equal to 10 Hz. Assuming that the impacts on the test specimen are purely elastic, this kind of source will give a force inside a frequency band Δf that may be expressed as

$$\tilde{F}_{\Delta f}^2 = 2 f_s \, I^2 \cdot \Delta f,$$

$$\text{where} \quad I = m_h v_0 = m_h \sqrt{2gH}. \tag{6.70}$$

v_0 is here the speed of the hammer at the moment of impact giving an impulse I. The quantity g is the acceleration due to gravity. Measuring the force using one-third-octave bands, i.e. $\Delta f \approx 0.23 \cdot f_0$, where f_0 is the centre frequency in the band, we get

$$\tilde{F}_{1/3\,\text{octave}}^2 \approx 0.90 \cdot f_0 \qquad (\text{N}^2). \tag{6.71}$$

The whole of this force is not necessarily transmitted to the floor under test; the ratio of the point impedance to the internal impedance of the source will be a determining factor. Instead of Equation (6.52), we shall have to express the mechanical power input to the floor by

$$W = \tilde{F}^2 \cdot \text{Re}\left\{ \frac{1}{Z + Z_h} \right\} = \tilde{F}^2 \cdot \text{Re}\left\{ \frac{1}{Z + j\omega m_h} \right\}. \tag{6.72}$$

Assuming $Z \gg Z_h$, which will always be the case when dealing with heavy floors as concrete etc., we are able to calculate the radiated power by inserting Equation (6.71), the equation for the force, into Equation (6.67). As an example, we shall use a homogeneous floor slab of thickness h. Restricting our discussion to the frequency range above the critical frequency enables us to set $\sigma \approx 1$ giving

$$W_{\text{ac}} = \text{const.} \cdot \frac{1}{\eta h^3}. \tag{6.73}$$

The quantities contained in the constant are the material data for the floor and the surrounding air. This implies that the radiated power will decease by the third power of the thickness in the given frequency range or, in other words, the sound power level will decrease by 9 dB for each doubling of the floor thickness. In addition, assuming that the loss factor is frequency independent, the impact sound pressure level L_n in the receiving room (see section 6.2.2), will be frequency independent.

Using these presumptions, we shall use Equations (6.21), (6.58) and (6.67) to calculate the normalized impact sound pressure level giving

$$L_n \approx 10 \cdot \lg\left(\frac{f_c}{m^2 \eta}\right) + 82 \text{ dB}. \tag{6.74}$$

Measurements of the loss factor in practice have shown that it is slightly frequency dependent. Following Craik (1996), we may, as a rough estimate for heavy floors of concrete etc., write

$$\eta = \frac{1}{\sqrt{f}} + 0.015. \tag{6.75}$$

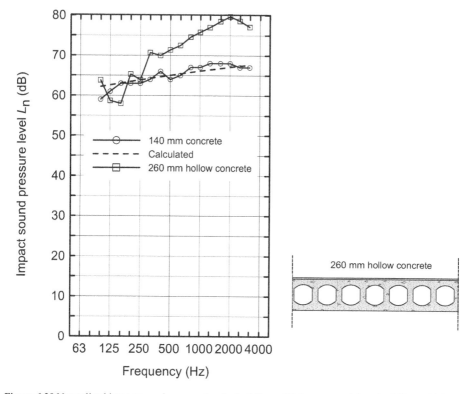

Figure 6.20 Normalized impact sound pressure level of a 140 mm thick concrete slab and a 260 mm thick hollow core concrete slab. Laboratory measurement data together with calculated results for the concrete slab.

An example is given in Figure 6.20 showing results of laboratory measurements on a concrete slab of thickness 140 mm. Comparing with calculated results we have used Equation (6.74) together with Equation (6.75) for the loss factor. The mass per unit area of the slab is 320 kg/m² and the critical frequency is 120 Hz. As seen from the figure measured and calculated results compares very well. The standard EN 12354–2 gives an alternative expression for the normalized impact sound pressure level, which is

$$L_n = 155 - 30 \lg m + 10 \lg T_s + 10 \log \sigma + 10 \lg \frac{f}{f_{ref}} \qquad \text{(dB)}. \qquad (6.76)$$

However, using this equation we will obtain the same results as shown in the figure. The quantities T_S and f_{ref} are the structural reverberation time given by Equation (6.59) and a reference frequency of 1000 Hz, respectively.

In Figure 6.20, we have also included a result measured on a thicker concrete slab of the hollow core type. A vertical section of such a slab is shown in the sketch in the same figure. These kinds of element exist in different thicknesses, also having hollows of different shapes. Statically considered they are equivalent to the massive slabs but, as seen from the figure, the curve shape of the impact level is quite different. To our knowledge, a similar model for the impact sound pressure level, as given for the massive slab, is not known.

6.5 AIRBORNE SOUND TRANSMISSION. SOUND REDUCTION INDEX FOR SINGLE WALLS

To calculate airborne sound transmission we are presented with a more complicated problem than with impact sound. We are again forced to calculate the bending wave field induced by the excitation and thereafter find the resulting radiated power due to this field. In this case, however, the vibration pattern of the structure is more complex having two components:

- A forced vibration field; imparted to the wall due to the external sound field. This is also called the non-resonant field.
- A resonant field; a vibration field due to the natural modes excited by reflections from the boundaries.

The radiated sound power may now be expressed as

$$W_{ac} = \rho_0 c_0 S \left\{ \left\langle \tilde{u}_f^2 \right\rangle \cdot \sigma_f + \left\langle \tilde{u}_r^2 \right\rangle \cdot \sigma_r \right\}, \qquad (6.77)$$

where the indices f and r indicate "forced" and "resonant", respectively. An exact theoretical treatment of this case will be rather involved, partly due to these two different mechanisms, partly due to a complicated dependency of the angle of sound incidence. We shall choose to give an overview of the physical background for these phenomena, followed by a calculation procedure covering the case of most interest, the airborne sound transmission by a diffuse field.

By analogy with the treatment of impact sound, it is useful to start considering a single wall or floor modelled as an infinitely large thin plate excited by a single plane wave. Such a simplification may be justified by the fact that several of these predicted results also apply for plates of finite dimensions. The reasoning behind this fact will be treated in section 6.5.2.

6.5.1 Sound transmitted through an infinitely large plate

We shall assume that the plate lies in the xz-plane and that it is driven by an external force per unit area, a sound pressure $p(x, z, t)$, due to the incident plane wave (see Figure 6.21).

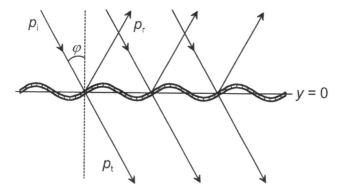

Figure 6.21 Plane wave incidence on a thin, infinitely large plate.

The first task will be, analogous to the case of point force excitation, to find an expression for the plate velocity as a function of the sound pressure driving the plate. To solve this we have to use a differential equation, a wave equation, where the driving pressure is represented on the right side of the equation. Hence, to calculate the sound reduction index we shall have to find the sound pressure in the transmitted wave. We shall use this procedure but as an introduction we shall treat a special case neglecting the bending stiffness of the plate, i.e. characterizing the plate by its mass impedance only.

6.5.1.1 Sound reduction index of a plate characterized by its mass impedance

We may visualize such a wall or plate as a membrane (without tensional forces) or a collection of loosely connected point masses. A plastic curtain or something comparable will in practice behave, acoustically speaking, in such a way. For simplicity, we shall also assume normal sound incidence. The resulting input impedance Z_g in this case (see section 3.5) will be

$$Z_g = \rho_0 c_0 + j\omega m = Z_0 + j\omega m. \tag{6.78}$$

This is a series connection of the mass impedance of the plate and the characteristic impedance of the air behind the plate. Seen from the side of the incident wave the plate will represent a boundary surface giving an absorption factor α that we may calculate using the following equation, derived in section 3.5.1

$$\alpha = \frac{4\,\mathrm{Re}\left\{\dfrac{Z_g}{Z_0}\right\}}{\left|\dfrac{Z_g}{Z_0}\right|^2 + 2\,\mathrm{Re}\left\{\dfrac{Z_g}{Z_0}\right\} + 1}. \tag{6.79}$$

Having characterized the plate by its mass impedance, also having no internal energy losses, the transmission factor τ of the plate must be equal to the absorption factor α. Inserting for Z_g after Equation (6.78), we get

$$\tau = \frac{1}{1 + \left(\dfrac{\omega m}{2Z_0}\right)^2}, \tag{6.80}$$

giving the sound reduction index

$$R_0 = 10 \cdot \lg \frac{1}{\tau} = 10 \cdot \lg\left[1 + \left(\frac{\omega m}{2Z_0}\right)^2\right] \approx 20 \cdot \lg\left(\frac{\pi f m}{Z_0}\right). \tag{6.81}$$

This is the so-called *mass law* in its simplest form; the sound reduction index increases by 6 dB by each doubling of frequency and/or mass per unit area. The approximation given by the last expression, however presuppose that the mass impedance is much larger than the characteristic impedance of air. This condition is normally fulfilled for panels used in buildings. Inserting the characteristic impedance of air at 20°C we get:

$$R_0 \approx 20 \cdot \lg(m f) - 42.5 \quad \text{(dB)}. \tag{6.82}$$

6.5.1.2 Bending wave field on plate. Wall impedance

Taking the bending stiffness into account we have, as mentioned above, to solve a wave equation where the sound pressure of the incoming wave is the driving force. The wave equation may be written as

$$B \nabla^2 \nabla^2 \xi + m \frac{\partial^2 \xi}{\partial t^2} = p(x, z, t), \tag{6.83}$$

where B and m are the bending stiffness per unit length and the mass per unit area, respectively. The quantity ξ is the particle displacement, the deflection of the plate surface. Assuming an harmonic time function $e^{j\omega t}$ and furthermore, using the velocity u as a variable we get

$$\nabla^2 \nabla^2 u - k_B^4 u = \frac{j\omega}{m} p(x, z). \tag{6.84}$$

For plane wave incidence we can cast $p(x,z)$ in the form

$$p(x, z) = \hat{p}(k_x, k_z) \cdot e^{jk_x x} \cdot e^{jk_z z}, \tag{6.85}$$

and by rotating the coordinate system one of the partial wave numbers k_x and k_z may be set equal to zero. We shall set k_z equal to zero and assume that the solution for the velocity u have the same form as the one for the pressure. Inserting into Equation (6.84) we obtain the following relation between the amplitudes of the pressure and the velocity

$$\hat{u}(k_x) = \frac{j\omega \, \hat{p}(k_x)}{B\left(k_x^4 - k_B^4\right)}. \tag{6.86}$$

We observe again, as when discussing the radiation factor of plates, the important relationship between the acoustic wave number and the bending wave number. This becomes more evident when we calculate the velocity, having a situation as sketched in Figure 6.21. For the sound pressure in the incident, reflected and transmitted wave, respectively, we shall write:

$$
\begin{aligned}
p_i &= \hat{p}_i \, e^{-jky\cos\varphi} e^{-jkx\sin\varphi} & y > 0, \\
p_r &= \hat{p}_r \, e^{jky\cos\varphi} \, e^{-jkx\sin\varphi} & y > 0, \\
p_t &= \hat{p}_t \, e^{-jky\cos\varphi} \, e^{-jkx\sin\varphi} & y < 0.
\end{aligned}
\tag{6.87}
$$

Hence, the total pressure on the plate is ($y = 0$):

$$p(x,z) = \left(\hat{p}_i + \hat{p}_r - \hat{p}_t\right)e^{-jkx\sin\varphi}. \tag{6.88}$$

Inserting this expression into Equation (6.86) with k_x equal $k \cdot \sin\varphi$, we arrive at the equation giving the relationship between the driving sound pressure and the resulting velocity:

$$\hat{u} = \frac{j\omega\left(\hat{p}_i + \hat{p}_r - \hat{p}_t\right)}{B\left(k^4 \sin^4 \varphi - k_B^4\right)}. \tag{6.89}$$

The ratio of the driving pressure to the velocity is generally known as *wall impedance*. This quantity, for which we shall use the symbol Z_w, will be given by

$$Z_w = \frac{\hat{p}_i + \hat{p}_r - \hat{p}_t}{\hat{u}} = \frac{B}{j\omega}\left(k^4 \sin^4 \varphi - k_B^4\right). \tag{6.90}$$

Under the condition $k > k_B$ we shall always find an incident angle φ where Z_w is equal to zero, making the velocity "infinitely" large. The plate will not present any obstacle for the sound wave! The conditions determining this *trace matching* were discussed in section 6.3.2 concerning sound radiation from a plate. The important point in this connection is that the angle giving a maximum radiation is also the one giving maximum excitation. This is an example of a general principle in acoustics, the so-called *reciprocity principle*, which we shall address in section 6.6.1.

A further discussion on Equation (6.90) will be easier when introducing the critical frequency f_c and also some energy losses by way of a complex bending stiffness $B(1+j\cdot\eta)$. We may then write

$$Z_w = j\omega m\left[1 - \left(\frac{f}{f_c}\right)^2 \cdot (1+j\eta)\sin^4 \varphi\right]. \tag{6.91}$$

The equation clearly shows the presumption for the derivation in the preceding section; the wall impedance will become a pure mass impedance at frequencies far below the critical frequency.

6.5.1.3 Sound reduction index of an infinitely large plate. Incidence angle dependence

The transmission factor τ and the sound reduction index R are calculated from the ratio of the sound pressure amplitudes in the transmitted and incident wave. By definition:

$$\tau = \frac{W_t}{W_i} = \left| \frac{\hat{p}_t}{\hat{p}_i} \right|^2 . \tag{6.92}$$

We may by using Equation (6.89) express the velocity as

$$u = \hat{u} e^{-jkx\sin\varphi} = \frac{\left(\hat{p}_i + \hat{p}_r - \hat{p}_t \right)}{Z_w} \cdot e^{-jkx\sin\varphi} . \tag{6.93}$$

The normal component of the acoustic particle velocity v on both sides of the plate must be equal to the plate velocity u. Hence, the following relationship must apply,

$$\hat{v}_i + \hat{v}_r = \hat{u} = \hat{v}_t . \tag{6.94}$$

The relationship between these velocity amplitudes and the corresponding pressure amplitudes is easily found by applying the force equation (Euler equation),

$$v_{y=0} = -\frac{1}{j\omega\rho_0} \left(\frac{\partial p}{\partial y} \right)_{y=0} . \tag{6.95}$$

Applying this to Equations (6.87), we get

$$\hat{v}_i = \frac{\hat{p}_i}{Z_0} \cos\varphi, \quad \hat{v}_r = -\frac{\hat{p}_r}{Z_0} \cos\varphi \quad \text{and} \quad \hat{v}_t = \frac{\hat{p}_t}{Z_0} \cos\varphi. \tag{6.96}$$

The Equations (6.93), (6.94) and (6.96) give us the relationship between the pressure amplitudes we are looking for as we find

$$\hat{u} = \frac{2\hat{p}_i}{Z_w + \dfrac{2Z_0}{\cos\varphi}} = \frac{\hat{p}_t \cos\varphi}{Z_0} . \tag{6.97}$$

The transmission factor and the reduction index will then be given by

$$\tau = \left| \frac{1}{1 + \dfrac{Z_{\mathrm{w}} \cos \varphi}{2 Z_0}} \right|^2, \qquad R = 10 \cdot \lg \frac{1}{\tau} = 10 \cdot \lg \left[\left| 1 + \frac{Z_{\mathrm{w}} \cos \varphi}{2 Z_0} \right|^2 \right]. \qquad (6.98)$$

The wall impedance Z_{w} will here be given by Equation (6.91). An example using these equations is shown in Figure 6.22, where the sound reduction index of a plate of surface weight 10 kg/m^2 and critical frequency 1000 Hz is given for a number of incident angles.

The low values around the critical frequency should be noted, furthermore, how this "dip" approaches the critical frequency by increasing the angle should also be observed. The determining factor in this frequency range is the damping of the plate characterized by the loss factor, which in this example is rather high. We shall further note that the mass law gives an appropriate description at frequencies somewhat lower than the critical frequency. Above coincidence we observe that the dependence on frequency is much greater than the corresponding one in the mass law range. The bending stiffness will be the determining factor, and far above coincidence there will be an increase of 18 dB per octave.

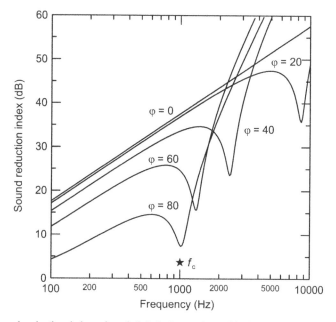

Figure 6.22 Sound reduction index of an infinitely large plate with the angle of incidence as parameter. Material data: $m - 10$ kg/m^2, $\eta - 0.1, f_{\mathrm{c}} - 1000$ Hz.

6.5.1.4 Sound reduction index by diffuse sound incidence

On real partitions in buildings we normally have sound incidence from many angles at the same time. To calculate the sound insulation we could in principle use Equations (6.98) and (6.91), make a weighting according to the given distribution of incident angles and sum up the contributions. In practice, however, the actual distribution is seldom known. As mentioned in the introduction, the only viable solution is then to carry out the calculatation assuming an ideal diffuse incident sound field; i.e. assuming sound incidence evenly distributed over all angles and with random phase. We shall approach the practical problem by again going back to our infinitely large plate.

In Chapter 5 (section 5.5.3.2) we calculated the statistical absorption factor for a sound-absorbing surface. The same type of integral may be used to calculate a statistical or diffuse field transmission factor τ_d. We shall write

$$\tau_d = 2 \int_0^{\pi/2} \tau(\varphi) \sin \varphi \cos \varphi \, d\varphi. \tag{6.99}$$

Inserting the transmission factor from Equation (6.98) with the wall impedance according to Equation (6.91) we shall not be able to give an analytical solution to the integral. Limiting the solution to low frequencies by using the approximation given by Equation (6.80), the result may be written

$$R_d = 10 \cdot \lg \frac{1}{\tau_d} = R_0 - 10 \cdot \lg \left[0.23 \, R_0 \right] \qquad \text{(dB)}. \tag{6.100}$$

The expression is commonly referred to as the *diffuse field mass law*. Cremer presented a similar expression as far back as 1942, valid for frequencies above the critical frequency, which we shall write

$$R_d = 10 \lg \left[\frac{\pi f m}{Z_0} \right] + 10 \lg \left(\frac{2\eta f}{f_c} \right) - 5 \text{ dB} \qquad \text{for} \qquad f \gg f_c. \tag{6.101}$$

We shall return to these expressions below when we address the problem of transmission through real walls or panels, taking the finite size into consideration.

6.5.2 Sound transmission through a homogeneous single wall

Starting out from the observations on sound transmitted through an infinitely large plate, we shall move on to the practical case of sound transmission from one room to another by way of a single homogeneous wall or floor. Several options are available for calculations. One possibility is an analytical solution starting out from a description of the sound field in the rooms coupled to the structural wave field in the wall, all of them expressed by a sum of the natural modes. The task is then to calculate the coupling between each of these modes, which is a relatively complex task (see e.g. Josse and Lamure (1964) or Nilsson (1974)). Using finite element methods (FEM) may be seen as a modern version of such procedures (see e.g. Pietrzyk (1997)). The power of such methods lies in the ability to investigate specific situations, preferably in the lower frequency range.

Statistical energy analysis (SEA) is also a powerful method under the condition that the modal fields have a sufficient number of eigenfrequencies inside the actual frequency band (see e.g. Craik (1996)). The strength of this method lies therefore in treating problems in the mid and high frequency range. We shall therefore give an introduction, backed up by some examples, for this method in the next chapter.

Being pragmatic, we could investigate the applicability of the infinite plate model to real situations and in the preceding section we gave an expression for the sound reduction index for diffuse field incidence at low frequencies (see Equation (6.100)). In practice this equation will give values a little too low, a slightly better model for the field situation is

$$R_\mathrm{d} = R_0 - 5 \text{ dB} \approx 20 \cdot \lg(f \cdot m) - 47 \text{ dB}, \qquad (6.102)$$

a result which will come close to the one obtained by performing the integration of Equation (6.99), using an upper limit of approximately 78°. This is explained by the fact that wave components near to grazing incidence will be of less importance for finite size partitions. It does not explain, however, why such a simple expression gives quite a good prediction at low frequencies, i.e. for frequencies below the critical frequency.

This is connected to the phenomena outlined in the introduction to this section. The wave field in a finite plate will have the following two components: 1) A forced field set up by the sound field in the same way as in the "infinite" case and 2) a free field originating at the boundaries due to the impact of the forced field. The forced field cannot by itself satisfy the boundary conditions. The point now is, as we demonstrated in section 6.3.4, that the radiation from the free field or resonant modes is very inefficient at frequencies below the critical frequency. The forced field due to its longer wavelengths therefore mainly determines the transmission. This is the reason behind the fact that results calculated for an infinitely large wall with considerable success are transferable to one having a finite size.

The simple expression given above needs, however, some modification. The finite sized area influences the forced transmission. Sewell (1970) has calculated this effect based on calculating the transmission by diffuse field incidence of a plate surrounded by an infinite baffle. In the expressions given below for the transmission factor (see Equation (6.103)), the mentioned effect shows up in the radiation factor for forced transmission.

The importance of the loss factor should also be noted. The resonant modes will certainly be reduced in amplitude by increasing the loss factor. However, these modes are of minor importance in the acoustic radiation below the critical frequency. The effect of increasing the loss factor will therefore be very small. However, in the frequency range around the critical frequency and upwards, where the resonant transmission is dominant, any increase in the loss factor will be beneficial.

It is also worth noting that below a certain frequency, below the fundamental natural frequency, a plate will pass from the mass-controlled area to the stiffness-controlled one. Ideally, the sound reduction index will then increase with decreasing frequency. This effect is normally not observed in measurement data for walls in buildings. The reason is partly that this frequency range in normally below the one used for measurement, partly that the coupling to the resonant room modes makes the sound reduction index vary in an irregular and not very transparent manner.

This, however, should not make us believe that low frequency and stiffness-controlled transmission cannot be important in design of sound insulating devices. Enclosures designed for noise control of various machines and equipment will often include small size panels. The offending noise will often contain frequencies below the fundamental frequency of these panels, maybe even below the fundamental frequency of the air cavity of the enclosure. The stiffness of the panels, not their mass, is therefore of vital importance. This case is not covered by the formulae given below.

6.5.2.1 Formulas for calculation. Examples

One should not be surprised, taken the complexity of the task to even calculate the sound reduction index of a single homogeneous wall between two rooms, that a number of formulas are cited in the literature. A number of these are found in EN 12354–1, in which Equation (6.103) is given for the transmission factor. The dimensions of the wall are given by the quantities a and b, η_{tot} is the total loss factor, σ and σ_f are the radiation factor for resonant and non-resonant transmission, respectively. The latter is expressed by the formula by Sewell (Equation (6.49)), whereas the corresponding one for σ is a little more involved than the one given in section 6.3.4.2.

$$\tau = \left(\frac{Z_0}{\pi f m}\right)^2 \left[2\sigma_f + \frac{(a+b)^2}{a^2+b^2}\sqrt{\frac{f_c}{f}}\cdot\frac{\sigma^2}{\eta_{tot}}\right] \quad f < f_c,$$

$$\tau = \left(\frac{Z_0}{\pi f m}\right)^2 \left[\frac{\pi\sigma^2}{2\eta_{tot}}\right] \quad\quad\quad\quad f = f_c, \quad\quad (6.103)$$

$$\tau = \left(\frac{Z_0}{\pi f m}\right)^2 \left[\frac{\pi f_c\sigma^2}{2 f\,\eta_{tot}}\right] \quad\quad\quad f > f_c.$$

For rough estimates one may make some simplifications. For forced transmission ($f < f_c$), the simple mass law, given in Equation (6.102), is often sufficient. A slightly better alternative is to neglect the contribution from the resonant transmission but to include a slightly simplified area effect. Fahy (1987) has suggested that for the range $f < f_c$ one should use the following expression:

$$R_f = R_0 - 10\cdot\lg\left[\ln\left(\frac{2\pi f}{c_0}\cdot\sqrt{ab}\right)\right] + 20\cdot\lg\left[1 - \left(\frac{f}{f_c}\right)^2\right], \quad\quad (6.104)$$

where index f on the reduction index indicates that we are dealing with forced transmission. Inserting for R_0 according to Equation (6.82) we get

$$R_f \approx 20\cdot\lg(m\cdot f) - 10\cdot\lg\left[\ln\left(\frac{2\pi f}{c_0}\cdot\sqrt{ab}\right)\right] + 20\cdot\lg\left[1 - \left(\frac{f}{f_c}\right)^2\right] - 42\,\text{dB}. \quad (6.105)$$

In the frequency range above the critical frequency we may set $\sigma \approx 1$, and when using the last entry in (6.103), we obtain

$$R = 20\cdot\lg(m\cdot f) + 10\cdot\lg\left[2\eta_{tot}\frac{f}{f_c}\right] - 47\,\text{dB} \quad f > f_c. \quad\quad (6.106)$$

This expression is identical to the one given for a plate of infinite size (see Equation (6.101)). Below, we shall present several examples where we compare measured and calculated data. In all cases we shall use the complete set of equations given in Equation (6.103) and where the radiation factors are taken from Equations (6.48) and (6.49).

A major problem performing these comparisons is the availability of a complete set of specifications for the measured object. Material data as well as dimensions may not be completely described. Presumably, the loss factor is the most critical parameter. A prediction of the sound reduction index in the frequency range around the critical frequency and above will be quite uncertain without this information.

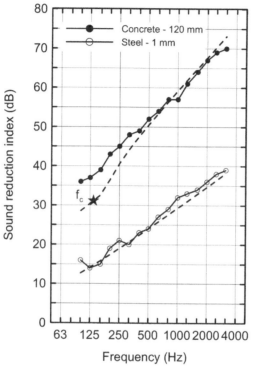

Figure 6.23 Sound reduction index of a 1 mm steel panel and a 120 mm concrete wall. Dashed lines: calculated data from Equations (6.103). The critical frequency for the concrete wall is indicated. Measured data from Homb et al. (1983).

Two examples on such a comparison are shown in Figure 6.23, measured and predicted sound reduction index of a 1 mm thick steel panel and a 120 mm thick concrete wall. In the first case we find that the critical frequency is approximately 12 kHz, making the panel mass controlled in the whole measuring range. The fit between measured and calculated data is very good. As for the 120 mm concrete, the fit around the critical frequency between these data is rather poor. However, no measured data for the loss factor were available making it necessary to use an estimate from Equation (6.75).

Results from reproducibility tests, comparing results from different laboratories measured on the same specimen, have shown that the reproducibility standard deviation becomes very large if the loss factor is not properly controlled. The prediction accuracy is also not very good around the critical frequency. Measured data on thick and massive walls normally exhibit a more or less constant "plateau" in the reduction index curve, as opposed to thin panels where there is normally a distinct "dip" in the curve.

An example of the latter effect is given in Figure 6.24, which shows measurement data and calculated results on single glazing. This example is used due to the complete set of data given, both for the material and the geometry. Measurements are performed using three separate sheets, each of area (560 x 1680) mm², mounted together in a frame making a total measuring area of (2020 x 1800) mm². Measurements were conducted on samples of thickness 3, 4 and 6 mm, of which we are presenting the last two.

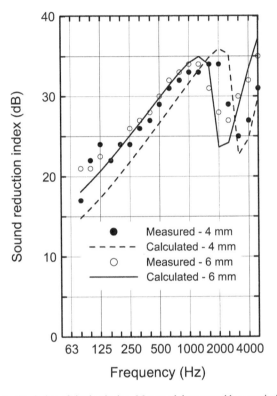

Figure 6.24 Sound reduction index of single glazing. Measured data reused by permission from Quirt (1982).

The fit between measured and calculated results is very good for the 6 mm glazing, whereas the expected lower reduction index for the 4 mm glazing does not show up in the mass-controlled region. This discrepancy could be due to several factors; presumably it is caused by the influence of the frame. It should be noted that the scale used for the ordinate is different from the one used in Figure 6.23.

More recently, Callister et al. (1999) have reported on measurements and calculations on single glazing using a test area much smaller than by Quirt, specifically 0.61m x 0.91 m. Their calculated results are based on Sewell's expression for the reduction index at low frequencies and Cremer's Equation (6.101) for frequencies above coincidence, together by a certain interpolation around the latter, following Sharp (1978). Doing this, they obtain a very good agreement between measured and calculated results.

Finally, we shall use the example shown in Figure 6.24 to illustrate the relationship between the resonant and non-resonant transmission in a specific case. Figure 6.25 again shows measured data for 6 mm glazing plotted together with the predicted ones. In addition we have also plotted the resonant part separately. As evident from the figure, the resonant part is nearly negligible below the critical frequency, whereas it makes up the dominant contribution around and above this frequency.

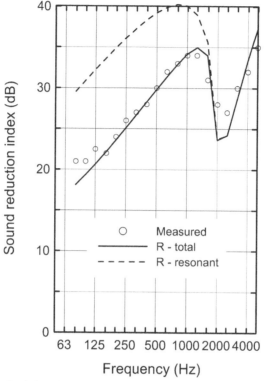

Figure 6.25 Sound reduction index of 6 mm glazing. Calculated result for resonant and total (resonant plus non-resonant) transmission. Measured data reused by permission from Quirt (1982).

6.5.3 Sound transmission for inhomogeneous materials. Orthotropic panels

As distinct from e.g. glass, a number of other building construction elements cannot reasonably be characterized as homogeneous and isotropic. The latter term means that the element is composed of a material whose properties are independent of direction. The material properties of commonly used panels, however, may vary in the transverse direction as well as laterally. A typical example of the former type is sandwich elements, commonly an assembly of three layers but there may also be more layers. We shall defer the treatment of such elements to Chapter 8, which concerns sound transmission through partitions having several layers, double walls etc.

In this section we are concerned with *orthotropic* plates, a commonly used term for plates where the elastic properties are different in two axial directions. For plane plates this is caused by material anisotropy, this is typical for wooden materials where the bending stiffness depends on the direction of the fibres. Fibre-reinforced materials are another example. Another large class of building elements, which are orthotropic, are corrugated plates. Plates having a "wavy" corrugation are well known but more common in industrial buildings are plates where the corrugations have a trapezoidal shape, commonly called cladding. We have previously (see section 3.7.3.3) referred to the literature giving the equivalent components of stiffness for such plates, enabling us to apply the general theory for plane orthotropic plates. We also performed a calculation of natural frequencies for a "wave"-corrugated plate.

The advantage of corrugated plates is great strength as compared to weight. They are more lightweight and cheaper than plane plates having equal strength. The disadvantage, however, is that the sound reduction index may become much less than for plane plates of equal thickness. We shall use the symbol B_1 to denote the bending stiffness (per unit length) about an axis lying in the plane of the plate normal to the corrugations, i.e. about the z-axis of Figure 6.26. Correspondingly, B_2 is the bending stiffness about the x-axis. We shall then find two critical frequencies given as

$$f_{c1} = \frac{c_0^2}{2\pi} \sqrt{\frac{m}{B_1}} \qquad \text{og} \qquad f_{c2} = \frac{c_0^2}{2\pi} \sqrt{\frac{m}{B_2}}, \qquad (6.107)$$

where m as usual represents the mass per unit area. The corrugations may increase the bending stiffness considerably, which certainly is the purpose, but it is followed by a low value of f_{c1}. This implies that the resonant transmission may become the dominant factor over a large part of the useful frequency range.

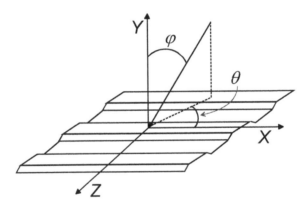

Figure 6.26 Corrugated plate with incident sound wave in direction (φ, θ).

Certainly, there are a multitude of such plates in use. With some there is a relatively small difference between the stiffness components, which makes the difference small between the critical frequencies. The coincidence range with the typical dip in the sound reduction curve will then just be a little broader. Commonly, however, the difference in stiffness is much larger. Whereas f_{c1} could be in the range of some hundred hertz, the corresponding f_{c2} could be 15–30 kHz.

Instead of Equation (6.91) we get an expression for the wall impedance Z_w dependent on two critical frequencies, at the same time dependent on two angles. In addition to the angle of incidence φ, the angle to the plate normal, the impedance will be a function of the azimuth angle θ as well. We get (see e.g. Hansen (1993)

$$Z_{\mathrm{w}} = \mathrm{j}\omega m \left[1 - \left(\frac{f}{f_{c1}} \cos^2 \theta + \frac{f}{f_{c2}} \sin^2 \theta \right)^2 (1 + \mathrm{j}\eta) \sin^4 \varphi \right]. \qquad (6.108)$$

The expression, however without the inclusion of the loss factor, was given back in 1960 by Heckl. The transmission factor for a given angle of incidence, according to Equation (6.98), hence becomes

$$\tau(\varphi, \theta) = \left| 1 + \frac{Z_{\mathrm{w}} \cos\varphi}{2 Z_0} \right|^{-2}. \qquad (6.109)$$

Setting out to calculate the transmission factor for diffuse field incidence we have to integrate this expression over all angles of incidence (see Equation (6.99)). Hence

$$\tau_{\mathrm{d}} = \frac{2}{\pi} \int_0^{\pi/2} \left[2 \int_0^{\pi/2} \tau(\varphi, \theta) \cos\varphi \sin\varphi \, \mathrm{d}\varphi \right] \mathrm{d}\theta. \qquad (6.110)$$

Inserting τ from Equation (6.109), we may write

$$\tau_{\mathrm{d}} = \frac{2}{\pi} \int_0^{\pi/2} \int_0^1 \frac{\mathrm{d}\left(\sin^2 \varphi\right) \mathrm{d}\theta}{\left| 1 + \dfrac{Z_{\mathrm{w}}}{2 Z_0} \cos\varphi \right|^2}. \qquad (6.111)$$

The evaluation of this expression must be performed numerically. However, Heckl (1960) also gives some approximate expressions (for η equal zero). In the frequency range below the lowest critical frequency, we may use ordinary mass law. In two other frequency ranges, being the range between the critical frequencies and above the highest one, respectively, Heckl gives the following expressions:

$$\tau_{\mathrm{d}} \approx \frac{Z_0}{2\pi^2 m} \cdot \frac{f_{c1}}{f^2} \left(\ln \frac{4f}{f_{c1}} \right)^2 \quad \text{for} \quad f_{c1} < f < f_{c2}$$

$$\qquad (6.112)$$

and $\qquad \tau_{\mathrm{d}} \approx \frac{Z_0}{2m} \cdot \frac{\sqrt{f_{c1} f_{c2}}}{f^2} \quad \text{for} \quad f > f_{c2}.$

It should be noted that the expressions above only apply to an infinitely large plate. Taking the finite dimensions into account, we may as before (see section 6.5.2.1) introduce the correction factor after Sewell (1970). Hansen (1993) introduces a correction by substituting the upper limit one (1.0) in the integral (6.111) over $\sin^2\varphi$ by a variable limit

$$\left(\sin^2 \varphi \right)_{\text{upper limit}} = 1 - \frac{c_0}{2\pi f \sqrt{S}}, \qquad (6.113)$$

where S is the area of the panel and where we recognise the last term from the expressions given in section 6.5.2.1. In the examples we shall use, we have for simplicity fixed the upper limit at 0.96, which corresponds to a maximum angle of incidence of approximately 78°.

The first example, given in Figure 6.27, shows the sound reduction index for a specific case where the panel has a weight of 7.5 kg/m² and where the critical frequencies are 400 Hz and 4000 Hz, respectively. The results shown are calculated by a numerical integration of the integral in Equation (6.111), where the integration is performed for loss factors of 0.01 and 0.1 (1% and 10%). In addition, Heckl's approximation for the frequency range between the two critical frequencies is also included. As the loss factor is not included in this approximation one could expect that the best fit would be obtain for a low loss factor. The most important thing to note is, however, the far lower results one get as compared by the mass law.

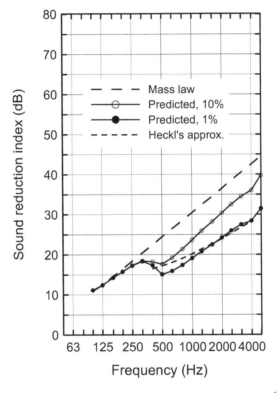

Figure 6.27 Predicted sound reduction index of corrugated panel of weight 7.5 kg/m² and critical frequencies (f_{c1}, f_{c2}) equal (400,4000) Hz. Calculated using loss factor 1% and 10%, additional data using Heckl's approximation in the range $f_{c1} < f < f_{c2}$.

The second example is taken from an extensive series of measurements performed by Hansen (1993). The series comprised 10 different types of corrugated panel, where the panels partly had the 10 m² size normally used in laboratory tests, some very small; approximately 1.5 m². Experiments were also conducted by additional damping of the panels.

We shall show results from measurements on the type denoted Hi Span '800' (see Figure 6.28). This has a large measuring area, and in Hansen's own calculations he uses a loss factor of 0.011 (1.1%) and the critical frequencies are 378 Hz and 30 400 Hz, respectively. The measured results in Figure 6.28 are reproduced from Hansen's data but the calculated results are performed using Equation (6.111), setting a fixed upper limit for the incidence angle of 78°. The differences between these data and Hansen's own calculation are negligible when applied to the large measurement area.

Other results does not fit equally well with the calculated ones as in this example. Hansen does point out that when high accuracy is demanded one has to rely on measured results, which may be unnecessary to point out. A typical feature that often is observed in measurement results of corrugated panels is "dips" in the curve not attributed to coincidence phenomena. This may be seen in the measured result around 4000 Hz. Two explanations are suggested: acoustic wave standing wave resonances between the ribs and panel vibration resonances. By panel resonances we shall not understand the ones due to the eigenmodes of the entire panel but to the sub-panels of the cladding, i.e. vibration modes of particular sections of the profile. A later series of measurements combined with predictions using FEM analysis indeed have substantiated the latter explanation (see Lam and Windle (1995 a, b)).

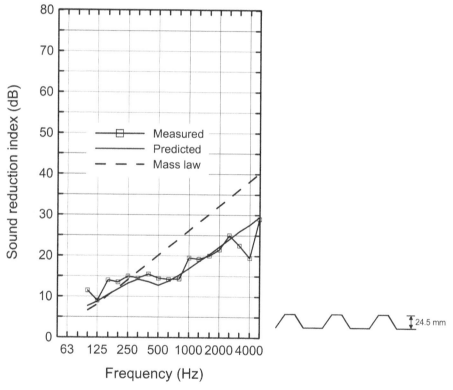

Figure 6.28 Sound reduction index of corrugated panel of weight 4.6 kg/m² and critical frequencies (f_{c1}, f_{c2}) equal (378, 30400) Hz. Measurement data reproduced from Hansen (1993). Predicted data from Equation (6.111).

6.5.4 Transmission through porous materials

As pointed out in the introduction to this chapter, high sound insulation is based on high reflection from a dividing partition, not on dissipating the sound energy in the partition itself. Applying a porous material for good sound insulation is therefore not appropriate. This does not imply, however, that estimating the sound reduction index for such materials is not relevant. It is certainly of interest to be able to estimate the added insulation when mounting a porous absorber on to a wall or below a ceiling.

Measured data for the sound reduction index of mineral wool of different densities are available; e.g. by Homb et al. (1983). An example is given in Figure 6.29 where measured data for rock wool samples of density 50 kg/m³ are compared with calculations. In these calculations we obviously cannot use the formulae in the present chapter, apart from the one performing the averaging over the incidence angle (Equation (6.99)). Calculating the transmission factor of the porous material we shall have to use the equations given in Chapter 5 (section 5.7.1), where we calculated the impedance and absorption factor for such materials. It should be noted that we assume that the material is of infinite extent in the lateral direction, but we have reasons to believe that the boundary conditions are of minor importance in this case. (Set up an expression for the transmission factor e.g. by normal incidence using these equations.)

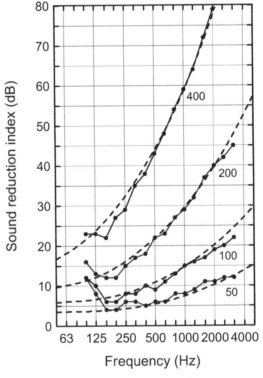

Figure 6.29 Sound reduction index of rock wool of density 50 kg/m³. The thickness in mm is indicated on the curves. Measured data from Homb et al. (1983). Dashed curves are predicted results using the model of Mechel for describing a porous material having flow resistivity 12 kPa·s/m² and porosity 95 %.

Apart from the very low frequency range we get a very good fit between measured and calculated results. The discrepancies at low frequencies may be caused partly by the limited specimen size, partly by a certain stiffness making the mats act like a plate. A more probable explanation is found in the measurement conditions; very low reduction indexes imply feedback between the sending and receiving room. When deriving Equation (6.6), which gives the sound reduction index, we implicitly assumed that there was no coupling between these rooms. At the same time one may get problems concerning the diffusivity when the partition is strongly absorbing.

6.6 A RELATION BETWEEN AIRBORNE AND IMPACT SOUND INSULATION

We have in general treated our two cases, a construction's behaviour by point impacts versus a distributed excitation as a sound field, separately. Under certain conditions, however, we are able to derive a direct relationship between these two ways to characterize the sound insulation properties of a partition, i.e. a relation between the sound reduction index and the impact sound pressure level. One then makes use of a very important principle in vibroacoustics, the *principle of reciprocity*, a principle we have referred to several times in the preceding chapters. Reciprocity implies a mutual relationship and in its most general form tells us that the response at a certain point in a linear elastic system that is caused by exciting another point in the system is invariant by interchanging the source and receiver.

This was postulated by Lord Rayleigh as far back as 1873, but one had to wait nearly 100 years before a formal proof was put forward by Lyamshev (1957). This laid the foundation for many practical applications of the principle. Among these are many connected to sound radiation from vibrating structures, response of structures excited by sound fields and sound transmission. In all cases, reciprocal measurements of transfer functions may often be simpler and less time consuming than the equivalent direct measurements.

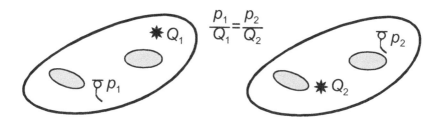

Figure 6.30 Acoustic reciprocity. Exchanging the position of a monopole source and the position of the sound pressure measurement.

More recently Fahy (1995) has presented an extensive review of theory and practical applications. We shall not repeat it here, only to give a short overview and present an expression for the relationship between the radiated acoustic power from a structure excited by a point force and, conversely, the velocity when the structure is excited by a diffuse sound field. Next, we shall use this relationship to derive an expression linking the sound reduction index and the impact sound pressure level.

6.6.1 Vibroacoustic reciprocity, background and applications

Generally, presenting the reciprocity theorem one uses the purely acoustical case as depicted in Figure 6.30. The sound pressure at a given frequency measured at a certain point in a fluid, this caused by an acoustic monopole situated at another position is independent of an interchange of source and receiver. This is true even if the transmission path comprises different media, boundary surfaces giving diffraction etc. The only basic requirement is that the boundary surfaces react linearly. Questions related to the influence of the dynamic behaviour of boundaries have engaged many scientists, e.g. will the principle work when including porous materials in the system? Does it require the boundaries to be locally reacting?

In fact, Rayleigh's general principle of reciprocity implicitly implies that all types of component may take part in the dynamic process, provided that their kinetic, potential or dissipative energy is finite and positive functions of the velocity. Vibrating structures may then take part without invalidating the principle, this being formally proved by Lyamshev (1957). From this it follows that the transfer function between a mechanical point force applied to a structure (e.g. a plate or a shell) and the sound pressure in a point (see Figure 6.31) may be determined by exciting the structure by sound emitted by an acoustic monopole.

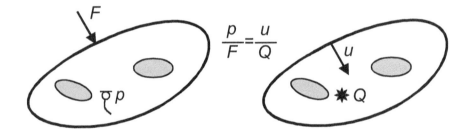

Figure 6.31 Application of the principle of reciprocity. Mechanical forces and sound radiation.

An extension of this point-to-point connection was to prove that a similar reciprocal relationship exists between the sound radiation from a mechanical structure vibrating in a given mode and the response of this mode for incident sound. This leads to the question we shall be concerned with; the relationship between the point response of a structure to diffuse field incidence and the radiated power from the structure excited in the same point. The derivation is given in Cremer et al. (1988), where the following thought experiment is presented:

A plate, comprising a part of the wall in a room, is driven by a point force F (see Figure 6.32). We shall assume that the sound field set up in the room is diffuse and that the sound power emitted may be expressed by

$$W = a \cdot \tilde{F}^2, \tag{6.114}$$

where a is a factor of proportionality. The power sets up an acoustic field in the room, with a resulting sound pressure

$$\tilde{p}^2 = \frac{4\rho_0 c_0}{A} \cdot W = a \cdot \frac{4\rho_0 c_0}{A} \tilde{F}^2, \tag{6.115}$$

where A is the total absorption area of the room. We further assume that, imbedded into another wall of the room, we have a small (in relation to the wavelength) mass-controlled piston of mass m and area S. The resulting force F_p on the piston, caused by the sound field in the room, induces a piston velocity u_{ps} given by

$$\tilde{u}_{ps} = \tilde{F}_p \cdot \frac{1}{\omega m} = S\sqrt{2\tilde{p}^2} \cdot \frac{1}{\omega m}. \tag{6.116}$$

This enables us to write

$$\frac{\tilde{u}_{ps}^2}{\tilde{F}^2} = a \cdot \frac{8\rho_0 c_0 S^2}{A m^2 \omega^2}, \tag{6.117}$$

where the angular frequency ω is understood to be the centre frequency of a band broad enough to give diffuse field conditions.

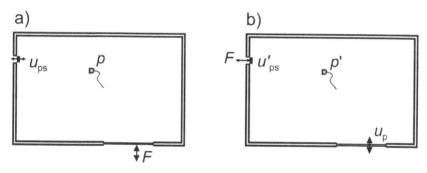

Figure 6.32 Sketch of a room used for a thought experiment. a) A force F is driving a plate being part of a wall, b) A monopole source drives the plate via the sound field in the room.

In the next part of the thought experiment (see Figure 6.32), we shall drive this piston by the same point force used to drive the plate. The piston then gets a velocity u'_{ps}, thereby radiating a power W' into the room equal to the power from a piston in a baffle. At low frequencies, the piston will act like a monopole source and the power may be written (see sections 3.4.1 and 3.4.4)

$$W' = \frac{\rho_0 c_0 k^2}{2\pi} \cdot Q'^2 = \frac{\rho_0 c_0 k^2}{2\pi} \cdot \left(S\tilde{u}_{ps}'\right)^2 = \frac{\rho_0 c_0 k^2}{2\pi} \cdot \left(S\frac{\tilde{F}}{\omega m}\right)^2. \tag{6.118}$$

In the last expression we have inserted the relationship between the force and the resulting velocity of the piston. This power will again set up a sound field in the room having a sound pressure p' given by

$$\left(\tilde{p}'\right)^2 = \frac{4\rho_0 c_0}{A} \cdot W' = \frac{2\left(\rho_0 c_0 k\, S\, \tilde{F}\right)^2}{\pi\, A\omega^2 m^2}. \tag{6.119}$$

The sound pressure p' will then drive the plate, resulting in a velocity u_P that in the linear case must be proportional to the applied pressure. Hence, we may write

$$\tilde{u}_P^2 = b \cdot \left(\tilde{p}'\right)^2, \tag{6.120}$$

where b is a second factor of proportionality. Using Equation (6.119) we get

$$\frac{\tilde{u}_P^2}{\tilde{F}^2} = b \cdot \frac{2\left(\rho_0 c_0 k\, S\right)^2}{\pi\, A\omega^2 m^2}. \tag{6.121}$$

Looking at Equations (6.117) and (6.121) we observe that we have just interchanged the source and receiving point. From the principle of reciprocity, these equations shall then be identical resulting in

$$\frac{a}{b} = \frac{2\left(\rho_0 c_0 k\, S\right)^2}{\pi\, A\omega^2 m^2} \cdot \frac{A m^2 \omega^2}{8\rho_0 c_0 S^2} = \frac{\rho_0 c_0 k^2}{4\pi}. \tag{6.122}$$

From this follows that we may generally find the velocity of a structure placed in a diffuse field (see Equation (6.120)), if we know the power radiated from the structure by a point force excitation given by Equation (6.114). It should be noted that this specifically applies to a given point-to-point relationship. The proportionality factors a and b will generally be space dependent.

6.6.2 Sound reduction index and impact sound pressure level: a relationship

Finally, we shall follow Cremer et al. (1988), giving an example on how to use Equation (6.122) to derive a simple functional relationship between the impact sound pressure level of a massive floor construction and its sound reduction index. We shall assume that the frequency is above the critical frequency, i.e. setting the radiation factor $\sigma \approx 1$ is applicable both for airborne and impact sound. We shall cast the impact sound pressure level L_n, given by Equation (6.21), into the form

$$L_n = 10 \cdot \lg\left\{\frac{\tilde{p}^2}{p_0^2} \cdot \frac{A}{A_0}\right\}, \tag{6.123}$$

where p_0 and A_0 are the reference values $2 \cdot 10^{-5}$ Pa and 10 m^2, respectively. This is equivalent to a radiated power W_n to the receiving room

$$W_n = \frac{\tilde{p}^2}{4\rho_0 c_0} \cdot A = a \cdot \tilde{F}^2. \tag{6.124}$$

The last expression is introduced from Equation (6.114). We may then write for L_n

$$L_n = 10 \cdot \lg \left\{ \frac{4 \rho_0 c_0 \, a \, \tilde{F}^2}{p_0^2 A_0} \right\}. \tag{6.125}$$

For the sound reduction index we shall be looking for an expression for the proportionality factor b, the relationship between the velocity of the floor and the driving sound pressure. By definition, the sound reduction index is given by

$$R = 10 \cdot \lg \frac{W_i}{W_t} = 10 \cdot \lg \frac{\dfrac{\tilde{p}_i^2}{4 \rho_0 c_0} S}{\rho_0 c_0 S \, \tilde{u}^2 \sigma} = 10 \cdot \lg \frac{\tilde{p}_i^2}{4 \rho_0^2 c_0^2 \, \tilde{u}^2 \sigma}, \tag{6.126}$$

where the radiation factor is included in the transmitted power W_t. Setting this factor equal to one, and furthermore, using Equation (6.120) we get

$$R = 10 \cdot \lg \frac{1}{4 \rho_0^2 c_0^2 \cdot b}. \tag{6.127}$$

Combining this expression with Equation (6.125) and using Equation (6.122), we obtain

$$L_n + R = 10 \cdot \lg \left[\frac{k^2 \tilde{F}^2}{4 \pi \, p_0^2 A_0} \right] = 10 \cdot \lg \left[\frac{\pi \, f^2 \tilde{F}^2}{c_0^2 p_0^2 A_0} \right]. \tag{6.128}$$

Inserting for the reference values together with the force from the tapping machine, the latter given by Equation (6.71), we arrive at a very simple expression applying to measurements in one-third-octave band

$$L_n + R \approx 30 \cdot \lg f + 38 \, \text{dB}. \tag{6.129}$$

It should be noted that we have applied the point-to-point relationship according to the principle of reciprocity in a situation where both sound pressure and velocity are mean values taken over a room and a surface, respectively.

As seen from the examples shown in Figure 6.33, which are results from laboratory measurements, Equation (6.129) gives a very good prediction in the case of a massive concrete floor with the top cover. However, the fit is not particularly good for the hollow concrete floor, which may be caused by the non homogeneity of this type of floor. As shown earlier in Figure 6.20, we found that the impact sound pressure level of this type of floor exhibited quite another frequency characteristic than the homogeneous one.

We shall also call attention to two other conditions that must be fulfilled applying Equation (6.129). First, the transmission must take place only through the floor, i.e. the flanking transmission must be negligible. Second, we have assumed when deriving Equation (6.129) from Equation (6.128) that the force from the tapping machine is not reduced by any kind of elastic layer, a floating floor etc. Constructions intended for reducing impact sound will be treated in Chapter 8; here we just wanted to illustrate the effect of such measures.

In fact, we could have used expressions derived earlier to arrive at this relationship between the sound reduction index and the impact sound level. If we sum the expressions given for R and L_n by Equations (6.74) and (6.106) we get, as expected, the same result as in Equation (6.129).

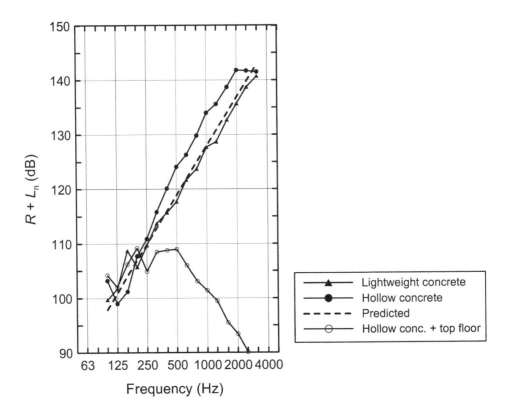

Figure 6.33 The sum of impact sound pressure level and sound reduction index, calculated for two types of floor. 1) Lightweight concrete floor, 175 mm thick with 60 mm concrete top cover and 2) Hollow concrete floor, 260 mm thick. The latter is also measured with an added lightweight floating floor (chipboard on elastic layer).

6.7 REFERENCES

EN 12354–1: 2000, Building acoustics – Estimation of acoustic performance of buildings from the performance of elements. Part 1: Airborne sound insulation between rooms.

EN 12354–2: 2000, Building acoustics – Estimation of acoustic performance of buildings from the performance of elements. Part 2: Impact sound insulation between rooms.

ISO 140–3: 1995, Acoustics – Measurements of sound insulation in buildings and of building elements. Part 3: Laboratory measurements of airborne sound insulation of building elements.

ISO 717–1: 1996, Acoustics – Rating of sound insulation in buildings and of building elements. Part 1: Airborne sound insulation.

ISO 717–2: 1996, Acoustics – Rating of sound insulation in buildings and of building elements. Part 2: Impact sound insulation.

ISO 140–1: 1997, Acoustics – Measurements of sound insulation in buildings and of building elements. Part 1: Requirements for laboratory test facilities with suppressed flanking transmission.

ISO 140–4: 1998, Acoustics – Measurements of sound insulation in buildings and of building elements. Part 4: Field measurements of airborne sound insulation between rooms.

ISO 140–6: 1998, Acoustics – Measurements of sound insulation in buildings and of building elements. Part 6: Laboratory measurements of impact sound insulation of floors.

ISO 140–7: 1998, Acoustics – Measurements of sound insulation in buildings and of building elements. Part 7: Field measurements of impact sound insulation of floors.

ISO 15186–1: 2000, Acoustics – Measurements of sound insulation in buildings and of building elements using sound intensity. Part 1: Laboratory measurements.

ISO/TR 7849: 1987, Acoustics – Estimation of airborne noise emitted by machinery using vibration measurement.

Callister, J. R., George, A. R. and Freeman, G. E. (1999) An empirical scheme to predict the sound transmission loss of single-thickness panels. *J. Sound and Vibration*, 222, 145–151.

Craik, R. J. M. (1996) *Sound transmission through buildings using statistical energy analysis*. Gower Publishing Limited, Aldershot.

Cremer, L. (1942) Theorie der Schalldämmung dünner Wände bei schrägem Einfall. *Akustische Zeitschrift*, 7, 81–102.

Cremer, L., Heckl, M. and Ungar, E. (1988) *Structure-borne sound*, 2nd edn. Springer-Verlag, Berlin.

Fahy, F. (1987) *Sound and structural vibration*. Academic Press, London.

Fahy, F. (1995) The vibroacoustic reciprocity principle and applications to noise control. *Acustica*, 81, 544–558.

Hansen, C. H. (1993) Sound transmission loss of corrugated panels. *Noise Control Eng. J.*, 40, 187–197.

Heckl, M. (1960) Untersuchungen an Orthotropen Platten. *Acustica*, 10, 109–115.

Homb, A., Hveem, S. and Strøm, S. (1983) Sound insulating constructions (in Norwegian). Report 28, Norwegian Building Research Institute (NBI), Oslo.

Josse, R. and Lamure, C. (1964) Transmission du son par une paroi simple. *Acustica*, 14, 266–280.

Lam, Y. W. and Windle, R. M. (1995a) Noise transmission through profiled metal cladding. Part I: Single skin SRI prediction. *Building Acoustics*, 2, 341–356.

Lam, Y. W. and Windle, R. M. (1995b) Noise transmission through profiled metal cladding. Part II: Single skin measurements. *Building Acoustics*, 2, 357–376.

Leppington, F. G., Broadbent, E. G. and Heron, K. H. (1982) The acoustic radiation efficiency of rectangular panels. *Proceedings of the Royal Society of London*, A392, 245–271.

Ljunggren, S. (1991) Airborne sound insulation of thin walls. *J. Acoust. Soc. Am.*, 89, 2324–2337.

Lyamshev, L. M. (1957) A method for solving the problem of sound radiation by thin elastic shells and plates. *Soviet Physics Acoustics*, 5, 122–124.

Macadam, J. A. (1976) The measurement of sound radiation from room surfaces in light-weight buildings. *Applied Acoustics*, 9, 103–118.

Nilsson, A. C. (1974) Sound transmission through single leaf panels. Report 74-01, Chalmers Tekniska Högskola, Gothenburg, Sweden.

Novak, A. (1995) Studies of sound insulation in buildings. PhD thesis KTH, TRITA-BYT 95/0171, Stockholm.

Pietrzyk, A. (1997) Sound insulation at low frequencies. Report F 97-01, Chalmers Tekniska Högskola, Gothenburg, Sweden.

Quirt, J. D. (1982) Sound transmission through windows I. Single and double glazing. *J. Acoust. Soc. Am.*, 72, 834–844.

Sewell, E. C. (1970) Transmission of reverberant sound through a single leaf partition surrounded by in infinite baffle. *J. Sound and Vibration*, 12, 21–32.

Sharp, B. H. (1978) Prediction methods for the sound transmission of building elements. *Noise Control Eng.*, 11, 53–63.

Venzke, G., Dämmig, P. and Fischer, H. W. (1973) Der Einfluss von Versteifungen auf die Schallabstrahlung Schalldämmung von Metallwänden. *Acustica*, 29, 29–40.

Wallace, C.E. (1972) Radiation resistance of a rectangular panel. *J. Acoust. Soc. Am.*, 51, 946–952.

CHAPTER 7

Statistical energy analysis (SEA)

7.1 INTRODUCTION

We have several times referred to statistical energy analysis, SEA for short, which is a general prediction model for complex continuous systems comprising both acoustic and structural members. It has found many applications in building acoustics (see e.g. Craik (1996)), which warrants an introductory chapter giving a background for the examples being presented in the remaining two chapters.

One could argue against the notion of a "method" due to the fact that the user has the choice on how SEA should be applied to a specific system, but we shall not go into that discussion here. SEA originated in connection with the US space programme in the 1960. The problem addressed was the prediction of the response, both of the complete structure and of single components, to the enormous sound and vibration forces released during takeoff. Later applications include the transmission of sound and vibration onboard ships, airplanes and other means of transport and also, as mentioned above, in buildings. The literature covering the field is quite extensive, and the list of references is by no means complete. We shall, however, give reference to a couple of general books on the subject, Lyon (1975) and Craik (1996), together with a review article, Fahy (1994). The last few years have seen a lot of work into the subject of estimating the uncertainty of the method. This, together with the advent of quite a number of commercially available computer programs (see section 2.5.3.1) has opened up SEA for more general use.

SEA is used to model complex *resonant* systems, which may contain structural members such as beams, plates and shells together with acoustical members such as air ducts and rooms. The response, represented by vibration levels (of velocity, acceleration or displacement) and sound pressure levels are calculated for the given excitation (mechanical force, acoustic pressure). The term *statistical* implies that the analysis, contrary to finite element methods (FEM, BEM), does not give any exact information on the behaviour of the system, e.g. how the system responds to an excitation of a single frequency. The calculated data will represent averaged values, not only over given frequency bands, but which also represents averaged values for an ensemble of systems which are nominally identical to the actual one but with a certain statistical spread. The latter is easy to forget because one normally observes, let alone makes measurements on, a given single system.

In the context of building acoustics the aforementioned consideration represents a strength due to the fact that the building components themselves and how they are interconnected, are not in every detail the same for nominally identical systems. Also, one is seldom interested in a detailed frequency description. Rough estimates on how the sound or vibration levels vary with frequency, i.e. in octave or one-third-octave bands, are normally sufficient.

Sound and/or vibration energy are the primary variables in SEA and it may therefore be classed as an energy flow method. In that respect it is analogous to methods calculating heat flow in a system. We shall therefore use this analogy to illustrate the method.

Finally, we shall refer to some additional features important in practical applications of the method:

- The most important choice is specifying the model, which is solely dependent on experience. The calculations themselves are relatively trivial.
- SEA normally works at its best on systems having many cross couplings between the elements (subsystems). The accuracy will normally be less when elements have a series connection only.
- SEA is most suitable for investigation of the type "what happens if…?", i.e. in situations searching for the effects of modifications.
- Experiments are just as important as the analysis.
- Tools for estimating the accuracy of the results may be difficult to obtain.

7.2 SYSTEM DESCRIPTION

As pointed out in the introduction, the primary variable is the energy in the system. The other dynamic variables are deduced from the energy. Specifically, the energy is the *modal energy*, the energy per mode in the separate elements or subsystems. The modal density of these subsystems is therefore an important parameter. Furthermore, the energy loss mechanisms of each single subsystem are characterized by a loss factor and, analogously, we shall use *coupling loss factors* to characterize the power flow between subsystems. Fahy (1998) has pointed out that the latter may not be the best alternative to describe the energy transport between subsystems, and he suggested using a *power transfer coefficient*. We shall not, however, treat this development here.

7.2.1 Thermal–acoustic analogy

To illustrate the terms, we shall as a starting point use a simple model of a thermal system depicted in Figure 7.1. Two identical components (subsystems) are coupled together, one of them connected to a heat supply. We shall assume that the thermal conductivity for both subsystems is high enough making the temperature the same inside each subsystem.

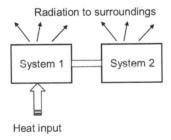

Figure 7.1 Energy transport in a thermal system having two components (subsystems).

Energy will be lost by radiation to the surroundings and by the exchange of energy by a thermal coupling. We shall look at the possible combinations when the radiation losses and the conductivity in the coupling, respectively, either has a low or high value. The number of squares in Table 7.1 represents the energy (temperature) inside the subsystems for the different combinations, certainly giving just a qualitative picture. A situation which we shall look into further on is when the radiation losses are small combined with a high conductivity, i.e. strong coupling. In this case the subsystems are "sharing" the energy; there will be a so-called *equipartition* of energy.

Table 7.1 A qualitative picture of the energy (temperature) in the thermal system shown in Figure 7.1.

Radiation loss	High		Low	
Conductivity	System 1	System 2	System 1	System 2
High	■■■	■■	■■■■■■■■	■■■■■■
Low	■■■■	■	■■■■■■■■■■	■■■■■

Transferring this model into a vibroacoustic one having resonant acoustic volumes (rooms etc.) and resonant solid structures, we may establish the following analogous quantities:

- Thermal radiation losses ⇔ Losses due to absorption, internal losses in materials characterized by the reverberation time T_{60} or loss factor η.
- Conductivity ⇔ Measure of the coupling strength, coupling loss factor η_{ij} (may also be given by an impedance).
- Temperature (energy) ⇔ Sound pressure level in room, vibration level (velocity etc.) of solid structures.
- Thermal capacity ⇔ Modal density.

The last item is not self-evident but it is connected to one of the most important assumptions for SEA.

7.2.2 Basic assumptions

The most important assumptions behind the method are the following:
1. The loss of energy within a subsystem is proportional to the total energy of the subsystem.
2. The energy transmitted from one subsystem to another is proportional to the modal energy difference.
3. The forces driving the different subsystems are independent, statistically speaking. We may add the energy response resulting from these forces to arrive at the total (modal) energy of each subsystem.

To give an illustration of these assumptions and express them in a mathematical form, we shall again start with a system having two components or subsystems, marked 1 and 2 in Figure 7.2. The total energy is E_1 and E_2, respectively, of the subsystems having modal densities n_1 and n_2. The symbol W represents the power; W^{in} for input power, W' for transmitted power between subsystems and W^{diss} for the energy dissipated or lost inside a subsystem. Implicitly; all quantities E, n and W are functions of frequency and generally we shall use the angular frequency ω in the equations.

A simple practical example is depicted in Figure 7.3, a freely suspended plate forced into vibrations by the sound field set up in the room by a loudspeaker. In section 7.3.1 below we shall apply the equations, which follow from the assumptions above, to calculate the amount of vibration resulting from a given input power to the room.

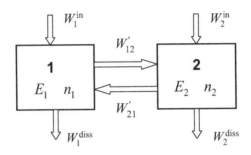

Figure 7.2 System with two components (subsystems).

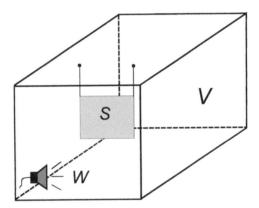

Figure 7.3 Example of a system with two components; a room being the acoustic component having volume V and a free hanging panel of area S as the solid structure.

Assumption no. 1 above gives

$$W^{\text{diss}} = a \cdot E, \qquad (7.1)$$

where a is factor of proportionality. As shown earlier (see section 6.4.1), we found for a plate, having an area S and a mass m per unit area, that the relationship between the its mean square velocity and mechanical power W was given by

$$W = m\, S\omega\eta \cdot \left\langle \tilde{u}^2 \right\rangle_{\Delta\omega} = E\omega\eta, \qquad (7.2)$$

where η is the loss factor. It is assumed that the RMS-value is taken as a mean value over the area of the plate and also over a frequency band $\Delta\omega$ to include a number of modes around the frequency ω. The model energy E_{modal} is expressed as

$$E_{\text{modal}} = \frac{E}{n \cdot \Delta\omega},$$
(7.3)

and assumption no. 2 tells us that the net power W_{12} transmitted between the two subsystems may be written

$$W_{12} = W_{12}' - W_{21}' = b \cdot \left(\frac{E_1}{n_1 \cdot \Delta\omega} - \frac{E_2}{n_2 \cdot \Delta\omega} \right) = b' \cdot \left(\frac{E_1}{n_1} - \frac{E_2}{n_2} \right).$$
(7.4)

The quantities b and $b' = b/\Delta\omega$ are factors of proportionality. The equation may be cast into a more suitable form by introducing the aforementioned coupling loss factors η_{ij}. We then have

$$W_{12}' = E_1 \omega \eta_{12}$$
$$\text{and} \qquad W_{21}' = E_2 \omega \eta_{21}.$$
(7.5)

Hence, $\dfrac{b'}{n_1} = \omega\eta_{12}$ and $\dfrac{b'}{n_2} = \omega\eta_{21}$, which gives

$$n_1 \cdot \eta_{12} = n_2 \cdot \eta_{21},$$
(7.6)

which is called the *reciprocity* or the *consistency relation*. Equation (7.4) may then be written as

$$W_{12} = \omega\eta_{12}\left(E_1 - \frac{n_1}{n_2} E_2 \right).$$
(7.7)

The last assumption given was that the forces driving the various subsystems were statistically independent. This implies that we may calculate the total modal energy of each subsystem by adding the responses resulting from each force input. For system containing k subsystems we get the following set of equations:

$$\omega \cdot \begin{bmatrix} \sum_{i \neq 1} \eta_{1i} + \eta_1 & -\eta_{21} & \cdot & -\eta_{k1} \\ -\eta_{12} & \sum_{i \neq 2} \eta_{2i} + \eta_2 & \cdot & \cdot \\ \cdot & \cdot & \cdot & \cdot \\ \cdot & \cdot & \cdot & \cdot \\ \cdot & \cdot & \cdot & \cdot \\ -\eta_{1k} & \cdot & \cdot & \sum_{i \neq k} \eta_{ki} + \eta_k \end{bmatrix} \cdot \begin{bmatrix} E_1 \\ E_2 \\ \cdot \\ \cdot \\ \cdot \\ E_k \end{bmatrix} = \begin{bmatrix} W_1^{\text{in}} \\ W_2^{\text{in}} \\ \cdot \\ \cdot \\ \cdot \\ W_k^{\text{in}} \end{bmatrix}.$$
(7.8)

The terms on the diagonal represent the total loss factor of each subsystem. It should be noted that in the literature some are using double indices on the symbol for the internal loss factors, i.e. η_{11} for η_1 etc.

7.3 SYSTEM WITH TWO SUBSYSTEMS

For the system shown in Figure 7.2, using Equations (7.8), we get

$$W_1^{in} = \omega E_1 \eta_1 + \omega E_1 \eta_{12} - \omega E_2 \eta_{21}$$

and $\quad W_2^{in} = \omega E_2 \eta_2 + \omega E_2 \eta_{21} - \omega E_1 \eta_{12}.$ $\quad\quad$ (7.9)

Driving one subsystem only, e.g. setting W_2^{in} equal zero, the last equation may be written

$$\frac{E_2}{E_1} = \frac{\eta_{12}}{\eta_2 + \eta_{21}} = \frac{n_2}{n_1} \cdot \frac{\eta_{21}}{\eta_2 + \eta_{21}}. \quad\quad (7.10)$$

Two comments should be made about this equation and here we may compare with the simple thermal model that was our starting point:

a) Assuming that the internal energy losses in subsystem no. 2 are small in comparison with the energy transmitted back to subsystem no. 1, equipartition of modal energy will occur. This implies

$$\eta_{21} \gg \eta_2 \text{ resulting in } \quad \frac{\dfrac{E_2}{n_2}}{\dfrac{E_1}{n_1}} \Rightarrow 1.$$

In such a case there is no point in adding damping to subsystem no. 2, i.e. increasing η_2, unless it could be increased in size of the order of η_{21}.

b) E_2/n_2 will always be smaller than E_1/n_1, which means that η_{21} always will have a positive value.

7.3.1 Free hanging plate in a room

We shall give an example, used by Vér (1992), applying Equation (7.10) to the situation depicted in Figure 7.3. A loudspeaker is creating a sound field in a room of volume V and we shall assume that the field is a diffuse one. The plate, having an area S and a mass per unit area m is forced into vibration by the sound field. The task is to calculate the mean velocity amplitude of the plate. Representing the room and the plate by subsystem 1 and 2, respectively, will be a suitable choice. The total energies may be written

$$E_1 = w \cdot V = \frac{\left\langle \tilde{p}^2 \right\rangle}{\rho_0 c_0^2} \cdot V \quad\quad \text{and} \quad\quad E_2 = mS \cdot \left\langle \tilde{u}^2 \right\rangle, \quad\quad (7.11)$$

where p represents the acoustic pressure in the room and u is the velocity we are seeking. The quantity w is the energy density in the room, for which we have used the simple expression valid for plane waves. For the corresponding modal densities we shall write

$$n_1 = \frac{\omega^2 V}{2\pi^2 c_0^3} \qquad \text{and} \qquad n_2 = \frac{\sqrt{12} \cdot S}{4\pi\, c_L h}. \tag{7.12}$$

The quantities h and c_L are the plate thickness and longitudinal wave speed, respectively. The equation for n_1 is derived from the classical expression giving the modal density in a room of rectangular shape (see section 4.4.1), which is

$$n(f) = \frac{\Delta N}{\Delta f} \approx \frac{4\pi\, f^2}{c_0^3} \cdot V + \frac{\pi\, f}{2 c_0^2} \cdot S + \frac{L}{8 c_0}. \tag{7.13}$$

The quantity ΔN is the number of natural frequencies inside the bandwidth Δf, S and L are here the total surface area of the room and the sum of all the edges in the room, respectively. Going to higher frequencies, the first term will become dominant and is therefore used in Equation (7.12) besides the transformation using the angular frequency.

It now remains an expression for η_{21}, representing the radiated power from the plate, i.e. the power radiated back into the room resulting from the movement of the plate. This output power, radiated from both sides of the plate may be written as

$$W_{\text{out}} = 2 \cdot \rho_0 c_0 S \cdot \langle \tilde{u}^2 \rangle \cdot \sigma \equiv E_2 \omega \eta_{21} = mS \cdot \omega \cdot \eta_{21} \langle \tilde{u}^2 \rangle. \tag{7.14}$$

Hence, the relationship between the coupling loss factor η_{21} and the radiation factor σ will be

$$\eta_{21} = \frac{2\rho_0 c_0}{\omega m} \cdot \sigma. \tag{7.15}$$

Inserting Equations (7.11), (7.12) and (7.15) into Equation (7.10), we arrive at the relation between the velocity of the plate and the acoustic pressure in the room:

$$\frac{\langle \tilde{u}^2 \rangle}{\langle \tilde{p}^2 \rangle} = \frac{\pi\sqrt{12}\, c_0^2}{2\rho_0 c_0 h c_L \omega^2 m} \cdot \frac{1}{1 + \dfrac{\eta_2 \omega m}{2\rho_0 c_0 \cdot \sigma}}. \tag{7.16}$$

Example To put in some numbers we shall use a steel plate of 5 mm thickness being driven by a sound field where the sound pressure level in the room is 100 dB inside a one-third-octave band centred on 1000 Hz. The loss factor η_2 will be quite small as the plate is freely hanging in the room; we may estimate $\eta_{21} \approx 10^{-4}$. Using the following data: $h - 5$ mm, $m - 13.5$ kg/m^3, $c_L - 5200$ m/s, $\rho_0 c_0 - 415$ kg/m^{2}s and $c_0 - 340$ m/s we obtain the term

$$\frac{\eta_2 \omega m}{2\rho_0 c_0 \cdot \sigma} \approx \frac{0.01}{\sigma}.$$

An estimate for the radiation factor σ could be 0.05–0.1 in the chosen frequency band as the critical frequency is approximately equal to 2500 Hz. In effect, we may neglect this term and transform Equation (7.16) into

$$\frac{\left\langle \tilde{u}^2 \right\rangle \omega^2}{\left\langle \tilde{p}^2 \right\rangle} = \frac{\left\langle \tilde{a}^2 \right\rangle}{\left\langle \tilde{p}^2 \right\rangle} \approx \frac{\pi \sqrt{12}\, c_0^2}{2\rho_0 c_0 h c_L m}, \tag{7.17}$$

where a denotes the acceleration. A sound pressure level of 100 dB corresponds to a sound pressure of 2 Pa, which gives

$$\left\langle \tilde{a}^2 \right\rangle = 17.3\, \frac{\text{m}^2}{\text{s}^4}\,, \text{ or expressed as an acceleration level } L_a = 10 \cdot \lg\left(\frac{\tilde{a}^2}{a_0^2}\right) \approx 132\,\text{dB}.$$

The reference value a_0 is 10^{-6} m/s^2. Expressed as a RMS-value, the acceleration is thereby 4.2 m/s^2 (or approximately half the acceleration of gravity).

It should, however, be no problem to increase the internal loss factor by a factor of 100 by applying a visco-elastic layer to the plate, resulting in a η_2 of 10^{-2}. (What will be the effect on the acceleration level?)

7.4 SEA APPLICATIONS IN BUILDING ACOUSTICS

Based on the early applications of SEA, in particular on calculating the response of panel constructions in reverberant sound fields (see e.g. Maidanik (1962)), the method gained acceptance for solving building acoustic problems such as transmission through single and double wall constructions (Crocker and Price (1969); Price and Crocker (1970)). Transmission through constructions containing several layers such as double walls will be treated in the next chapter and we shall then include examples where SEA models are used.

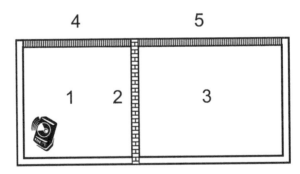

Figure 7.4 Sound transmission between two rooms by way of a separating wall and flanking walls.

It will be appropriate when presenting this short overview to use another building acoustic problem as an example, namely the sound transmission between two rooms including sound transmission by way of flanking walls. As we shall see later (see Chapter 9), there will be classical models to estimate both the direct transmission and the flanking transmission. These models will, however, become quite complicated for multiple layer wall constructions and complex junctions. In such cases, modelling by SEA may be appropriate. A hint on how such problems could be treated will be outlined using the example shown in Figure 7.4, where the rooms are modelled as subsystems 1 and 3. The separating wall constitutes subsystem 2, and we shall include only two flanking walls in the model, one wall in each room being subsystems 4 and 5.

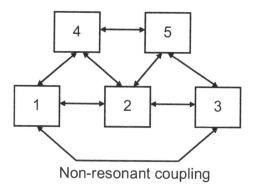

Non-resonant coupling

Figure 7.5 SEA model for the situation shown in Figure 7.4.

One very important aspect, which should be borne in mind, is that the whole concept of SEA presupposes resonant systems and resonant transmission. The non-resonant part of the sound transmission through the wall at frequencies below the critical frequency, the part calculated by the mass law, must therefore be included in a separate way. In Figure 7.5, this part of the transmission is shown by the direct connection between subsystems 1 and 3, which are the rooms.

From the simple example above, comprising two subsystems, we have already some of the expressions needed to fill into the matrix Equation (7.8) to solve for the sound transmission problem. However, it is easily seen that the key terms will be the coupling loss factors, which may be hard to estimate. Determining these factors by experiments may in practice be the only solution. (Set up the complete matrix equation for the system given in Figure 7.5 and try to pick out the most important components.)

Finally, we shall present an example on measured and predicted sound pressure level difference between two rooms caused by the different transmission paths using the above model. The results are laboratory measurement results from Building Research Establishment, UK, reproduced in the book by Craik (1996). The laboratory is constructed to measure the influence of flanking transmission, in the specific case the partition wall, subsystem 2, is concrete block cavity wall. The external flanking walls, subsystems 4 and 5 are lightweight block and brick cavity walls.

Figure 7.6 shows both the overall measured and predicted sound pressure level difference together with the predicted dominant transmission paths. The symbols included in parenthesis are taken from the calculation models in the standard EN 12354 Part 1. The symbols D and F refer to the partition and flanking wall, respectively, seen from the sending room and the small letters, d and f, correspondingly refer to the ones seen from the receiving room. Further treatment of these models is postponed until Chapter 9.

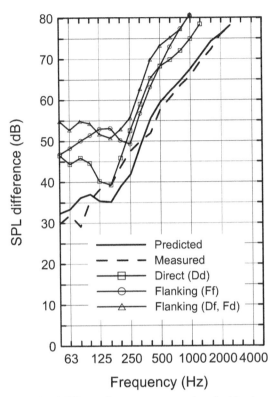

Figure 7.6 Sound pressure level difference between two rooms in a flanking transmission laboratory (Building Research Laboratory, UK). Overall predicted and measured results together with predicted dominant transmission paths. For symbols in parentheses; see EN 12354–1. Data reproduced from Craik (1996).

7.5 REFERENCES

EN 12354–1: 2000, Building acoustics – Estimation of acoustic performance of buildings from the performance of elements. Part 1: Airborne sound insulation between rooms.

Craik, R. J. M. (1996) *Sound transmission through buildings using statistical energy analysis*. Gower Publishing Limited, Aldershot.

Crocker, M. J. and Price, A. J. (1969) Sound transmission using statistical energy analysis. *J. Sound and Vibration*, 9, 469–486.

Fahy, F. J. (1994) Statistical energy analysis: A critical overview. *Philosophical Transactions of the Royal Society of London*, Series A, 346, 431–447.

Fahy, F. J. (1998) An alternative to the SEA coupling loss factor: Rationale and method for experimental determination. *J. Sound and Vibration*, 214, 261–267.

Lyon, R. H. (1975) *Statistical analysis of dynamical systems. Theory and applications*. MIT Press, Cambridge, MA.

Maidanik, G. (1962) Response of ribbed panels to reverberant acoustic fields. *J. Acoust. Soc. Am.*, 34, 809–826.

Price, A. J. and Crocker, M. J. (1970) Sound transmission through double panels using statistical energy analysis. *J. Acoust. Soc. Am.*, 47, 683–693.

Vér, I. L. (1992) Interaction of sound waves with solid structures. In L. L. Beranek and I. L. Vér (eds) *Noise and vibration engineering*. John Wiley & Sons, New York.

CHAPTER 8

Sound transmission through multilayer elements

8.1 INTRODUCTION

Using the methods presented in Chapter 6, we may, with reasonable accuracy, predict the sound transmission properties of walls and floors characterized as single homogeneous elements. The complexity increases quickly when the elements are composed of several layers, also mechanically coupled in ways that are difficult to specify.

This chapter deals with these types of multilayer element but will, as in Chapter 6, treat the specific elements as being disconnected from their boundary elements. Certainly, this imposes a limitation as other connected structures certainly will influence the vibration wave field of the element in question, and more so if the mechanical couplings are strong. We shall postpone further discussion of this question to Chapter 9, which deals with the theme of flanking transmission, i.e. the prediction of sound transmission between rooms with several interacting structures.

We shall use the term multilayer to characterize a building element made up of two or more homogeneous layers. These may contain mechanical coupling elements, such as ties, studs or elastic layers. A double wall is a pertinent example, where the leaves are mechanically coupled or uncoupled depending on a common or separate system of studs. Another example is a floating floor construction where the top layer is coupled to the base floor through either a continuous elastic layer or by elastic point supports.

The level of coupling may be difficult to ascertain. Two common boards (plasterboard, chipboard etc.) placed in contact with another or even screwed together are, acoustically speaking, "weakly" coupled. With boards of identical thickness, the critical frequency of the combination will be approximately equal to the one for a single board and the sound reduction index will ideally increase by 6 dB due to the increase in weight by a factor of two. Gluing the boards together, however, we end up with a *sandwich element*, where the coupling depends on the type of glue and the element itself acquires some quite different properties than in the first mentioned case.

We shall, according to the terms used in the standard EN 12354, use the term element when speaking of a building component like a partition wall, a floor, a door etc., even when these are composite structures. In many cases one should perhaps better use the term "construction", a collection of elements. Hopefully, the reader should not be confused finding both terms being used throughout the book.

8.2 DOUBLE WALLS

With the term double walls we are to understand constructions having two independent wall elements separated by an air-filled cavity that may contain a porous absorber. As mentioned in the introduction, the wall elements may be mechanically coupled in different ways. We shall not put any restrictions on the type of elements; these may be massive heavy elements of concrete, brick etc. as well as lightweight elements such as plasterboard, chipboard, glazing etc. There is a great demand for lightweight constructions offering high sound insulation and we shall therefore put some emphasis on lightweight board constructions. These have at least two layers, certainly because single lightweight layers will never be able to fulfil the requirements e.g. imposed in building codes for dwellings. Combinations, such as a massive heavy element and an additional lightweight lining are practical constructions. Such combinations may achieve a considerably higher sound reduction index than the primary heavy one.

Searching for literature on the sound insulation of double walls, one quickly discovers that it is rather extensive, certainly on lightweight constructions. The theoretical analysis is, however, less developed than the one for single leaf constructions, which should not come as a surprise in view of the complexity involved. It is not only the mechanical couplings between the leaves that is difficult to quantify, the importance of the different, and at the same time distributed, energy loss mechanisms is difficult to ascertain. An illustration is given in Figure 8.1, indicating how energy may be transmitted between rooms by a lightweight double wall. A direct path coupling the two leaves across the cavity is indicated together with the paths across structural stud connections and along the outer boundaries.

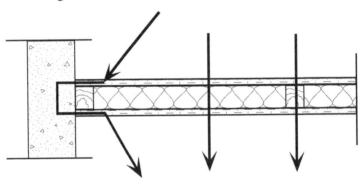

Figure 8.1 Transmission paths for a lightweight double wall.

We do not intend to give a general overview of the literature dealing with double walls but to render an understanding of the physical variables involved by presenting a number of idealized examples. Prediction models using infinite size elements are appropriate also in this case. One cannot, however, avoid presenting results based on statistical energy analysis (SEA), a method applied to these questions as early as 1970 onwards (see Chapter 7). Using SEA, suitable software may be developed for calculating transmission factors and sound reduction indices. In practice, it may also be advantageous to develop formulae giving rough estimates of the performance. Such expressions, developed on an empirical basis will also be given here.

8.2.1 Double wall without mechanical connections

Assuming that there are no structural connections between the two leaves of the wall, the energy transmission from one leaf to the other must take place by a forced excitation by way of the cavity. As a first approximation to this case, which may well be found in practice, we shall deal with a situation where we assume diffuse sound fields, not only in the sending and receiving rooms but in the cavity as well. This certainly presupposes the cavity depth is larger than the wavelength. Such a situation is depicted in Figure 8.2 where the sending and receiving rooms are separated by a double wall represented by two single partitions having sound reduction indexes R_1 and R_2.

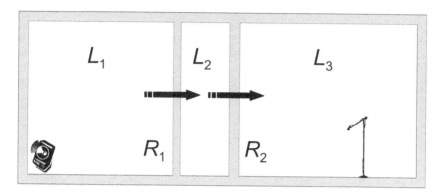

Figure 8.2 Rooms separated by a double wall with a large cavity.

We may express these sound reduction indexes as:

$$R_1 = L_1 - L_2 + 10 \cdot \lg \frac{S}{A_2}$$

$$\text{and} \qquad R_2 = L_2 - L_3 + 10 \cdot \lg \frac{S}{A_3}, \qquad (8.1)$$

where S is the area of the partitions. L and A denote the sound pressure level and the total absorption area, respectively, having indices according to the figure. The sound reduction index R_d of the "double" wall is expressed as

$$R_d = L_1 - L_3 + 10 \cdot \lg \frac{S}{A_3}, \qquad (8.2)$$

which, by inserting the Equations (8.1), gives

$$R_d = R_1 + R_2 + 10 \cdot \lg \frac{A_2}{S}. \qquad (8.3)$$

In spite of the assumption, having a diffuse field in the cavity, the importance of the cavity damping is a general one. A porous absorber inside the cavity will generally attenuate waves running parallel to the wall leaves, and we know that obliquely incident waves are easily transmitted.

As an example we shall perform a calculation on an unbounded double wall of two 9 mm plasterboards with a 50 mm cavity filled with a porous absorber of mineral wool type (see Figure 8.3). These calculations are performed using a transfer matrix model described in Chapter 5 (section 5.7.1.1). The wall impedance of each board, see Chapter 6, section 6.5.1.2, is given by

$$Z_{\mathrm{w}} = \mathrm{j}\omega m \left[1 - \left(\frac{f}{f_{\mathrm{c}}} \right)^2 \cdot (1 + \mathrm{j}\eta) \sin^4 \varphi \right]. \tag{8.4}$$

The model of Mechel (1976) is used to describe the porous absorber in the same way as when calculating the transmission through the absorber alone; see Chapter 6 (section 6.5.4). The flow resistivity of the porous layer is varied in steps to simulate various degrees of cavity damping, starting from a value of 10 kPa·s/m², which corresponds to the value found for common products of mineral wool.

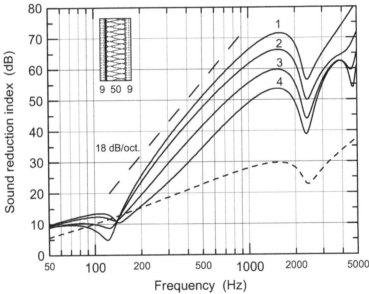

Figure 8.3 Sound reduction index of an unbounded double wall. Diffuse field incidence. Two 9 mm plasterboards, 7.2 kg/m² with critical frequency 2250 Hz. 50 mm cavity with porous material. Parameter is the flow resistivity: curve 1 having r = 10 kPa.s/m² and for each curve 2–4 the resistivity is reduced by a factor of five. Lower dashed curve – reduction index for a single board.

Figure 8.3 exhibits some typical features found for the sound reduction index of double walls with no structural connections between the leaves. At low frequencies, where the distance between the leaves is much smaller than the wavelength, the leaves will be strongly coupled by the acoustic stiffness of the air in the cavity, this in spite of the presence of the porous absorber. If the vibrations of the leaves are mass controlled, the wall will behave like a single leaf, having a mass equal to the sum of the masses of the leaves; compare with the dashed curve giving the reduction index of each leaf.

The acoustic coupling across the cavity will give a *double wall resonance*, resulting in a minimum value of R at a frequency expressed as

$$f_0 = \frac{1}{2\pi} \sqrt{\frac{s}{m_1} + \frac{s}{m_2}}, \tag{8.5}$$

where m_1 and m_2 are the mass per unit area of the two leaves and where s is the stiffness per unit area represented by the cavity. As evident from the expression, this resonance corresponds to a symmetrical movement of two masses connected by a spring. Assuming that the spring stiffness is solely determined by the air in the cavity, it will have a value in the range between $\rho_0 c_0^2/d$ and $P_0/(\sigma \cdot d)$. Here the quantities d, P_0 and σ are the distance between the leaves, the barometric pressure and the porosity of the porous material, respectively.

In the latter expression we have assumed that the sound propagation takes place isothermally, whereas the former will apply for an empty (air-filled) cavity. In practice, it is not very important which one to use as the difference is given by the factor $\gamma \cdot \sigma$, where γ is the adiabatic constant. The latter is approximately equal to 1.4 for air and the porosity is normally above 0.9, which leaves us with a difference in the range of 10–15 %. For a rough estimate one normally finds expressions that applies to an air-filled cavity, such as

$$f_0 = \frac{c_0}{2\pi} \sqrt{\frac{\rho_0 (m_1 + m_2)}{m_1 \cdot m_2 \cdot d}} \approx 60 \sqrt{\frac{m_1 + m_2}{m_1 \cdot m_2 \cdot d}}. \tag{8.6}$$

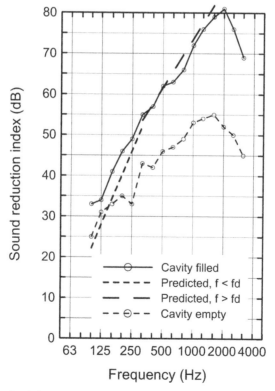

Figure 8.4 Sound reduction index of a double wall, 13 mm plasterboards mounted on separate studs, 150 mm cavity depth. Measured data from Homb et al. (1983). Predicted data from model by Sharp (1978).

In our example (see Figure 8.3), we get $f_0 \approx 140$ Hz when using this expression. It should be noted that we presuppose that there is no elastic coupling of the leaves due to a possible elasticity in the porous material.

In the frequency range above the resonance frequency the sound reduction index will increase by 18 dB per octave. It is not evident from the figure that above a given frequency $f_d \approx 55/d$ (see Equation (8.7), the increase is less strong and according to empirical data estimated as 12 dB per octave. In our example f_d will be approximately 1100 Hz, thus the effect will be partly masked by the "dip" due to coincidence.

These phenomena are all to be found in results from laboratory measurements. An example is given in Figure 8.4, presenting the sound reduction index of a double wall, two leaves of 13 mm plasterboard mounted on separate studs. The cavity depth is 150 mm, and measurements were performed both leaving the cavity empty and also being completely filled up by rock wool of density 20 kg/m^3. One cannot detect the double wall resonance as the measurements were limited downwards to 100 Hz, whereas the resonance should be around 65 Hz. The frequency f_d, marking the transition from a 18 dB per octave to a 12 dB per octave should be approximately equal to 370 Hz.

The predicted results are based on an empirical model by Sharp (1978). Using classical expressions and a large measurement database, he presented the following simple set of equations to predict the sound reduction index for double walls without structural connections, however having the cavity filled with a porous absorber:

$$R = \begin{cases} R_M & f < f_0, \\ R_1 + R_2 + 20 \cdot \lg(f \cdot d) - 29 \text{ dB} & f_0 < f < f_d, \\ R_1 + R_2 + 6 \text{ dB} & f > f_d, \end{cases} \tag{8.7}$$

where f_d, as given above, is equal to $55/d$. The index M indicate that the reduction index is to be calculated from the total mass of the leaves, $M = m_1 + m_2$. The predicted results shown in Figure 8.4 for the frequency range $f < f_d$ does not fit too well to the measured ones but due to lack of accurate specifications only the simple mass law is applied for calculating R_1 and R_2, i.e. the one given in Chapter 6 (section 6.5.2):

$$R = 20 \cdot \lg(f\,m) - 47 \text{ dB}. \tag{8.8}$$

It may seem odd that no specifications as to the porous material, filling the cavity, enter into Equations (8.7). The attenuation caused by this material certainly depends on parameters such as flow resistance etc. Brekke (1979) suggested the following expression to be used for the frequency range above f_d :

$$R = R_1 + R_2 + ATT - 20 \cdot \lg \frac{Z_i}{Z_{\text{ref}}}, \tag{8.9}$$

where Z_i and ATT is the input impedance of the absorber (as seen from the first leaf) and the attenuation offered by it. The reference impedance Z_{ref} is equal to the impedance of a porous material having a flow resistivity of 7 kPa·s/m^2.

In spite of the influence of the cavity material there are several reasons for not gaining much by using more complicated expressions for the high frequency range than the one given in Equation (8.7). In practice, the reduction indices are normally much larger than the ones in the lower frequency range, even when taking the weighting curve into account (see Figure 6.4). The calculation accuracy which is certainly interesting from a theoretical point of view, then becomes of less interest in practice. Furthermore, judging from Figure 8.3, there has to be large variations in the flow resistivity to really affect the reduction index, much larger variations than between the products normally used in lightweight double walls. The big difference lies in the amount of filling, either just a part or completely. There may be differences in the reduction index of 4–5 dB from a percentage of filling being 30–50% as compared with 90–100%.

8.2.1.1 Lightly damped cavity

Double walls, where the cavity is not efficiently damped by a porous absorber, will necessarily give a sound reduction index somewhere between the one found for an empty cavity and a completely filled one (see Figure 8.4). A case where one will benefit from a filling of the cavity, but for obvious reasons cannot fill it completely, is by window constructions applying a lining inside the window frame. In practice, these are cases having a reasonably large cavity depth, at least more than 50 mm, excluding common compact double (or triple) glazed units. For a double construction equipped with a frame absorber one could use Equation (8.3) as a first approximation, writing

$$R = R_1 + R_2 + 10 \cdot \lg \frac{A}{S} = R_1 + R_2 + 10 \cdot \lg \frac{\alpha U d}{S}, \qquad (8.10)$$

where α is the absorption factor for the absorber along the frame, and U is the circumference or total length of the frame. Most correctly, one should use the absorption factor for normal incidence. (Why is that?)

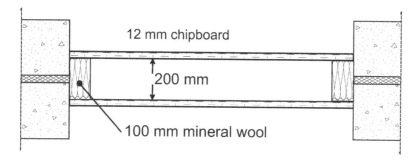

Figure 8.5 Arrangement for measurements on a double leaf construction (2.25 x 1.25 m) with no structural connection between the leaves. Adapted from Brekke (1979).

Even without an absorber in the cavity there will be a certain surface absorption due to viscous effects but the magnitude is difficult to estimate. In a relatively early phase of applying SEA models on problems in building acoustics the cases of an empty cavity and a cavity with a frame absorber were treated (see e.g. Crocker et al. (1971); Brekke (1979)). Both also treated the case when the leaves are structurally connected, a case recently taken up by Craik and Smith (2000), also within the framework of SEA. In parallel with the SEA type of modelling, analytical models for incorporating the effect of structural connections as studs or other types of mechanical links have been developed. A number of these efforts are mentioned by Wang et al. (2005), who themselves are treating the case of a double leaf lightweight partition with periodically placed studs. We shall postpone the treatment of such mechanical connections to the next section.

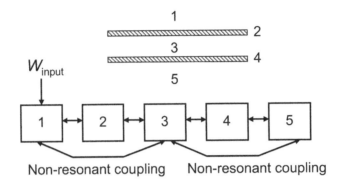

Figure 8.6 SEA model for the set-up shown in Figure 8.5. Subsystems 1 and 5 represent the sending and receiving room, respectively. Subsystem 3 is the cavity between the leaves.

      Referring back to the discussion in Chapter 7, we should expect that predictions using SEA would be accurate when the cavity is lightly damped. A cavity completely filled with a porous material may hardly be characterized as a resonant subsystem. Brekke (1979) used a set-up as shown in Figure 8.5, a double leaf construction of 12 mm chipboard with an absorber lining that was placed in a measuring opening of dimensions 2.25 x 1.25 metres. The system was modeled using SEA according to the scheme shown in Figure 8.6. The system of equations is easy to formulate, estimating the loss factors is harder. By determining these factors mainly from independent measurements, the fit between measured and calculated results were reasonably good (see Figure 8.7).

      Crocker et al. used a double leaf construction, which they called a double panel, aluminum panels of 3.2 mm thickness placed in an opening of 1.55 x 1.97 meters. The cavity, having a depth of 71 mm, was empty. The fit between measured and calculated results was good but based on estimated data for the internal material losses (see Figure 8.7).

8.2.2 Double walls with structural connections

      An accurate prediction of sound insulation indexes of double wall constructions with different types of structural connection between the leaves (see Figure 8.1) has been and still is a challenge. In the cases cited above, which were using SEA modelling, the inclusion of structural connections has been an obvious extension. This includes coupling along a line, i.e. ribs or studs in lightweight partitions as well as discrete point connections, the latter being binders in heavy walls as brick or concrete. These connections may be modelled as separate modal subsystems or purely as coupling elements.

      Sharp (1978) introduced an extension to the simple set of calculations in Equations (8.7) covering point and line connections for lightweight walls, assuming that these connections were infinitely stiff. An extension of the work of Sharp, taking the stiffness of the connections into account, has been suggested by Davy (1991). Later developments have been on models, partly of the type "smeared" model by representing the studs by uniformly distributed elastic springs as well as a more accurate model taking the discrete placing of the studs into account by treating the wall as a periodic structure, see for example Lee and Kim (2002), who treated single panels with stiffeners and Wang et al. (2005), who extended this model to double leaf constructions. The "smeared" type of model follows the approach used for floating floors (see section 8.4.1) and in practice, there certainly are double wall constructions where the layers are connected in this manner, e.g. by a continuous elastic layer such as stiff mineral wool. A model for such cases has been presented by Kropp and Rebillard (1999).

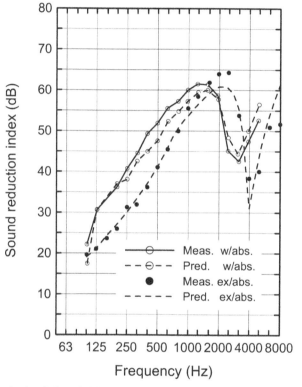

Figure 8.7 Sound reduction index of lightweight double leaf constructions without structural connections. Measured and predicted results by Brekke (1979), using a frame absorber (w/abs.). Measured and predicted results by Crocker et al. (1971), using an empty cavity (ex/abs).

By lightweight double leaf partitions of plasterboard, chipboard etc. mounted on common studs, forces or moments transmitted through these studs normally are the determining factor in the frequency range above the double wall resonance. The development of studs having reduced stiffness is an important task as specially profiled steel studs have shown to give large improvements. Certainly, there will normally always be direct structural connections along the perimeter of a partition (edge coupling) but these are not equally important for lightweight constructions as for heavy, massive systems.

We shall treat all these methods of couplings starting with a double construction of a special type: a heavy wall or ceiling covered by an additional so-called *acoustical lining*. This implies a lining having a low bending stiffness and a high critical frequency as compared with the primary heavy construction. This is a common method of improving the sound reduction index of a wall or the impact sound insulation of a ceiling (floor). The reason behind starting out with this example is that we may assume that the movement of the primary construction is not affected by the lining; i.e. we assume that there is no "feedback" in the system, which would certainly be the case of lightweight walls. A laboratory standard for measuring the improvement of such linings has recently become available (see ISO 140 Part 16).

8.2.2.1 Acoustical lining

The treatment of this case may be found in the book by Cremer et al. (1988), but was presented by Heckl as early as in 1959. As seen from Figure 8.8, we shall assume that the basic or primary construction is a heavy, massive wall (or floor) for which the critical frequency lies below the observed frequency range. Furthermore, we assume that the radiation factor is approximately equal to 1.0 in this frequency range. The lining, on the other hand, has such a high critical frequency that the bending wave near field, caused by vibrations transmitted through the studs or ties, will dominate in the radiated sound. This is the basic idea behind such additional acoustical linings; even if the lining is firmly connected to the basic wall, and thus obtains the same velocity as the latter at the connections, the total radiation will be reduced.

The sound reduction index for the combination (see Figure 8.8 b)), may be expressed as

$$R = 10 \cdot \lg \frac{W_i}{W_2} = 10 \cdot \lg \left(\frac{W_i}{W_{2,P} + W_{2,B}} \right), \tag{8.11}$$

where W_i is the incident power on the primary wall. The radiated power from the lining is divided into two parts: the power $W_{2,B}$ radiated from the bending wave near field and the power $W_{2,P}$ due to the transmission through the cavity. The latter may, at frequencies above the double wall resonance (see Equation (8.6)), be written

$$\frac{W_{2,P}}{W_i} = \frac{W_1}{W_i} \cdot \left(\frac{f_0}{f} \right)^4. \tag{8.12}$$

The quantity W_1 is the power transmitted through the primary wall. This equation tells us that the relationship between the mean velocity amplitude of the primary wall and the lining is proportional to the frequency squared. The cavity acts as a pure spring situated between the wall and the lining, these being represented by two masses. The radiated power from *one* of the structural bridges, which may be a stud (line connection) or a tie (point connection), may be written

$$\Delta W_{2,B} = \rho_0 c_0 \tilde{u}_{2,B}^2 \cdot S \cdot \sigma_B. \tag{8.13}$$

The quantity σ_B is the radiation factor, however here defined by the velocity of the bridge, not as earlier by a mean velocity of the plate. Having a number n bridges distributed over wall area S we may just multiply by n to arrive at the total radiated power. We have assumed that the radiation factor of the basic wall is approximately equal to 1.0, hence

$$W_1 = \rho_0 c_0 \left\langle \tilde{u}_1^2 \right\rangle S. \tag{8.14}$$

Equations (8.12) through (8.14) then give

$$\frac{W_i}{W_{2,P} + W_{2,B}} = \frac{W_i}{W_{2,P} + n \cdot \Delta W_{2,B}} = \frac{W_i}{W_1} \left[\left(\frac{f_0}{f} \right)^4 + n \cdot \frac{\tilde{u}_{2,B}^2}{\left\langle \tilde{u}_1^2 \right\rangle} \sigma_B \right]^{-1}. \tag{8.15}$$

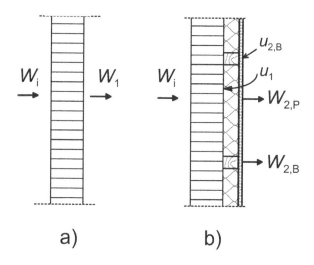

Figure 8.8 Heavy basic wall with additional acoustical lining.

Inserting this equation into Equation (8.11), we get

$$R = R_1 + \Delta R,$$

$$\text{where} \quad \Delta R = -10 \cdot \lg \left[\left(\frac{f_0}{f} \right)^4 + n \cdot \frac{\tilde{u}_{2,B}^2}{\left\langle \tilde{u}_1^2 \right\rangle} \sigma_B \right]. \tag{8.16}$$

The quantity ΔR is thereby the improvement of the reduction index due to the additional lining. To calculate this improvement we shall need an expression for the radiation factor σ_B. However, we have shown in Chapter 6 (section 6.4.2.1) that the radiated power from a bending wave near field on a plate driven at a point and along a line, respectively, is given by

$$W_{\text{point}} = \frac{8\rho_0 c_0^3}{\pi^3} \cdot \frac{\tilde{u}_0^2}{f_c^2},$$

$$W_{\text{line}} = \frac{2\rho_0 c_0^2 \tilde{u}_0^2}{\pi f_c} \cdot \ell, \tag{8.17}$$

where u_0 is velocity in the point or on the line, and ℓ is the length of the line. The radiation factor for these two cases may then be expressed as

$$\sigma_{\text{B,point}} = \frac{8c_0^2}{\pi^3} \cdot \frac{1}{S} \cdot \frac{1}{f_c^2}$$

$$\text{and} \quad \sigma_{\text{B,line}} = \frac{2c_0}{\pi} \cdot \frac{\ell}{S} \cdot \frac{1}{f_c}. \tag{8.18}$$

In the normal case using a set n of studs distributed evenly over the surface area S, where we shall apply the last expression, the centre-to-centre distance between the studs will be $S/(n \cdot \ell)$.

If we look at the expression for the sound reduction index, Equation (8.16), and initially assume that the ratio between the velocity of the primary wall and the velocity of the bridges are frequency independent, the improvement will increase by 12 dB per octave until it reaches a maximum, a plateau. This maximum will be determined by the critical frequency of the lining and the degree of mechanical contact between the wall and the lining. It should be noted that the improvement will go to zero towards the critical frequency. The lining will then became just as good a radiator as the primary wall and no improvement, except for the one caused by a small increase in mass, is to be expected. A sketch showing the improvement in principle is presented in Figure 8.9.

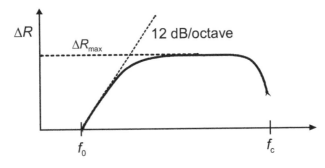

Figure 8.9 Sound reduction improvement caused by an additional lining.

The maximum improvement offered by a lining connected to the primary wall by studs is shown in Figure 8.10. It should be noted that the data assume infinitely stiff connections between the primary construction and the lining. For the case of a lining detached from the primary wall, i.e. having contact along the edges only, one may in practice set the c-c distance to be equal to the smallest lining dimension.

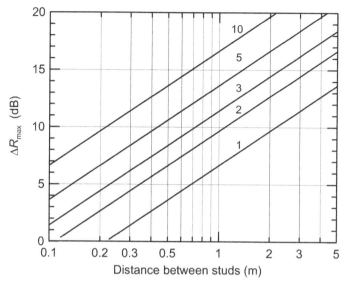

Figure 8.10 Acoustic lining attached to the primary construction by infinitely stiff studs. The maximum improvement in the sound reduction index is given as a function of c-c distance of studs. The parameter shown on the curves is the critical frequency of the lining in kHz.

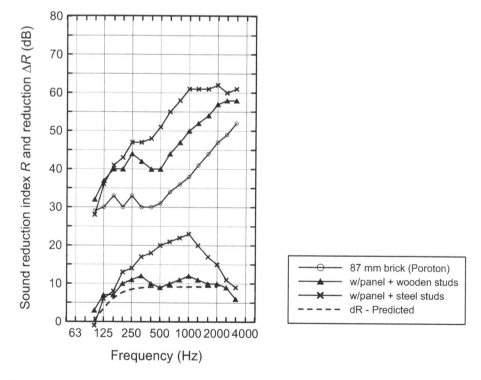

Figure 8.11 Brick wall with lining. Sound reduction index and improvement by lining of 12 mm chipboard directly attached to the wall. Measured data from Homb et al. (1983).

The effect of such a lining of 12 mm chipboard attached to a lightweight brick wall of 87 mm thickness in shown in Figure 8.11. The cavity depth is 50 mm and filled with a porous absorber. The uppermost three curves show the sound reduction for the brick wall alone together with the result when adding the lining attached to the wall with wooden studs and steel profiles, respectively.

The lower curves show the improvement by the lining when using the two types of studs together with predicted results using Equation (8.16), assuming the studs are infinitely stiff. Obviously, using the steel profiles one has not got a perfectly stiff connection and furthermore, the ratio u_2/u_1 is frequency dependent.

Finally, dealing with acoustical linings we shall call attention to the achievable improvement in practice, being normally limited by flanking transmission. In most cases, the flanking transmission will have a greater influence by airborne sound transmission than by impact sound transmission. The improvement offered by a lining is therefore often higher for impact sound than for airborne sound.

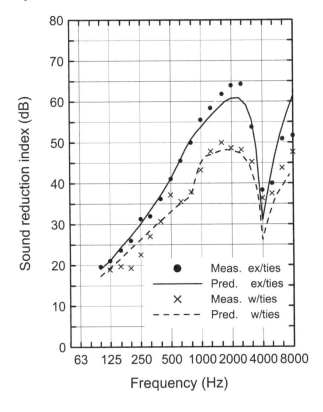

Figure 8.12 Sound reduction index of double aluminium panel with (w/ties) and without tie beams (ex/ties). Predicted results using SEA modelling. Adapted from Crocker et al. (1971).

8.2.2.2 Lightweight double leaf partitions with structural connections

A double leaf partition with structural connections, in the form of point or line connection between the leaves, should be suited to modelling by SEA. In the model shown in Figure 8.6 we shall have to include either a coupling element or a subsystem between the subsystems 2 and 4. An early example of such calculations, mentioned above, were performed by Crocker et al. (1971), where two aluminum panels of dimensions 1.55 x 1.97 m were interconnected by a total of 50 "ties", these being short beams of aluminium of thickness 0.7 mm and width 25 mm. The comparison between measured and predicted results is shown in Figure 8.12. The fit between the two sets of data is very good but it should be noted that it is based on estimated values for internal loss factors of the material, η equal to 0.005 in the frequency range below 800 Hz and η equal to 0.02 above 800 Hz.

Structural connections in the form of studs are more relevant in building partitions. In spite of being a special laboratory model, the set-up used by Brekke (1979) (see Figure 8.5) gives results that are typical in practice. Brekke performed measurements and predictions where the panels were interconnected by different types of stud. In this case, the depth of the cavity was reduced to approximately 95 mm to accommodate the various types of studs. Measured sound reduction indexes of the double panel with and without two types of stud are depicted in Figure 8.13. Calculations for the set-up with wooden studs were performed using a SEA model determining the coupling loss factor for the studs by measurement. The fit between measured and predicted results were reasonably good with a maximum deviation of 5 dB but predicted results are omitted here.

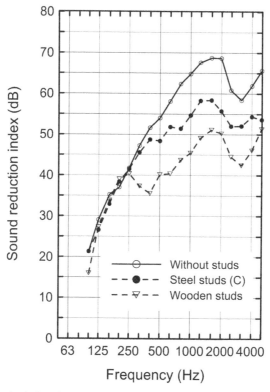

Figure 8.13 Sound reduction index of experimental double panel, after Brekke (1979). 12 mm chipboards with a cavity of depth 95 mm filled with mineral wool. See also Figure 8.5.

There is certainly a need for performing simple estimations of the sound reduction index of double constructions on common studs. We shall therefore include the prediction model by Sharp (1978), a commonly cited reference. The model that assumes infinitely stiff connections, either point or line connections, again uses Equation (8.11) as a base. We may write it in the following way

$$R = 10 \cdot \lg \frac{W_i}{W_2} = 10 \cdot \lg \left(\frac{W_i}{W_{2,P} + W_{2,B}} \cdot \frac{W_{2,P}}{W_{2,P}} \right)$$

(8.19)

or $\quad R = 10 \cdot \lg \left(\frac{W_i}{W_{2,P}} \right) - 10 \cdot \lg \left(1 + \frac{W_{2,B}}{W_{2,P}} \right) = R_{\text{without}} - 10 \cdot \lg \left(1 + \frac{W_{2,B}}{W_{2,P}} \right).$

We have then got an expression for the sound reduction index as a difference between the reduction index for the partition *without* the structural connections and a term due to these connections. Assuming that the sound radiation caused by these connections or bridges is dominant, i.e. $W_{2,B} \gg W_{2,P}$, Sharp shows that in the frequency range $f_0 < f < f_d$, where R_{without} increases by 18 dB per octave the last term will increase by 12 dB per octave. Similarly, this term will increase by 6 dB per octave where R_{without} increases by 12 dB per octave, that is to say when $f > f_d$. Without going into detail, the resulting reduction index will in effect have a shape as sketched in Figure 8.14. We end up with a term ΔR added to the reduction index R, the latter determined by the total mass $M = m_1 + m_2$ of the partition:

$$R = R_M + \Delta R,$$

where $\quad \Delta R = -10 \cdot \lg \left(n \cdot \sigma_B \right) + 20 \cdot \lg \left[\frac{m_1}{m_1 + m_2} \cdot \frac{|Z_1 + Z_2|}{|Z_1|} \right].$

(8.20)

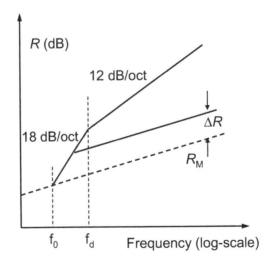

Figure 8.14 Principal shape of the sound reduction index of a lightweight double leaf partition with and with and without infinitely stiff structural connections. Sketch according to Sharp (1978).

The radiation factor σ_B of each of the n bridges is given by Equation (8.18). As distinct from the case of the acoustical lining treated in section 8.2.2.1, where we assumed that the primary wall was not influenced by the lining, we have now got an added term depending on the mass and input impedance of the leaves. From these impedances we shall understand the input impedance of the leaves seen from the bridges. For point connections we may use the expression valid for the point impedance of an infinitely large plate. As explained earlier (see Chapter 6, section 6.4.1), this expression also gives the space averaged mean value for a plate of finite dimensions. We shall write

$$Z_{\text{point}} = 8\sqrt{B \cdot m} = \frac{4c_0^2 m}{\pi f_c}, \tag{8.21}$$

where the critical frequency is introduced in the last expression. Inserting this expression into the term giving ΔR together with the expression for $\sigma_{B,\text{point}}$ in Equation (8.18), we get

$$\Delta R_{\text{point}} \approx 20 \cdot \lg\left\{a \cdot f_{g,\text{point}}\right\} - 45 \text{ dB},$$
$$\text{where} \quad f_{g,\text{point}} = \frac{m_1 f_{g,2} + m_2 f_{g,1}}{m_1 + m_2}. \tag{8.22}$$

Here we have assumed that the point connections are arranged in a square pattern, the quantity a being the centre-to-centre distance between points.

In a similar way we shall make use of the expression for the point impedance of an infinitely long beam to calculate the input impedance of a plate driven along a line. The point impedance of an infinite beam having a mass m_ℓ per unit length is, according to Cremer et al. (1988):

$$Z_{\text{beam}} = 2(1 + j)m_\ell \cdot c_B, \tag{8.23}$$

where c_B is the bending wave speed. This impedance is, as opposed to Z_{point} in Equation (8.21), a complex quantity. Driving the plate along a line by a force F distributed over a length L_y (see Figure 8.15), we may envisage that the plate is built up from a set of beams having cross-section $\Delta L_y \cdot h$. The impedance of each of these beams is

$$\left(Z_{\text{line}}\right)_\Delta = \frac{\Delta F}{u} = 2(1 + j)\Delta L_y h \cdot \rho \cdot c_B. \tag{8.24}$$

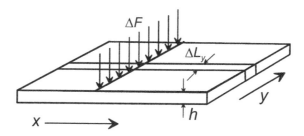

Figure 8.15 A plate driven along a line.

As we are driving all these "beams" in parallel, we get

$$Z_{\text{line}} = \frac{F}{u} = \frac{\sum \Delta F}{u} = 2(1+\text{j})L_y \cdot m \cdot c_\text{B}$$

or $\qquad Z_{\text{line}} = 2(1+\text{j})L_y \cdot c_0 \cdot m\sqrt{\dfrac{f}{f_c}}.$ (8.25)

The critical frequency is introduced here in the last expression. Inserting this expression into Equation (8.20), which gives ΔR, together with $\sigma_{\text{B,line}}$ from Equation (8.18), we arrive at the approximate equation

$$\Delta R_{\text{line}} \approx 10\lg(b \cdot f_{\text{c,line}}) - 23 \text{ dB},$$

where $\qquad f_{\text{c,line}} = \left[\dfrac{m_1\sqrt{f_{c,2}} + m_2\sqrt{f_{c,1}}}{m_1 + m_2}\right]^2.$ (8.26)

The quantity b is the centre-to-centre distance between the line connectors (studs). It is apparent that this expression is identical to the one depicted in Figure 8.10. Here, however, the modified critical frequency $f_{\text{c,line}}$ takes the position of the critical frequency of the lining.

Before showing examples based on this expression, we shall also refer to a work by Davy (1991), having extended the work by Sharp by taking into account the elasticity of the connections. He uses the case of stud connections and introduces the *compliance* (inverse stiffness) C_M of these connectors. The expressions, which are also cited in Bies and Hansen (1996), become relatively complicated and may be a little difficult to follow. However, we shall repeat them here and also make a comparison with the equations given above.

In the same manner as above, the energy transmission is divided into two parts, one part transmitted by way of the cavity, the other by way of the connections. In the frequency range between the double wall resonance f_0 and $f_{c,1}$, where the latter is the lowest of the critical frequencies of the two leaves, the transmission factor for the part caused by the connections is

$$\tau_{\text{B,line}} = \frac{64\rho_0^2 c_0^3}{\left\{g^2 + \left(4c_0(2\pi f)^{3/2} m_1 m_2 \cdot C_M - g\right)^2\right\} \beta b(2\pi f)^2},$$ (8.27)

where

$$g = m_1\sqrt{2\pi f_{c,2}} + m_2\sqrt{2\pi f_{c,1}}$$

and

$$\beta = \left[1 - \left(\frac{f}{f_{c,1}}\right)^2\right]\left[1 - \left(\frac{f}{f_{c,2}}\right)^2\right].$$

The quantity b denotes, as before, the centre-to-centre distance between the studs. For "commonly used" steel studs, a compliance C_M equal to 10^{-6} m$^2 \cdot$N^{-1} is indicated and for wooden studs, C_M is zero. However, common experience shows that steel studs may have quite different elastic properties; the elasticity is not frequency independent. Estimating this property must therefore be based on experience.

The transmission factor for the part being transmitted across the cavity is

$$\tau_P = \frac{1 - \dfrac{c_0}{2\pi f \sqrt{S}}}{\left[\dfrac{m_1^2 + m_2^2}{2m_1 m_2} + a_1 a_2 \bar{\alpha} \dfrac{c_0}{2\pi f \sqrt{S}} \right] \left[\dfrac{m_1^2 + m_2^2}{2m_1 m_2} + a_1 a_2 \bar{\alpha} \right]}, \qquad (8.28)$$

where

$$a_i = \left[\frac{\pi f m_i}{\rho_0 c_0} \right] \cdot \left[1 - \left(\frac{f}{f_{c,1}} \right)^2 \right].$$

As apparent from the expression, the influence of a finite area S is taken into account. Further, a mean absorption factor $\bar{\alpha}$ for the cavity is introduced. This may be put equal to 1.0 having a completely filled cavity but in other cases it may be difficult to estimate. The sound reduction index which includes both contributions according to Equations (8.27) and (8.28) is then

$$R = -10 \lg(\tau_{B,\text{line}} + \tau_P) \qquad \text{for} \qquad f_0 < f < f_{c,1}. \qquad (8.29)$$

Davy (1991) also gives an estimate for the frequency range above $f_{c,1}$, applied to the case of infinitely stiff studs but we shall not quote that here.

An example on the use of Equations (8.20) and (8.26) after Sharp and Equation (8.29) after Davy is shown in Figure 8.16. The specimen is a double leaf partition of 13 mm plasterboards having a cavity depth of 70 mm, which was filled with mineral wool of nominal thickness of 60 mm. The boards are mounted on common studs, either wooden or steel. The predicted results are nearly identical when it comes to the case of wooden studs and the frequency is sufficiently below the critical frequency. In this case, however, none of the predictions fits the measured data particularly well.

Setting C_M equal to $5.0 \cdot 10^{-6}$ m$^2 \cdot$ N^{-1} as the compliance for the steel studs, the fit between the prediction using Davy's equation and the measured data is surprisingly good. The crucial question remains however: How to estimate C_M?

More recently, Hongisto et al. (2002) conducted a large experimental study on double walls, although on small size specimens (1105 x 2250 mm). The set-up was thus similar to the one used by Brekke (see Figure 8.5). However, here they used 2 mm thick steel panels and altogether four types of steel stud plus wooden studs were tested. Other variables were the cavity depth, the amount of absorber material filling the cavity and the flow resistivity of the absorber. Altogether 54 tests were conducted where the cavity conditions and the coupling between the panels were varied. No attempt to compare with prediction models was made.

For uncoupled panels the results show, as discussed in section 8.2.1, that the important parameters are the cavity thickness and the amount of filling. The flow resistivity played a minor role. For the coupled case, the stiffness certainly was the important factor, for wooden studs the spacing of the fastening screws also played an important part. Wang et al. (2005) used one of these measurement results, one with wooden studs and an empty cavity, to compare with their rather complex analytical model treating the double wall as a periodic structure. The periodic model gives quite an undulating sound reduction curve but followed the general trend of the measured data, although the model was strictly two-dimensional.

Another approach to such analytical modelling, also taking advantage of the periodicity of the studs, is offered by Brunskog (2005). The effect of the studs on the sound field in the cavity is also taken into account but again the cavity was empty. The agreement between measured and predicted results is quite good.

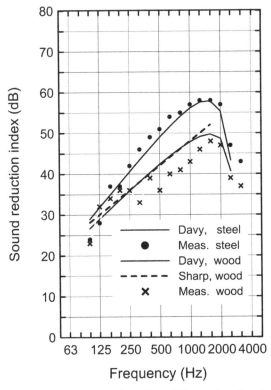

Figure 8.16 Sound reduction index of double leaf partition of 13 mm plasterboards on common wooden and steel studs, respectively. Measured data from Homb et al. (1983). Predictions according to Sharp (1978) and Davy (1991).

8.2.2.3 Heavy (massive) double walls

We have in the preceding section dealt with lightweight double leaf constructions where the leaves were connected by studs of various types. In the cases where the leaves are mounted on a separate system of studs there normally exists structural connections along the edges of the wall. In many cases such couplings are "weak" in the sense that the adjoining constructions have a much larger mass and stiffness, and the coupling is of little consequence unless a common, stiff frame is used.

This situation is quite distinct from the one experienced in double constructions involving brick or concrete, even including so-called lightweight concrete. The coupling along the boundary may completely determine the achieved result. In extreme cases there may be no improvements by using a double construction as compared with a single one.

Figure 8.17 gives an example, which shows that the difference in the reduction index, using a single wall of 150 mm thickness as compared with a double wall, 150 mm and 100 mm thick, is negligible as long as one uses a common frame. It should be noted that the main part of the curve lies above the critical frequency, which for the 150 mm thick lightweight concrete partition is approximately 250 Hz. Both partitions have, however, a thin layer of plaster added which probably makes the effective critical frequency slightly lower.

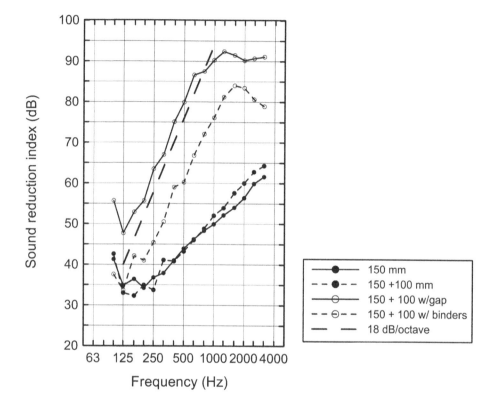

Figure 8.17 Sound reduction index of lightweight concrete walls showing the effect of coupling along the boundary. A double wall with a common vs. a separate frame (with gap). Effect of binders in the case of a separate frame. See Figure 8.18. Data from Homb et al. (1983).

In these laboratory measurements the elements were mounted on a foundation or frame in the form of a niche between the measuring rooms. In fact, there are two niches, one in each room structurally separated by a gap (see Figure 8.18). As shown in Figure 8.18 b) this makes it possible to mount the elements separately, each on its side of the gap. The distance between the elements gets larger (220 mm) but this is of minor importance as compared with the minimizing of the frame coupling. It should be noted that the dimensions of the boundary constructions is much larger than one normally finds in real buildings. As evident from Figure 8.17, we obtain a huge increase in the reduction index by separating the elements as compared with the case with the common frame. The frequency dependence is quite strong, approximately 18 dB per octave. The flattening of the curve in the higher frequency range is probably due to flanking transmission, which even with laboratory facilities will show up at extreme level differences.

Finally, all structural connections between the two elements will necessarily diminish the sound insulation. As indicated in Figure 8.18 b) the binders used are 3.5 mm thick steel bolts. There are distributed over the whole wall area with a centre-to-centre distance of 500 mm.

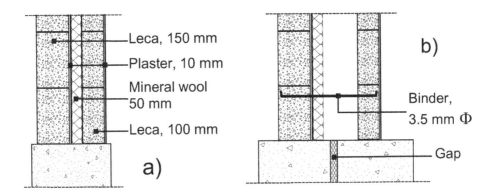

Figure 8.18 Lightweight concrete double wall. a) Mounted on a common frame; b) Mounted on separate frames.

8.3 SANDWICH ELEMENTS

The concept of a sandwich applies to a large group of multilayer elements with applications in building constructions, in transport systems and not least, in plane and ship constructions (high speed boats). The literature in this field is rather extensive, which stems from the fact that modelling the different layers may be quite different depending on the given application. In our context of building acoustics we shall define a sandwich element as a three-layer structure having two thin plates (face sheets) bounded by a lightweight core material. As opposed to conventional double leaf constructions, where the cavity has an infill of porous material, we now have a core as a continuous and solid coupling element between the outer sheets (see Figure 8.19).

The face sheets may not have the same material properties or have the same thickness but that is the normal case. In addition, we have characterized them as being "thin", which in practice implies that they are metal sheets, fibreboard, chipboard, plasterboard etc. As for the core material we find plastic foam as well as mineral wool, the latter having densities in the range of 100–150 kg/m^3 and also cut to make the direction of the fibres normal to the sheets. A much used type of core is the *honeycomb*, a beehive plate where the material may be metal or plastic. A very cheap variant, found in door leaves, is a core made up of cardboard rings, i.e. short cylinders of cardboard.

Elements with honeycomb cores offer lightweight and high stiffness elements. They become nearly incompressible in the crosswise direction, which implies that we may characterize the core by its shear stiffness only. Such a description cannot be used in a number of other core materials, e.g. plastic foams as polyurethane, polystyrene etc. In this case we could apply modelling by transfer matrices but a more advanced one than the equivalent fluid model used when calculating the data in Figure 8.3, now having to take the elastic properties into account. We have previously used a Biot model for a porous elastic material to calculate the absorption factor (see section 5.5.5). This model may thus be included in a transfer matrix calculation for the complete sandwich element by a procedure as e.g. used by Brouard et al. (1995). It should be noted, however, that these calculations presuppose infinite size layers. For finite size elements, one normally has to apply finite element methods (FEM) (see e.g. Vigran et al. (1997)).

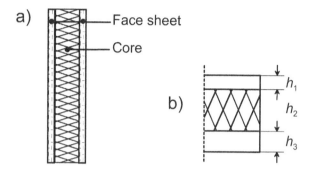

Figure 8.19 Sandwich element. a) Principal structure; b) Definition of layer thicknesses.

To illustrate the general features of sound transmission through sandwich elements we shall, however, use another approach. We shall start using the same assumption applicable for elements with a honeycomb core; the element being infinitely stiff in the normal direction having a core characterized by its shear stiffness only. As a second step we shall assume that the core is a general homogeneous elastic material. It may also be a porous material but we shall not have to model it using Biot theory as we will assume that the pores are closed.

8.3.1 Element with incompressible core material

The most pronounced feature of sandwich elements, as distinct from the partition elements treated up to now, is the frequency-dependent bending stiffness. In the static case, and also at sufficiently low frequencies, the core will act like an ideal spacer for the face sheets. For simplification, we may assume that the face sheets are identical, enabling us to express the low frequency bending stiffness of the element as

$$B_{\text{low}} \approx \frac{E_1 h_1 \left(h_1 + h_2\right)^2}{2} \qquad \text{when} \qquad h_1 = h_3 \quad \text{and} \quad E_1 = E_3. \qquad (8.30)$$

For the definitions of these quantities, see Figure 8.19. At a sufficiently high frequency, however, we get a decoupling of the face sheets and the bending stiffness will just be the sum of the bending stiffness of each sheet. We may then write

$$B_{\text{high}} \approx B_1 + B_3 = 2 \cdot B_1 \approx \frac{E_1 h_1^3}{6}. \tag{8.31}$$

Between these extremes the bending stiffness of the element will be determined by the properties of the core material. To find the effective bending stiffness we shall start with a differential equation for a sandwich element developed by Mead (1972). We shall assume that the element lies in the xz-plane having displacement ξ in the y-direction. The equation describing the free vibrations may be written

$$D_{\text{ys}} \nabla^6 \xi - g \left(D_{\text{ys}} + B \right) \nabla^4 \xi - m\omega^2 \nabla^2 \xi + m\omega^2 g(1-\upsilon^2)\xi = 0, \tag{8.32}$$

if we assume harmonic movements at an angular frequency ω. The quantity m is the total mass per unit area; D_{ys} and B are the total bending stiffness of the face sheets and the maximum bending stiffness of the complete element, respectively. The quantity g is a term that contains the bending stiffness of the core. These quantities are given by

$$D_{\text{ys}} = \frac{E_1 h_1^3 + E_3 h_3^3}{12\left(1-\upsilon^2\right)}, \qquad B = \left(\frac{h_1}{2} + h_2 + \frac{h_3}{2} \right)^2 \left\{ \frac{E_1 h_1 E_3 h_3}{E_1 h_1 + E_3 h_3} \right\},$$

$$m = \rho_1 h_1 + \rho_2 h_2 + \rho_3 h_3, \qquad g = \frac{\sqrt{G_x G_z}}{h_2} \left\{ \frac{1}{E_1 h_1} + \frac{1}{E_3 h_3} \right\}. \tag{8.33}$$

As seen from these equations, we may have different shear stiffness G in the x- and z-direction but for simplicity, we shall assume that these are equal. Expressing the shear stiffness of the core in the usual way by the modulus of elasticity and Poisson's ratio, we get

$$G_2 = \frac{E_2}{2(1+\upsilon_2)}. \tag{8.34}$$

Assuming a solution of Equation (8.32) of the form

$$\xi = \xi_0 e^{-jk_B x}, \tag{8.35}$$

we arrive at the following sixth order equation for the bending wave number:

$$k_B^6 + g\left(1 + \frac{B}{D_{\text{ys}}}\right) \cdot k_B^4 - \left(\frac{m\omega^2}{D_{\text{ys}}}\right) \cdot k_B^2 - \left(\frac{m\omega^2 g}{D_{\text{ys}}}\right) \cdot \left(1-\upsilon^2\right) = 0. \tag{8.36}$$

It should be mentioned that Ferguson (1986) uses this equation to arrive at an explicit expression for the critical frequency of the element. As mentioned above, we shall primarily use it to demonstrate the frequency dependence of the bending stiffness and, second, show how this affects the bending phase speed.

The polynomial Equation (8.36) may easily be solved numerically and we may then find the effective bending stiffness B_{eff} and the corresponding phase speed c_B from the following equations:

$$B_{\text{eff}} = \omega^2 \frac{m}{k_B^4} \quad \text{and} \quad c_B = \sqrt{\omega} \sqrt[4]{\frac{B_{\text{eff}}}{m}}. \tag{8.37}$$

Figure 8.20 shows a typical example of the frequency dependence of the bending stiffness. In this example, the face sheets are 9 mm chipboard and the core has properties corresponding to PVC foam (see Table 8.1). The bending stiffness is shown for two different thicknesses of the core, 50 mm and 100 mm. In the latter case, the dashed line shows the result of reducing the E-modulus and thereby also the shear stiffness by a factor of two.

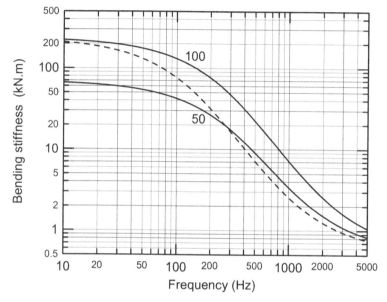

Figure 8.20 Bending stiffness of a sandwich element. Face sheets of 9 mm chipboard with foam core (PVC) of thickness 50 and 100 mm. Dashed curve indicate 100 mm core with reduced shear stiffness.

Table 8.1 Material data used in Figure 8.20.

	E-modulus (Mpa)	Density (kg/m^3)	Thickness (mm)	Poisson's ratio
Face sheets	4000	800	9	0.3
Core	50	60	50 - 100	0.3
Core (dashed curve)	25	60	100	0.3

What are then the consequences for the phase speed of the bending wave and further on for the sound reduction index of a sandwich element? Using the second calculation in Equation (8.37), the corresponding phase speed will be as depicted in Figure 8.21. Keeping the same bending stiffness as present at low frequencies would result in a very low critical frequency. However, depending on the core shear stiffness and thickness we obtain a more or less "flat" part on the curve before reaching coincidence.

It is also important to note that even if the critical frequency is relatively high, the radiation factor may be larger for a sandwich element as compared with a homogeneous one. The reason is that the phase speed c_B will have a value close to the speed of sound in air over a broad frequency range. This implies that resonant transmission will be a more important factor below coincidence than in the case of a homogeneous element.

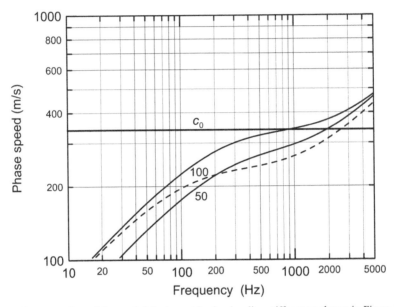

Figure 8.21 Phase speed c_B of the sandwich element having bending stiffness as shown in Figure 8.20. The speed c_0 is the phase speed in air.

How does one calculate the sound reduction index of a sandwich element having such frequency-dependent bending stiffness? This may be accomplished by using just the same expressions as given in Chapter 6 valid for a homogeneous single element, however, to calculate the reduction index at each frequency by using the radiation factor etc. appropriate for the bending stiffness in question.

An example is shown in Figure 8.22, where the reduction index is calculated for the element discussed above, however for three different core thicknesses. We have also assumed that the surface area is 10 m² and that the element has a total loss factor of 0.05. For comparison, a curve giving the simple mass law is included. It should be obvious that a poor design may produce quite dramatic failure in the sound insulation, added to the fact that one is starting out with a lower reduction index than predicted by the mass law. As opposed to the figures illustrating the bending stiffness and the phase speed, we have not calculated the reduction index corresponding to the dashed curve, where the shear stiffness of the 100 mm core is reduced by a factor of two. How would you expect the reduction index to be in that case?

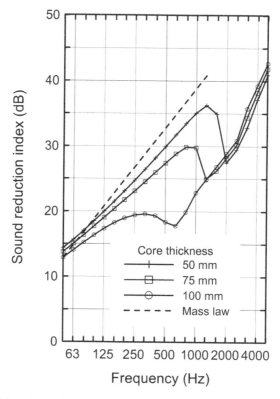

Figure 8.22 Sound reduction index of a sandwich element. Material data from Table 8.1.

8.3.2 Sandwich element with compressible core

Excluding sandwich cores of the honeycomb variety, we cannot in general neglect movements normal to the plane of the sandwich panel, i.e. there will be a transverse or dilatational movement where the face sheets move in opposite phase. There will be symmetric movements, dilatational modes, which may be considered as a generalized kind of double wall resonance, in addition to the anti-symmetric modes due to the bending waves as illustrated in Figure 8.23.

Moore and Lyon (1991) give analytical expression to calculate the sound reduction index of infinite size sandwich panels, for cases with isotropic as well as with orthotropic core materials. Concerning orthotropic materials, see Chapter 3 (section 3.7.3.3) and Chapter 6 (section 6.5.3). An interesting spin-off from their work is that a panel with an orthotropic core may in certain frequency ranges give a higher sound reduction index than predicted by the mass law.

The derivation of their expression is rather involved, and we shall not repeat it here. We shall look at the case of an isotropic core and show some calculated results where comparison with measured results is possible. Moore and Lyon's work is based on the transmission factor for plane wave incidence expressed by the wall impedances Z_s and Z_a for symmetric and anti-symmetric wave motion, respectively:

$$\tau\left(\varphi,\theta\right)=\left|\frac{\dfrac{\rho_0 c_0}{\cos\varphi}\left(Z_s-Z_a\right)}{\left(Z_s+\dfrac{\rho_0 c_0}{\cos\varphi}\right)\left(Z_a+\dfrac{\rho_0 c_0}{\cos\varphi}\right)}\right|^2.$$ (8.38)

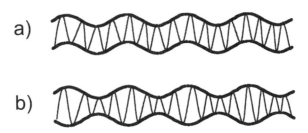

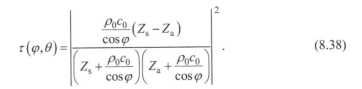

Figure 8.23 a) Anti-symmetric and b) symmetric wave motion in a sandwich element with compressible core.

To calculate the transmission factor for diffuse field incidence, we shall as usual integrate the expression; over a range 2π for the azimuth angle θ and up to an angle of approximately 80° for φ, the latter to simulate laboratory conditions.

The derivation of the expressions giving Z_s and Z_a starts from the equations of motion for the face sheets and the core. The equations for the face sheets allows for in-plane movements and for bending, whereas the core is described as a homogeneous and elastic material allowing for dilatational as well as shear wave motion. A 4x4 impedance matrix is set up to represent the core, relating both the normal and shear stress amplitudes to the velocity amplitudes of the face sheets, transverse as well as in-plane.

The next step is to link these impedance components, i.e. the matrix coefficients, to the equations of motion of the face sheets to arrive at the sought-after velocity amplitudes caused by the incident sound pressure. In this way, Moore and Lyon arrived at, given identical face sheets, two uncoupled equations describing the symmetric and nonsymmetric motion, respectively, and thereby to explicit expressions for Z_s and Z_a for direct input to Equation (8.38).

Measured and predicted results for a sandwich panel of 13 mm plasterboards and a 55 mm thick core of a polyurethane foam material (PUR) is shown in Figure 8.24. Taking account of the uncertainty of the material properties, the calculations are performed using two different values for the modulus of elasticity. The applied material data are given in Table 8.2.

As apparent from the figure, the fit between predicted and measured results is very good. The double wall resonance, caused by the symmetric or dilatational movement, is evident around 800 Hz in the same way as the effect of coincidence shows up in the frequency range 2500–3000 Hz. The frequency of the former one, the lowermost dilatational resonance, is given by

$$f_{\text{dil}} = \frac{1}{2\pi} \left[\frac{2K_c}{h_2 \left(\rho_1 h_1 + \frac{\rho_2 h_2}{6} \right)} \right]^{\frac{1}{2}} \quad \text{where} \quad K_c = \frac{E_2 \left(1 - \upsilon_2 \right)}{\left(1 - 2\upsilon_2 \right) \left(1 + \upsilon_2 \right)}. \quad (8.39)$$

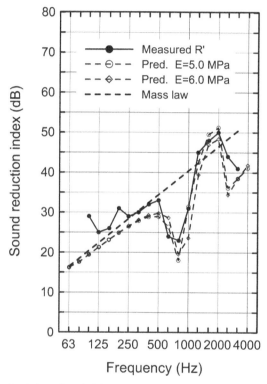

Figure 8.24 Sound reduction index of sandwich panel of 13 mm plasterboards and 55 mm polyurethane core. Predicted data using two different modulus of elasticity for the core. Measured data from Homb et al. (1983).

Table 8.2 Material data for sandwich panel in Figure 8.24.

	E-modulus (Mpa)	Density (kg/m³)	Loss factor	Poisson's number
Face sheets, 13 mm plasterboard	3200	800	0.05	0.30
Core, 55 mm polyurethane	5 – 6	50	0.05	0.35

Concerning the fit between predicted and measured results in the lower frequency range, one should be reminded that the prediction applies to an infinite size panel, i.e. no area effect is included. Furthermore, the panel area of 7.5 m² being used for the measurements was a little smaller than required for a standard laboratory measurement, which is 10 m². The predicted results are averaged values using data calculated for approximately 10 single frequencies inside each one-third-octave band.

Summing up sandwich elements: A critical factor in designing for good sound insulation is the core shear stiffness. By reducing the core thickness without changing any material parameters one may draw two important conclusions. The stiffness of the core will increase, moving the lowest dilatational resonance up in frequency. At the same time the bending stiffness of the element will decrease, which results in a reduced phase velocity for the anti-symmetric wave motion in the middle frequency range, thereby obtaining a higher reduction index.

8.4 IMPACT SOUND INSULATION IMPROVEMENTS

Adding a lining to the ceiling is *one* way of improving the airborne and/or the impact sound insulation of a floor, using a *floating floor* is another. The added floor "floats" on an elastic medium placed upon the primary floor construction. The elastic component is normally a continuous layer of mineral wool, plastic foam etc. but may be discrete load-bearing mounts, e.g. rubber mounts or rubber strips. A floating floor may then be considered as a generalised form of vibration isolation. Principal solutions for the case of a concrete floating floor slab are shown in Figure 8.25.

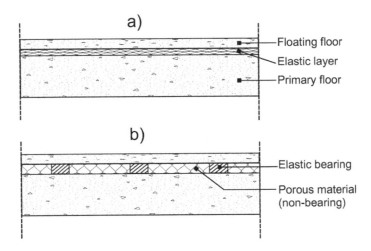

Figure 8.25 Principal types of concrete floating floors. a) On a continuous elastic layer, b) on elastic load-bearing unit mounts.

As far as the airborne sound insulation is concerned, such a construction is nothing less than a double wall construction without stiff structural connections. As for the improvements in the impact sound insulation, however, not only the properties of floating slab and the elastic layer are important, the properties of the ISO tapping machine may also influence the results. The latter applies in particular to cases where the floating floor is not a concrete slab but a lightweight construction, e.g. floorboards, parquet etc. We shall therefore mostly be concerned with the impact sound properties of such floating floors.

The top layer or floating floor may either be characterized as locally or resonantly reacting, which here means that the force from the tapping machine is either transmitted from the top layer to the primary floor just around the neighbourhood of the tapping point or a reverberating bending wave field is generated. For a very stiff top layer, a local reaction implies that the internal losses must be large; the wave field must be heavily attenuated before it arrives at the boundaries and furthermore; the free reflected waves created at the boundaries must also decay swiftly. This condition is not possible to realize when it comes to concrete slabs, except may be in combination with thick layers of asphalt etc.

A lightweight top layer of floorboards, parquet etc. may normally be characterized as be locally reacting, primarily due to their lower stiffness in combination with higher inner energy losses as compared with concrete slabs. The bending wavelength of lightweight top floors is also substantially less than for concrete slabs as the latter normally have a thickness 40–50 mm. These facts have, as will be shown below, implications for the design of the connections between the top floor and the primary floor at the boundaries.

Soft floor coverings, carpets etc. are, as opposed to floating floors, purely an agent for improving the impact sound insulation. Such elastic layers change the shape of the force impulse impacted by the tapping machine, thereby affecting the mechanical power transmitted to the primary floor.

The achieved improvement, whatever the top layer used, is certainly not independent of the type of primary floor. This is completely analogous to the assumptions we were allowed to use when dealing with the improvement of linings as opposed to the general case of lightweight double leaf partitions. In the former case we could assume that the primary construction was unaffected by the presence of the lining.

In the same way, we shall start with prediction models for the improvement offered by floating floors on heavy floor constructions. It should be mentioned that the basic concrete floor slab specified in ISO 140 Part 8, which deals with laboratory measurements of the improvement or reduction in transmitted sound by soft floor coverings, shall have a thickness in the range 100–160 mm, preferably 140 mm. It is, however, certainly of interest to know the reduction offered when placed on e.g. a lightweight wood joist floor and we shall give examples of this case when dealing with lightweight top layers. Lately, a laboratory standard for determining the reduction in impact sound by floor coverings on a lightweight basic floor has been issued. Altogether, three different lightweight floors have been specified (see ISO 140 Part 11).

8.4.1 Floating floors. Predicting improvements in impact sound insulation

Predicting the impact sound insulation improvement offered by a floating floor is not an easy task. Analogous to the prediction of airborne sound insulation we shall have to take account of both forced and resonant transmission, the latter being dependent on the boundary conditions for the floating as well as for the primary floor. The boundary conditions for these are not necessarily identical. Modelling the floating layer is also an important task. May we consider the layer to act like an ideal spring or are we forced to model it as medium supporting wave motion? We certainly cannot extensively deal with all these factors; we shall limit our treatment to a "classic" model for forced transmission in addition to an SEA model dealing with the resonant transmission.

The most well-known work, dealing with floating floor constructions, was performed by Cremer in 1952, to be found in Cremer et al. (1988). In Cremer's model the basic floor and the top floor are assumed to be two infinitely large and homogeneous slabs, characterized by their mass and bending stiffness and coupled together by an elastic layer characterized by its stiffness only. The system is driven by a point force, a falling hammer having a mass m_h. As the slabs are infinitely large there will only be a bending wave near field propagating outwards from the driving point. This means that there will be no reflections setting up a reverberant field, which implies that both slabs are locally reacting.

Surprisingly, even if the slabs are characterized both by their mass and stiffness, the latter property does not enter into the expression for the impact sound improvement ΔL_n. This quantity is defined as the difference in the radiated sound power from the primary construction applying the force directly on it and then to the floating slab. Cremer then showed that we get

$$\Delta L_n = 40 \cdot \lg\left(\frac{f}{f_0}\right) + 20 \cdot \lg\left|1 + \frac{j \cdot 2\pi f m_h}{Z_1}\right| \qquad \text{for} \qquad f > f_0, \qquad (8.40)$$

where f_0 is the double wall resonance given by Equation (8.5) and Z_1 is the mechanical point impedance of the floating slab. For heavy floating slabs such as concrete we may normally neglect the second term in the equation, thereby obtaining an improvement of 12 dB per octave. The assumption concerning local reaction for both slabs is however, as pointed out above, not valid in practice for such floating floor constructions. One therefore never experiences improvements as high as predicted by this equation. The standard EN 12354-1 proposes a modified version of Equation (8.40) where the constant 40 is substituted by 30, thereby reducing the frequency dependence to 9 dB per octave.

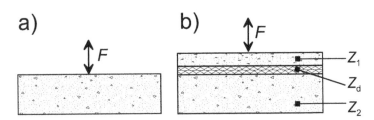

Figure 8.26 Model for calculating improvement in impact sound insulation.

It is also interesting to note that the Cremer equation is identical to the one arrived at using a simple one-dimensional model, where each layer is characterized by its mechanical impedance. Using the details found in Figure 8.26, we may calculate the velocity amplitudes of the primary floor, first, when being driven directly by a force F and, second, when the same force is driving the top floor. Letting these velocities equal u_{2a} and u_{2b}, respectively, we get

$$u_{2a} = \frac{F}{Z_2} \qquad \text{and} \qquad u_{2b} = \frac{F \cdot Z_d}{Z_1 \cdot Z_2 + Z_d\left(Z_1 + Z_2\right)}. \qquad (8.41)$$

The velocity ratio will be

$$\frac{u_{2a}}{u_{2b}} = 1 + \frac{Z_1}{Z_2} + \frac{Z_1}{Z_d} \xrightarrow{\quad Z_2 \gg Z_1 \gg Z_d \quad} \frac{Z_1}{Z_d}. \qquad (8.42)$$

With the last limit value we have assumed that the impedance of the primary slab is much larger than for the top slab, whereas the latter also is very much larger than the impedance of the elastic layer. Now assuming that the radiated power from the primary floor is proportional to the velocity squared, it follows

$$\Delta L_n = 10 \cdot \lg \left[\frac{(W_{rad})_a}{(W_{rad})_b} \right] = 10 \cdot \lg \left| \frac{u_{2a}}{u_{2b}} \right|^2 = 20 \cdot \lg \left| \frac{Z_1}{Z_d} \right|. \tag{8.43}$$

Assuming that the impedance of the top slab is an ideal mass impedance and the elastic layer is an ideal spring, i.e. $|Z_1| \sim \omega m_1$ and $|Z_d| \sim s_d/\omega$, we again obtain

$$\Delta L_n = 20 \cdot \lg \left(\frac{\omega^2 m_1}{s_d} \right) = 40 \cdot \lg \left(\frac{f}{f_0} \right). \tag{8.44}$$

An alternative to a continuous elastic layer is obtained by using elastic load-bearing unit mounts as shown in Figure 8.25 b). This type of connector may also be used to illustrate the influence of structural connections (sound bridges), normally unintentional, between the floating layer and the primary floor construction. Vér (1971) used a SEA model to calculate the improvement in the impact sound insulation by such floating floor constructions, assuming a reverberant bending wave field in the floating top slab. Other important assumptions were e.g. that the energy transmission from the top slab to the primary floor only takes place by way of the unit mounts, only transmitting forces and not moments. In other words, the coupling by way of the air stiffness in the cavity is disregarded. Furthermore, there is no correlation between the movements at the different mounts. Indicating the floating floor and the primary floor by the indices as above and having N mounts per unit area, each having a stiffness s, Vér gives the following result

$$\Delta L_n = 10 \cdot \lg \left[\frac{Z_1}{Z_2} + \frac{m_1 \eta_1}{m_2 \eta_2} + \frac{Z_1 \eta_1 N}{2\pi m_1} \cdot \frac{f^3}{f_0^4} \right], \quad \text{where} \quad f_0 = \frac{1}{2\pi} \sqrt{\frac{N \cdot s}{m_1}}. \tag{8.45}$$

Above a given frequency, when the last term inside the parenthesis becomes the dominating one, the frequency dependence of ΔL_n will be 9 dB per octave. Inserting for the impedance of the floating floor slab we may use the approximate expression

$$\Delta L_n \approx 10 \cdot \lg \left[\frac{2 c_{L_1} h_1 \eta_1 N}{\sqrt{3} \pi} \cdot \frac{f^3}{f_0^4} \right], \tag{8.46}$$

where h_1 and c_{L1} is the thickness and the longitudinal wave speed, respectively. It should be noted that the 9 dB per octave dependency presupposes that the loss factor of the floating slab as well as the stiffness of the elastic units are frequency independent.

An example of the measured improvement using a heavy floating floor, a 50 mm thick concrete slab on a 25 mm thick stiff mineral wool layer, is shown in Figure 8.27. Assuming that the total dynamic stiffness per unit area of the elastic layer is 8.0 MPa/m, we get a resonance frequency of approximately 40 Hz. This total stiffness represents the sum of the elastic stiffness of the mineral wool and the stiffness of the enclosed air (see section 8.4.4).

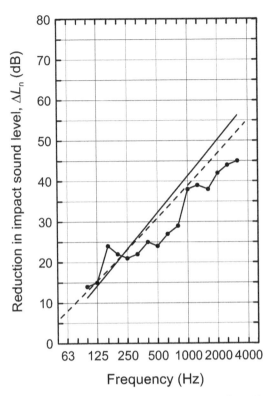

Figure 8.27 Improvement in the impact sound insulation by a heavy floating floor, 50 mm concrete slab, 25 mm mineral wool and 140 mm concrete basic floor. Measured data after Homb et al. (1983). Solid line – 9 dB per octave above resonance frequency. Dashed line – predicted for an equivalent elastic unit mounting.

The solid straight line represents the predicted improvement following a frequency dependence of 9 dB per octave above the resonance frequency, a curve that in this example gives a slightly better estimate than what is actually achieved (see comments below). It is also interesting to calculate the improvement in a thought experiment assuming that the top floor floats on mounting units having the same total stiffness as the mineral wool layer. The dashed line is calculated using Equation (8.45), where an empirical expression by Craik (1996) is used for the loss factor of concrete slabs (see also section 6.4.2.3),

$$\eta = \frac{1}{\sqrt{f}} + 0.015. \tag{8.47}$$

It should be stressed that the latter result is included just to give an illustration of the use of Equation (8.45), not because we expect that there should be a good fit to measured data applied to a continuous elastic layer.

There exists, however, a series of measurement data on heavy floating floors on continuous elastic layers that show a smaller frequency dependence than 9 dB per octave.

At very low frequencies one may see frequency dependence near to the Cremer model whereas at higher frequencies one may experience frequency dependence nearer to 6 dB than 9 dB per octave. This is probably caused by wave motion in the elastic layer, i.e. characterizing the layer by its compressional stiffness only is not appropriate. As opposed to this effect, measured results on floating floors on discrete mounts give a good fit to predicted data using Equation (8.46).

8.4.2 Lightweight floating floors

A couple of principal solutions when it comes to lightweight floating floors as floorboards, parquet etc. are sketched in Figure 8.28. In both cases there is a continuous elastic layer but in the first case the top floor is directly coupled to the elastic layer. We shall refer to this as a "surface mounted" case. In the second case, referred to as "line mounted", the top floor is mounted on beams or slats, the latter constituting the couplings to the elastic layer. This type of mounting may have the advantage that the dynamic stiffness of the enclosed air layer is diminished, at the same time also increasing the static stiffness of the top floor. In this line mounting case one may also replace the continuous elastic layer by unit mounts fastened below the beams, i.e. a mounting resembling the one shown in Figure 8.25 b).

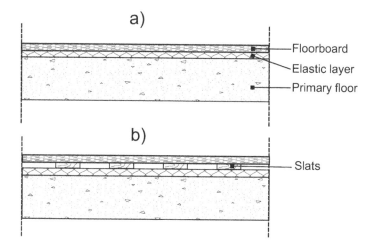

Figure 8.28 Lightweight floating floor on a heavy basic floor. a) "Surface" mounted, b) "Line" mounted.

The improvements gained by using such lightweight floors are distinctly different from the ones of concrete slab form. As mentioned above, this is partly due to a more local reaction by the lightweight floors. At the same time, however, the mass impedance of the tapping hammers may no longer be neglected in comparison with the input impedance of the floor. In this case, we shall use the complete Equation (8.40) which may be written

$$\Delta L_n = 40 \cdot \lg\left(\frac{f}{f_0}\right) + 10 \cdot \lg\left[1 + \left(\frac{f}{f_z}\right)^2\right], \quad \text{where} \quad f_z = \frac{4\sqrt{m_1 B_1}}{\pi \, m_h}. \quad (8.48)$$

The last term only applies to frequencies above f_z, a frequency that is determined by the ratio between the impedance of the top floor, specified by its mass and bending stiffness and the mass impedance of the hammer. At sufficiently high frequencies, the frequency dependence will be as high as 18 dB per octave, a result completely determined by the specific mass of the hammer. Testing such floors, however, is normally performed using additional loads of the order of 20–25 kg per unit area simulating the weight of furniture etc. This diminishes the effect of the hammer mass.

Two examples of the improvement gained by using lightweight floating floors are shown in Figure 8.29. In both cases the top floor is made of 22 mm thick chipboard, surface mounted on 15 mm plastic foam in the one case, line mounted by 22 x 95 mm beams on 25 mm stiff mineral wool in the second case. The primary floor is a 140 mm thick concrete slab in the first case and 200 mm lightweight concrete in the second. The difference in weight of the primary floor has negligible influence on the results as the mass in the latter case is just some 20 % less than in the case of the concrete floor.

The frequency f_z (see Equation (8.48)), will be approximately 600 Hz and we observe that the frequency dependence above f_z is very close to 18 dB per octave. We shall note that the measured data for the floor on plastic foam include the effect of a thin floor covering on top of the chipboard but this gives a contribution of maximum 5 dB (at 2000 Hz).

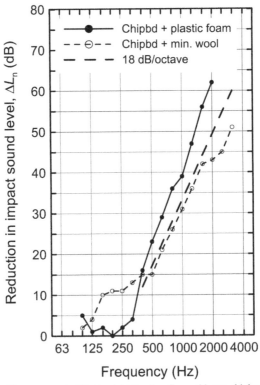

Figure 8.29 Impact sound improvement of lightweight floating floors, 22 mm chipboard surface mounted on 15 mm plastic foam (80 kg/m³) and line mounted on 25 mm mineral wool (100 kg/m³). Measured results from Homb et al. (1983).

8.4.2.1 Lightweight primary floor

In all examples shown relating to the improvement gained from floating floors, we have tacitly assumed that the primary floor is infinitely stiffer than the floating top floor. For lightweight primary floors such as wood joist floors, normally comprising lightweight panels both on top and on the underside, the impedance of the hammers will however influence the impact sound level of the primary construction as well, i.e. the measurement being the base for determining the reduction in transmitted sound. This problem, which was mentioned in the introduction to section 8.4, will be illustrated by measured data for the impact sound improvement of a lightweight floating floor combined with both a heavy and a lightweight primary floor. However, up to now we have not specifically looked into the impact sound insulation of such lightweight primary floor, partly due to the complexity in modelling, partly as they normally cannot offer sufficient impact sound insulation without being combined with a floating floor and/or a suspended ceiling.

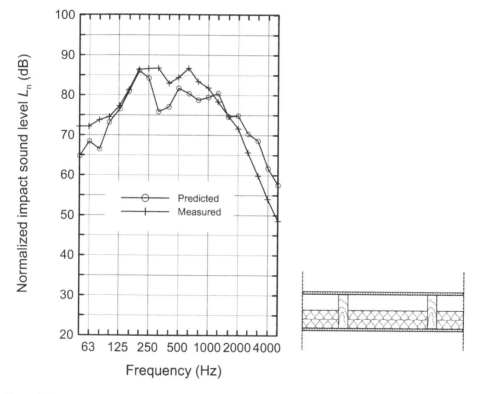

Figure 8.30 Normalized impact sound level of wood joist floor. Predicted results are mean values using 15 tapping positions. After Brunskog and Hammer (1999b).

Brunskog and Hammer (1999b) have presented a literature survey (see also Brunskog and Hammer (2000)), on the different approaches to the modelling and also presented their own prediction model taking the periodicity of the beam-plate system into account. Their approach is somewhat analogous to the one used by Lee and Kim (2002) treating airborne sound insulation (see section 8.2.2), as both start out from the governing equations for two plates connected and stiffened by beams. In several cases, Brunskog and Hammer's prediction model fits well to measured results. We shall present one example (see Figure 8.30), where the construction is depicted in the insert; a platform structure of two 22 mm matched boards connected by wooden beams of dimensions 67 x 220 mm, the cavity partly filled by mineral wool of thickness 120 mm and density 20 kg/m^3. The measured results are taken from an earlier work by Bodlund (1987). It is worth noting that the general frequency dependence is quite different from the ones found for heavy floor constructions (see Figure 6.20).

We shall now return to the case of the impact sound improvement of a lightweight floating floor combined with such a lightweight primary floor as opposed to a combination with a heavy floor. The floating floor is here a combination of 22 mm thick chipboards and 13 mm plasterboards surface mounted on 25 mm stiff mineral wool. It is combined with two different primary floors; one being a 200 mm thick lightweight concrete floor of density 1300 kg/m^3, the other a wood joist floor. The latter is a platform structure of 48 x 198 mm beams with 22 mm chipboard on top and combined with a ceiling of 2 x 13 mm plasterboards. As evident from Figure 8.31 we get, in the case of the heavy primary floor, a frequency dependency of 18 dB per octave just as shown in Figure 8.29. As expected, however, we do not see this effect using the wood joist floor.

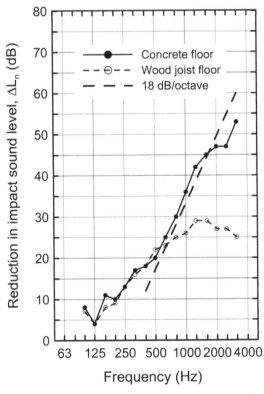

Figure 8.31 Impact sound improvement of floating floor, 22 mm chipboard and 13 mm plasterboard on 25 mm mineral wool. Measurements on two different primary floors, 200 mm lightweight concrete and a wood joist construction.

8.4.3 The influence of structural connections (sound bridges)

Structural connections, which we may refer to as sound bridges, may be of vital importance in the performance of a floating floor. This is certainly a case completely analogous to the effect of such structural connections in other types of multilayer constructions, e.g. when adding a lining to a basic construction. The conditions are however a little different for impact sound compared with airborne sound, as the influence of a sound bridge on the impact sound depends on whether the top floor is locally reacting or not, where there is no simple criterion to use. There is obviously a gradual transition between these extreme cases depending on material properties, thickness etc. making us dependent on experiments or experience to decide on a proper design. All investigations do, however, show that heavy floating floors such as concrete are extremely vulnerable to sound bridges, being in the form of point connections to the primary floor or as connections to the adjoining walls.

This was shown early on by Gösele (1964) in a laboratory experiment introducing point contacts to the primary floor in the form of cylinders of gypsum, 30 mm in diameter. The results are shown in Figure 8.32, where the parameter for the curves is the number of such bridges. It should be noted that just one of these bridges reduces the improvement to nearly half its value.

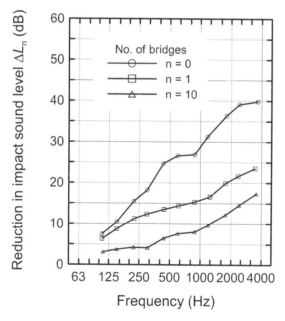

Figure 8.32 The influence of sound bridges, cylinders of gypsum of diameter 30 mm, on the measured impact sound improvement of a concrete floating floor. Data from Gösele (1964).

Correspondingly, Gösele (1964) performed a similar experiment where he made solid contact between the concrete top slab and one of the adjoining walls. This made in effect a direct structural connection between the floating floor and the primary floor. The result is shown in Figure 8.33 together with a sketch of the situation. The parameter on the curves is the length of the solid connection between the floating slab and the wall. It should be obvious that such stiff connections along the boundaries have a destructive effect on the working of a floating floor but the requirement as to the stiffness of such connections is not evident. Experience shows that such connections may have a stiffness that is a lot stiffer than the elastic layer before a reduction in the effect of the floating floor is detected.

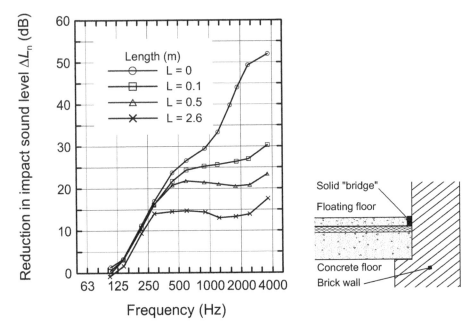

Figure 8.33 The effect of solid line contacts between a concrete floating floor and the adjoining wall. The length of the line bridges is the parameter on the curves. Measurements by Gösele (1964).

Extensive experiments (e.g. Holmås and Bjørklund (1979); Austnes and Hveem (1983)) on lightweight floating floors using floorboards, combinations of chipboard and plasterboard, parquet etc. show that stiff boundary connections have substantially less effect than in the case of heavy floating floors. Holmås and Bjørklund (1979) used a laboratory set-up with a wood joist primary floor combined with various types of lightweight floating floor. Fixed connections were established between the floating floor and a boundary wall in contact with the primary floor, the effect being an increase in the transmitted impact sound level at frequencies above approximately 300 Hz, this when placing the tapping machine at positions near to the boundary wall (< 1 metre) as compared to other tapping positions. This clearly indicates the local reaction of such lightweight floating floors, an effect making them less critical with regards to non-intentional structural connections.

8.4.4 Properties of elastic layers

Assuming that the mass *m* per unit area of the floating floor is substantially less than the mass of the primary floor, Equation (8.5) may be simplified to

$$f_0 = \frac{1}{2\pi}\sqrt{\frac{s}{m}}, \qquad (8.49)$$

where s is the total stiffness per unit area of the elastic layer. For porous layers the stiffness may be expressed as

$$s = s_{\text{skeleton}} + s_{\text{air}} = \frac{E_{\text{dyn}}}{d} + \frac{P_0}{d \cdot \sigma}, \qquad (8.50)$$

where s_{skeleton} and s_{air} denotes the stiffness of the solid frame and the stiffness of the enclosed air in the pores, respectively. As shown by the second expression the former stiffness is expressed by the dynamic modulus of elasticity and the stiffness due to the enclosed air is calculated by assuming isothermal motion. The quantity P_0 is the barometric pressure and d is the thickness of the porous layer of porosity σ. If the thickness of the layer is less than approximately 20 mm, we normally observe that the last term will be the dominating one.

The standard ISO 9052–1 specifies a method for determining the dynamic stiffness of materials intended for floating floors. A square specimen of dimension 200 mm is used, loaded by a given mass to make up a simple mass-spring system. Measuring the resonance frequency of this system determines the dynamic stiffness. Three principal arrangements are specified, of which two are sketched in Figure 8.34. In arrangement a) the loading mass is driven dynamically assuming the base is non-moving and in b) the base is driven and the differential motion between load and base is measured. A third possibility resembles the latter arrangement but now the load mass is driven. In all cases, only movements in the vertical direction are assumed.

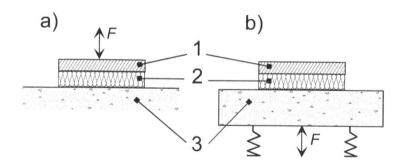

Figure 8.34 Test arrangement for determining dynamic stiffness according to ISO 9053–1 (some details are omitted). 1) Mass loading plate. 2) Test specimen. 3) Base (foundation). a) The mass load being driven, the base is fixed; b) Differential measurements of loading mass and base.

A complication concerning these arrangements is the necessary limited specimen area. For open-pore porous materials such as mineral wool, the flow resistivity will be a determining factor as it will determine whether or not the enclosed air will "escape" during the measurement. For high flow resistivity, $r \geq 100$ kPa·s/m^2, one will directly measure the total stiffness s. At intermediate values, 10 kPa·s/m$^2 < r < 100$ kPa·s/m^2, s_{skeleton} is determined by the measurement, and the total stiffness is calculated by Equation

(8.50). In cases where $r < 10$ kPa·s/m^2, moreover when we cannot assume that $s_{skeleton} \gg s_{air}$, the method is not suitable for determining s.

Some typical data of the dynamic elasticity modulus are given in Table 8.3, valid for a static load of 2 kPa. The dynamic stiffness and dynamic E-modulus are dependent on the static load and this dependency will not be the same for different materials. As an example we find that the E-modulus of rock wool, having a density in the range 150–175 kg/m^3, is approximately 0.3 MPa at a load of 2 kPa. Decreasing the load to 1 kPa the E-modulus will decrease by some 20%, whereas it will increase by some 30% with a load of 4 kPa.

Table 8.3 Dynamic modulus of elasticity.

Material	Density (kg/m^3)	Dynamic E-modulus (MPa) (static load $\approx$ 2 kPa)
Glass wool	approx. 125	0.11–0.13
Rock wool	150–175	0.27–0.33
Rock wool	110–135	0.25–0.30
Polystyrene foam	10–20	0.30–3.0
Polyurethane foam	33–72	7–19
Cork	120–250	10–30

Examples As pointed out above, the stiffness of the enclosed air may contribute substantially using thin elastic layers. Using a layer of thickness 10 mm only, assuming a porosity $\sigma \approx 1.0$, we get $s_{air} \approx 10$ MPa/m, which is in the same order of magnitude as $s_{skeleton}$ of a very elastic material.

A floating floor of 50 mm concrete on an elastic layer of 25 mm glass wool will have a resonance frequency of approximately 40 Hz. Using the same elastic layer together with a floating floor of type as shown in Figure 8.31, i.e. a combination of 22 mm chipboard and 13 mm plasterboard, the resonance frequency will now be approximately 90 Hz.

8.4.5 Floor coverings

Floor coverings such as carpets, vinyl, vinyl combined with felt and linoleum etc. are, as opposed to floating floors, purely a means of reducing the impact sound transmission. A soft covering changes the shape of the force impact from the tapping machine, thereby influencing the mechanical power transmitted to the floor below. The reduction in the impact sound power should therefore, in principle, be predicted from the difference in transmitted force with and without the covering. It has, however, been difficult to set up a good prediction model due to problems of characterizing the properties and behaviour of such coverings subjected to this kind of impact.

The main feature is that the speed of the hammer will decrease from its initial value v_0 when hitting the covering layer, to zero at the maximum compression of the layer. Thereafter the hammer will return. The time for this process is certainly dependent on the effective stiffness of the layer related to the area of the hammer(s) S_h, the latter being 7 cm^2. We may express this stiffness as

$$S_{\text{covering}} = \frac{E \cdot S_{\text{h}}}{d}. \tag{8.51}$$

The quantities E and d are the E-modulus and thickness of the covering, respectively. The questions to be raised here are what types of energy loss mechanism are involved and whether the process is linear or not. Here we shall use some measurement results and compare these with a simple linear model, a resistance in series with a stiffness given in Equation (8.51). The model, proposed by Lindblad (1968), was used by Brunskog and Hammer (1999a) to model the interaction between the tapping machine and lightweight floors (see section 8.4.2.1).

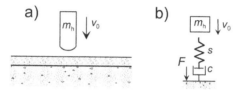

Figure 8.35 a) Floor covering hit by a hammer of speed v_0; b) Linear model of covering characterized by a spring stiffness in series with a resistance.

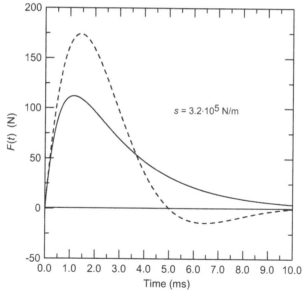

Figure 8.36 Time history of force pulses using elastic layer model given in Figure 8.35 b). Solid curve – overdamped case, approximately 20 % more than critical. Dashed curve – approximately 60 % less damped than critical.

The situation is depicted in Figure 8.35, where the hammer of mass m_{h} and speed v_0 is hitting the covering. We shall assume that the primary floor beneath the covering is infinitely stiff. Using the model in b) to calculate the improvement, we must, first, calculate the time function of the force F and thereafter compare this with the corresponding force without the covering.

The solution to the differential equation based on this model will depend on the damping of the system, i.e. the damping being less than critical or overdamped, which means we get an oscillation or just a positive pulse. The former case is assumed not to happen as the tapping machine has a mechanism for catching the hammers before they bounce back again after impact. It is, however, interesting to calculate the improvement assuming that the pulse becomes oscillatory.

Examples on calculated pulse forms are shown in Figure 8.36, using a covering of stiffness s equal to $3.2 \cdot 10^5$ N/m, giving a resonance frequency f_0 of approximately 130 Hz with a hammer mass of 0.5 kg. One of these pulse forms is slightly overdamped (≈ 20 %), while the other is less than critically damped (≈ 60 %). Critical damping is obtained when the damping coefficient c is equal to $\pi m f_0$.

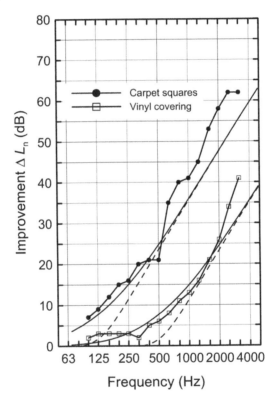

Figure 8.37 Impact sound improvement of two types of floor covering. Measured data from Homb et al. (1983). Predicted improvement with a linear model: stiffness of carpet squares $3.2 \cdot 10^6$ N/m, vinyl covering $5.2 \cdot 10^6$ N/m. Thin solid curves – overdamped case. Dashed curves – less than critically damped.

The reduction in the transmitted impact sound may now be determined by calculating the ratio of the Fourier transforms representing the actual force pulse and the corresponding one obtained without the covering. Figure 8.37 gives two examples of measured improvement data, one specimen being soft carpet squares and the other a vinyl covering with a felt backing. In the former case, we have assumed that the covering has the same stiffness as used in Figure 8.36, and the improvements are calculated using these two pulse forms; the thin solid curves predicted for the overdamped case and the dashed curves for the oscillatory case. For the former case, the fit between measured and predicted data, as far as the shape of the curves is concerned, is very good, certainly in view of the simple model used. The same comment may be made for the vinyl covering, where a stiffness of $5.2 \cdot 10^6$ N/m is used, equivalent to a resonance frequency of approximately 510 Hz.

We may therefore, at least for these two versions of floor covering, conclude that the simple linear model is satisfactory when assuming critical damping or slightly more. An absolute comparison is however not possible, due to lack of available data for the E-modulus of these coverings.

8.5 REFERENCES

EN 12354–2:2000, Building acoustics – Estimation of acoustic performance of buildings from the performance of elements. Part 2: Impact sound insulation between rooms. [Parts 1–3 of this standard are adopted by ISO with the number 15712 (from 2003).]

ISO 9052–1: 1989, Acoustics – Determination of dynamic stiffness. Part 1: Materials used under floating floors in dwellings.

ISO 140–8: 1997, Acoustics – Measurements of sound insulation in buildings and of building elements. Part 8: Laboratory measurements of the reduction of transmitted impact sound by floor coverings on a heavyweight floor.

ISO 140–11: 2005, Acoustics – Measurement of sound insulation in buildings and of building elements. Part 11: Laboratory measurements of the reduction of transmitted impact sound by floor coverings on lightweight reference floors.

ISO 140–16: 2006, Acoustics – Measurement of sound insulation in buildings and of building elements. Part 16: Laboratory measurements of the sound reduction index improvement by additional lining.

Austnes, J. and Hveem, S. (1983) Sound insulating lightweight floating floors (in Norwegian). Report No. 90. Norwegian Building Research Institute, Oslo, Norway.

Bies, D. A. and Hansen, C. H. (1996) *Engineering noise control*, 2nd edn. Spon, London.

Bodlund, K. (1987 Sound insulation of wood joist floors – rehabilitation project (in Swedish). Report No. 54, National Swedish Institute of Building Research, Stockholm.

Brekke, A. (1979) Sound transmission through single and double-leaf partitions (in Norwegian). DrEng thesis, Inst. for Husbyggingsteknikk, NTH, Trondheim, Norway.

Brouard, B., Lafarge, D. and Allard, J. F. (1995) A general method for modelling sound propagation in layered media. *J. Sound and Vibration*, 183, 129–142.

Brunskog, J. (2005) The influence of finite cavities on the sound insulation of double-plate structures. *J. Acoust. Soc. Am.*, 117, 3727–3739.

Brunskog, J. and Hammer, P. (1999a) The interaction between the ISO tapping machine and lightweight floors. Paper C in Licentiate thesis by Brunskog, J. Prediction of impact sound transmission of lightweight floors. Engineering Acoustics, LTH, Sweden. [Also published in *Acta Acustica/Acustica*, 89 (2003), 296–308.]

Brunskog, J. and Hammer, P. (1999b) A prediction model for the impact sound level of lightweight floors, incorporating periodicity. Paper D in Licentiate thesis by Brunskog, J. Prediction of impact sound transmission of lightweight floors. Engineering

Acoustics, LTH, Sweden. [Also published in *Acta Acustica/Acustica*, 89 (2003), 309–322.]

Brunskog, J. and Hammer, P. (2000) Prediction models for the impact sound level on timber floor structures: A literature survey. *J. of Building Acoustics*, 7, 89–112.

Craik, R. J. M. (1996) *Sound transmission through buildings using statistical energy analysis*. Gower Publishing Limited, Aldershot.

Craik, R. J. M. and Smith, R. S. (2000) Sound transmission through double leaf lightweight partitions. Part I: Airborne sound. Part II: Structure-borne sound. *Applied Acoustics*, 61, 223–269.

Cremer, L., Heckl, M. and Ungar, E. (1988) *Structure-borne sound*, 2nd edn. Springer-Verlag, Berlin.

Crocker, M. J., Battacharya, M. C. and Price, A. J. (1971) Sound and vibration transmission through panels and tie beams using statistical energy analysis. *Journal of Engineering for Industry*, 775–782.

Davy, J. L. (1991) Predicting the sound insulation of stud walls. *Proc. Internoise*, 91, 251–254.

Ferguson, N.S. (1986) A note on the critical frequency of sandwich panels. *J. Sound and Vibration*, 106, 171–172.

Gösele, K. (1964) Schallbrücken bei schwimmenden Estricken und anderen schwimmend   verlegten Belägen. *Berichte aus der Bauforschung*, Heft 35, Wilhelm Ernst & Sohn,   Berlin.

Heckl, M. (1959) Untersuchungen über die Luftschalldämmung von Doppelwänden mit Schallbrücken. Proceedings of the 3th ICA, Stuttgart. Elsevier, Amsterdam (1961).

Holmås, T. and Bjørklund, P. O. (1979) Wood joist floor with floating layer (in Norwegian). Siv.ing.-thesis in building acoustics, Inst. for Husbyggingsteknikk, NTH, Trondheim, Norway.

Homb, A., Hveem, S. and Strøm, S. (1983) Sound insulating constructions (in Norwegian), Report 28, NBI, Oslo, Norway.

Hongisto, V., Lindgren, M. and Helenius, R. (2002) Sound insulation of double walls – An experimental parametric study. *Acta Acustica/Acustica*, 88, 904–923.

Kristiansen, U. R. and Vigran, T. E. (1994) On the design of resonant absorbers using a slotted plate. *Applied Acoustics*, 43, 39–48.

Kropp, W. and Rebillard, E. (1999) On the air-borne sound insulation of double wall constructions. *Acustica*, 85, 707–720.

Lee, J.-H. and Kim, J. (2002) Analysis of sound transmission through periodically stiffened panels by space harmonic expansion method. *J. Sound and Vibration*, 251, 349–366.

Lindblad, L. (1968) Impact sound characteristics of resilient floor coverings. A study of linear and non-linear compliance. Bulletin 2, Division of Building Technology, LTH, Lund, Sweden.

Mead, D. J. (1972) The damping properties of elastically supported sandwich plates. *J. Sound and Vibration*, 24, 275–295.

Mechel, F. P. (1976), Ausweitung der Absorberformel von Delany und Bazley zu tiefen Frequenzen. *Acustica*, 35, 210–213.

Moore, J. A. and Lyon, R. H. (1991) Sound transmission loss characteristics of sandwich panel constructions. *J. Acoust. Soc. Am.*, 89, 2, 777–791.

Sharp, B. H. (1978) Prediction methods for the sound transmission of building elements. *Noise Control Eng.*, 11, 53–63.

Vér, I. L. (1971) Impact noise isolation of composite floors. *J. Acoust. Soc. Am.*, 50, 1043–1050.

Vigran, T. E., Kelders, L., Lauriks, W., Dhainaut, M. and Johansen, T. F. (1997) Forced response of a sandwich plate with a flexible core described by a Biot-model. *Acta Acustica/Acustica*, 83, 1024–1031.

Wang, J., Lu, T. J., Woodhouse, J., Langley, R. S. and Evans, J. (2005) Sound transmission through lightweight double-leaf partitions: Theoretical modelling. *J. Sound and Vibration*, 286, 817–847.

CHAPTER 9

Sound transmission in buildings. Flanking sound transmission

9.1 INTRODUCTION

With some exceptions, we have up to now treated sound transmission through a specific building element. A sound reduction index or an impact sound level is then ideally an element specification but as pointed out a number of times; the boundary conditions of an element may have considerable influence on the result. The type and properties of the connections to adjoining constructions are important factors when specifying the transmission properties of a given element. An example is the contribution to the total loss factor of an element by the vibration energy "leaked" to adjoining structures, making it advisable for laboratories following ISO 140 to determine the total loss factor of their test specimens. Another example of the importance of the couplings was presented in Chapter 8 when dealing with heavy double walls.

In this chapter we will deal with the interplay of building elements with the objective of predicting the airborne and impact sound transmission in real buildings, in which there are normally a number of transmission paths between the source and receiver. We shall look for models enabling us to predict the acoustic performance of buildings based on the acoustic performance of each element making up the complete structure. To prepare such models has been an important task for the European Standards Organization CEN and sound transmission inside buildings are covered by the standards EN 12354 Part 1 and Part 2. We have referred to these standards before concerning predictions of element performance. Here we shall show some examples of the full model.

The prediction accuracy of such models is obviously dependent on the types of element taking part and the complexity of the boundary conditions. Dealing with simple heavy constructions such as concrete, the accuracy will be good whereas combinations involving lightweight, multilayered elements are always difficult to handle. This should not prevent the use of these models in practical design cases, a use that contributes to the gathering of a larger information base for these standards.

By way of introduction, we shall refer back to Figure 6.3, which sketches a number of possible transmission paths between two rooms, one room being excited by a sound source. As shown, in addition to the airborne sound transmission through the partition wall, there will be energy transmitted by way of the flanking walls, through cracks and crannies, by way of the windows or a common ventilation or cable duct etc. We also pointed out that we shall reserve the concept of flanking transmission for the energy transport in the following way: the source excites the flanking constructions on the source side into vibration, thereby causing a part of this vibration energy to be transmitted to flanking constructions on the receiver side, which in turn radiates sound. Another important transmission path, which is not shown in Figure 6.3, is the transmission via a suspended ceiling common to two neighbouring rooms, i.e. the partition wall does not seal off the cavity above the ceiling. This certainly may result in some flanking transmission along the suspended ceiling panel, but the airborne sound transmission in the cavity between the basic ceiling and the suspended panel is normally more important.

We shall treat a number of these transmission paths before presenting a model where the flanking transmission is included. We shall start with a simple calculation of the sound transmission index of a partition made up of a combination of different parts, e.g. a wall including a window or door. We may use this result to exemplify the effect of non-intentional weaknesses of a partition such as badly sealed cable ducts or direct building defects such as cracks (slits) or apertures in the construction. Furthermore, sound transmission by way of common ventilation ducts will be looked at and, as an introduction to the theme of calculating the apparent sound reduction index, we shall treat the subject of sound transmission by way of a suspended ceiling.

9.2 SOUND REDUCTION INDEX COMBINING MULTIPLE SURFACES

When calculating the sound reduction index of a partition consisting of an assembly of two or more parts or surfaces, one normally assumes no interaction between the different parts; each part vibrates independently driven by the incident sound pressure. This is certainly a simplification but it may be justified by giving a rough and reasonable estimate. The total transmission factor τ_{total} of a number n of partial surfaces S_n having transmission factor τ_n will be given by

$$\tau_{\text{total}} = \frac{\tau_1 S_1 + \tau_2 S_2 + \cdots + \tau_n S_n}{S_1 + S_2 + \cdots + S_n} = \frac{\displaystyle\sum_{i=1}^{n} \tau_i S_i}{S_0}, \tag{9.1}$$

where S_0 is the total area. Expressed by the corresponding sound reduction indices, we find

$$R_{\text{total}} = 10 \cdot \lg \left[\frac{S_0}{\displaystyle\sum_{i=1}^{n} S_i \cdot 10^{\frac{R_i}{10}}} \right]. \tag{9.2}$$

An example showing the use of this expression is given in Figure 9.1, where there are just two components ($n = 2$) giving a diagram useful for dimensioning a partition containing a door or window. It should be noted that R_0 is the sound reduction index belonging to the total area S_0, i.e. the index *before* the smaller part of area S_1 with reduction index R_1 is inserted. The explicit expression, certainly assuming $S_1 \leq S_0$, is

$$R_0 - R_{\text{total}} = 10 \cdot \lg \left[1 - \frac{S_1}{S_0} + \frac{S_1}{S_0} \cdot 10^{\frac{R_0 - R_1}{10}} \right]. \tag{9.3}$$

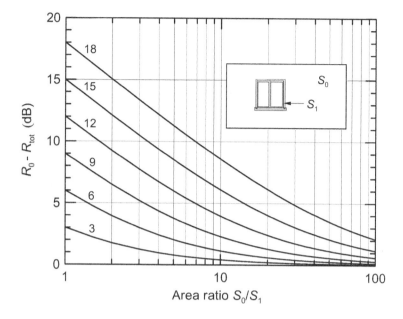

Figure 9.1 Calculation of the sound reduction index of a composite construction, e.g. a partition containing a door or window. The parameter is $R_0 - R_1$, the difference in the sound reduction indices.

9.2.1 Apertures in partitions, "sound leaks"

The results shown above may be used to estimate the influence of unintentional weaknesses in a building element, e.g. a partition, examples being badly sealed cable ducts or just building defects as slits or apertures. The problem is to find the effective transmission factor or the sound reduction index of such apertures, sealed or not. If the aperture is open one may, as a starting point, assume that the pertinent area has a reduction index equal to zero. However, as we shall see, this is too rough a simplification. Intuitively, one should expect that sound waves experience some difficulties in passing through apertures where the transverse dimensions are less than the wavelength. One should therefore expect such reduction indexes to be frequency dependent. A greater problem associated with such apertures is that one will encounter resonance phenomena that may in some frequency ranges give reduction indexes of the order −5 to −10 dB.

   Pioneering work for prediction tools in this field were performed by Wilson and Soroka (1965) and Gomperts and Kihlman (1967), the latter pair of authors also made comparison with full-scale measurements. These works are, however, limited to the treatment of apertures as being open, i.e. there is no kind of filling materials or sealing, an aspect taken up by Mechel (1986). Vigran (2004) extended the work on open apertures by including apertures having a variable transverse shape (conical apertures and wedge-shaped slits). The transmission aspects were, however, not the primary concern of this work but the design of resonator absorbers using panels having these types of perforation. The sound reduction index of an open aperture in a wall of thickness 100 mm is shown in Figure 9.2, illustrating the general frequency dependence of the transmission, the typical resonance phenomena (can you predict the frequency of these "dips"?) and the effect of shaping the aperture as a conical horn.

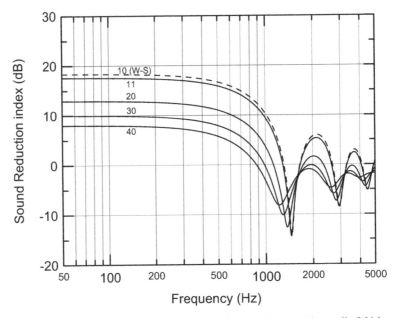

Figure 9.2 The sound reduction index at normal incidence of a conical aperture in a wall of thickness 100 mm as compared with a cylindrical aperture. The entrance radius is 10 mm and the exit radius (in mm) is indicated on the solid curves. Dashed curve gives results for a cylindrical aperture of radius 10 mm calculated from Wilson and Soroka (1965). After Vigran (2004).

For practical use in sound insulation in buildings, transmission through sealed apertures is more relevant and we shall therefore revert to the work by Mechel (1986). This is, however, a purely theoretical work but we shall make comparisons with results from other sources. The geometry used in calculating the transmission through a cylindrical aperture with radius a in a wall of thickness d is shown in Figure 9.3. The aperture is filled with a porous material characterized by propagation coefficient Γ and complex characteristic impedance Z_c. The aperture may also be sealed at one or both sides by ideal mass layers having a mass per unit area denoted m. The transmission factor for the aperture at plane wave incidence at an angle φ is given by

$$\tau(\varphi) = \frac{\rho_0 c_0}{\cos\varphi} \cdot \operatorname{Re}\{Z_{r2}\} \cdot \left| \frac{2Z_c}{Z_c(Z_1 + Z_2)\cosh\Gamma d + \left(Z_c^2 + Z_1 Z_2\right)\sinh\Gamma d} \right|^2 \quad (9.4)$$

with $Z_1 = j\omega m_1 + Z_{r1}$ and $Z_2 = j\omega m_2 + Z_{r2}$.

The quantity Z_r is the radiation impedance, here being the radiation impedance of a piston in an infinite baffle (see section 3.4.4). Having a diffuse incident field, Mechel gives a simple relationship between the transmission factor for diffuse sound incidence and normal incidence, as

$$\tau_{\text{diffuse}} = \pi \, \tau(0). \tag{9.5}$$

This implies that the sound reduction index for diffuse incidence will be approximately 5 dB smaller than for normal incidence.

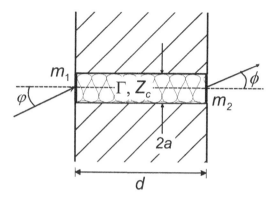

Figure 9.3 Cylindrical aperture in a wall filled with a porous material and sealed by mass layers.

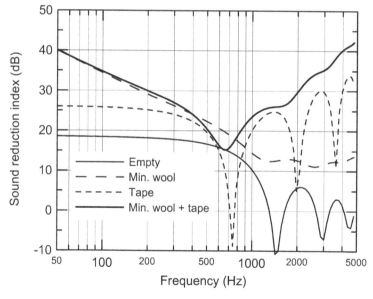

Figure 9.4 Sound reduction index at normal incidence. Cylindrical aperture of radius 10 mm in a wall of thickness 100 mm. Mineral wool of flow resistivity 5 kPa·s/m². Sealing tape of mass/area 100 g/m².

We shall illustrate the use of Equation (9.4) by again using a cylindrical aperture of radius 10 mm in a wall of thickness 100 mm (see Figure 9.4). The curve giving the result for a an open or empty aperture is the same as shown in Figure 9.2, however, now using

Mechel's Equation (9.4), which gives nearly identical results as predicted by Wilson and Soroka (1965).

Using sealing tape only, we get a mass-spring-mass resonance of a frequency which may also be easily calculated. We may also notice the effect of this resonance even if the aperture is filled with mineral wool. To calculate the propagation coefficient and complex impedance for the mineral wool a model by Mechel is used (see section 5.5.2). Calculating the reduction index for the case denoted "empty", we have introduced a small energy loss by setting the flow resistivity to 5 Pa·s/m^2, i.e. to 1/1000 of the one used above.

Calculation of the transmission factor for a slit shaped aperture becomes much more complicated than for a cylindrical aperture. There will be a dependency of the azimuth angle as well, in addition to an angle dependency of the impedances on both sides of the aperture. Calculating for a diffuse field incidence implies a complicated numerical integration. We shall therefore confine ourselves showing predicted results for normal incidence only, where we may use the same equations applicable for apertures, but where we have to exchange the radiation impedance for a circular piston with the proper one for an infinitely long and narrow slit. An analytical formula is available, expressed by Hankel and Struve functions (see e.g. Abramowitz and Stegun (1970)), but we shall not give it here. It should be noted, however, that the radiation impedance of a piston, of any shape (square, triangle etc.) and sitting in a baffle, may be calculated from the Fourier transform of the impulse response of the piston, i.e. the response when driven by a Dirac pulse (see Lindemann (1974)).

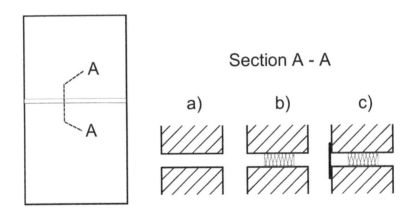

Figure 9.5 Test wall of dimensions 2250 mm x 1240 mm (2.8 m^2) for measuring the sound reduction index of a slit of depth 180 mm. a) Open slit; b) Slit containing 100 mm mineral wool; c) Slit with mineral wool and tape on one side.

We shall compare predicted and measured results for a slit, where the latter was a laboratory set-up to test the acoustic performance of different sealing methods, according to Alvestad and Cappelen (1982). In a test wall of surface area of 2.8 m^2 a slit was made across the aperture width of the wall, 1240 mm. The slit was formed by two steel UNP beams placed against one another, making a slit of adjustable width (see Figure 9.5).

In the experiments, a large number of combinations of mineral wool and sealing were tested on slit widths in the range 5 to 20 mm. We shall show some results for a slit width of 20 mm. When comparing with the predictions, we have in addition to the problem of sound incidence, two other deviations from the assumptions. The mineral wool did not fill the entire length of the slit, only approximately half of it. Also, the tape used for sealing was not properly specified in the written report. For our calculations, the first problem was solved by assuming a lower flow resistivity, approximately 1/6 of the one for the mineral wool used, whereas the second problem was eliminated by using a tape surface weight of 0.28 kg/m² measured on a similar brand of tape.

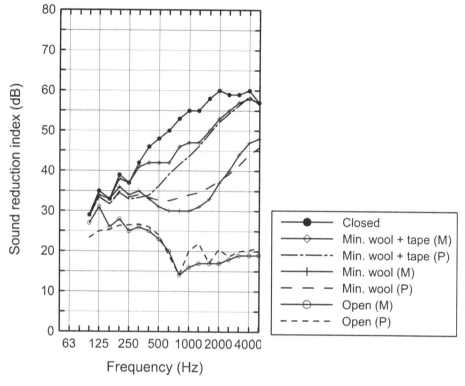

Figure 9.6 Sound reduction index of test wall of area 2.8 m² with slit of area 1240 mm x 20 mm and length 180 mm. Measuring situations as depicted in Figure 9.5. M – measurements in diffuse field. P – predicted for normal incidence.

As shown in Figure 9.6, we arrive at a reasonable fit between measured and predicted results using the assumptions mentioned above. The uppermost curve gives the reduction index in the case of the slit when completely closed, this reduction index being the one for a combined wall, a lower part being a 140 mm concrete slab and the upper part a 180 thick lightweight construction. Based on these data and the predicted reduction index for the slit, we may use Equation (9.2) to find the resulting reduction index for the cases a) to c) depicted in Figure 9.5. The largest discrepancy between measured and predicted results is found for case c), which may be attributed to resonance phenomena more predominant at normal incidence than when "smeared out" at diffuse incidence.

9.2.2 Sound transmission involving duct systems

Another example where we may assume that the sound transmission takes place by way of two or more independent surfaces, are when rooms are connected by a duct system. Two possible situations are depicted in Figure 9.7. In the first situation, the rooms are connected to the main ventilation duct system by way of the terminal units in the rooms. An open transmission path between the rooms is thereby established, indicated in the sketch as open-ended ducts.

The second situation is more complicated; the sound energy transmission between rooms takes place due to airborne sound transmitted through the duct walls in one room and a corresponding sound transmission out through the duct walls in the other room. This problem may arise when a duct passes through a very noisy room and the transmission may certainly affect areas other than the neighbouring room. One may also have situations where the neighbouring rooms have terminal units connected to the same system. It should also be noted, as indicated in b) that added to the airborne sound inside the duct there may also be flanking transmission by way of the solid duct walls.

The phenomena connected to transmission through duct walls, which is commonly referred to as acoustical *break-in* and *break-out*, will not be theoretically described here. We shall give a qualitative description only, together with a few examples from the literature. A fairly recent review paper is given by Cummings (2001).

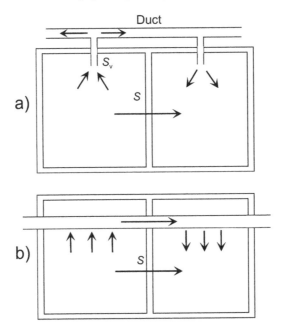

Figure 9.7 Sound transmission between rooms by way of a common duct system. a) transmission between terminal units; b) transmission through duct walls ("break-in" and "break-out").

As for the situation depicted in a) we are in a position to calculate the necessary attenuation, i.e. in the duct system between the terminal units, so as not to reduce the overall sound insulation between the rooms. This attenuation, which we shall denote D, may be defined as $10 \cdot \lg(1/\tau)$ where τ is the transmission factor for this transmission path;

the ratio of the radiated sound power from the terminal in the receiving room to the sound power entering the terminal in the source room. The attenuation includes a number of different sound energy losses; losses at branches and bends, energy reflected at terminal outlets, by silencers inserted in the duct etc. Prediction tools for all types of such losses are available in the literature (see e.g. ASHRAE (1999)).

Assuming that the area S of the partition is much larger than the area S_t of the terminal unit (grill, diffuser), we may use Equation (9.3) to calculate the necessary attenuation D. We may e.g. demand that the sound reduction index R (or R') of the partition should not be reduced by more than ΔR when the duct system is connected. This implies that

$$\Delta R \approx 10 \cdot \lg\left[1 + \frac{S_t}{S} \cdot 10^{\frac{R-D}{10}}\right] \qquad \text{or}$$

$$D \approx R - 10 \cdot \lg\left[\frac{S}{S_t}\left(10^{\frac{\Delta R}{10}} - 1\right)\right].$$

(9.6)

Example Having a partition of area 10 m^2 and sound reduction index 40 dB and demanding $\Delta R < 1$ dB, we need at least an attenuation D of 19 dB when there is a terminal unit of area (10 x 20) cm^2.

In principle, we should be able to handle the situation depicted in Figure 9.7 b), in the same way allocating a sound reduction index and an area to the duct walls. The problem is that reduction indexes of duct walls are difficult to predict with satisfactory accuracy, as they are dependent not only on the general shape but on details in the structure. Ducts with rectangular cross section have the lowest reduction indexes, whereas cylindrical ducts with an ideal circular cross section may exhibit very high reduction indexes, in particular at low frequencies. This is, however, only part of the aperture story. Details in the shape of cylindrical ducts, e.g. flanges, may dramatically decrease the reduction index. This may be explained, in the break-out case, by higher order vibration modes in the duct walls excited by the internal sound field, thereby increasing the radiated sound power.

Several definitions are in use concerning sound reduction indexes of duct walls. According to Cummings (2001), the most popular definition in the case of break-out is

$$R_{\text{duct}}^{\text{out}} = 10 \cdot \lg\left[\frac{W_i / S_i}{W_r / S_r}\right],$$

(9.7)

where W_i and W_r are the sound power inside the duct and the radiated power, respectively. The quantity S_i is the cross sectional area of the duct and S_r is the area of the sound radiating duct wall. Correspondingly, one may define a similar sound reduction index for sound transmission into the duct as

$$R_{\text{duct}}^{\text{in}} = 10 \cdot \lg\left[\frac{\tilde{p}^2 / 4\rho_0 c_0}{W / 2S}\right].$$

(9.8)

The sound pressure p is the pressure in the assumed reverberant field outside the duct, and W is the power transmitted into the duct having a cross-sectional area S. The rationale behind the number two in the denominator is that when the power is transported in the duct, one-half of it goes each way. Using the reciprocity relation (see section 6.6.1), assuming a point source being placed in a room outside the duct, thereafter inside the duct, one may show that there is a direct connection between these sound reduction indexes. Neglecting a possible attenuation of the wave in the axial direction over a duct length L, we get

$$R_{\text{duct}}^{\text{in}} = R_{\text{duct}}^{\text{out}} + 10 \lg \left[\frac{\pi S^2 f^2}{c_0^2 U L} \right], \qquad (9.9)$$

where f as usual is the frequency, and U is the outside perimeter of the duct.

There remains the problem of finding an expression for one of these reduction indexes. As mentioned above, this is not an easy task and it will be outside the scope of this book to provide a complete prediction model. We shall, however, give a couple of examples on measured and predicted results of the reduction index of ducts having rectangular and circular cross sections, respectively, both examples being taken from the paper by Cummings (2001).

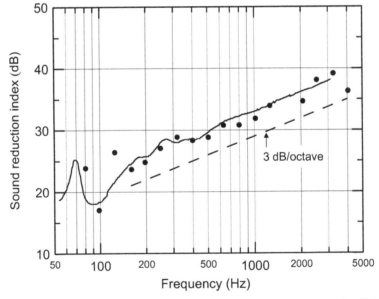

Figure 9.8 Sound reduction index for break-out. Duct with dimensions 457 x 229 mm and wall thickness 0.64 mm. Solid curve – predicted. Points – measured data in one-third-octave bands. Reproduced from Cummings (2001).

Figure 9.8 shows sound reduction index for break-out of a galvanized steel duct of cross section 457 x 229 mm and wall thickness of 0.64 mm. Measured results are as usual given in one-third-octave bands, showing a very good fit to predicted results using a wave solution, taking account of the coupling between the acoustic wave field inside the duct and the structural wave field in the duct walls. It should be noted that the reduction index increases only by approximately 3 dB per octave, as opposed to the 6 dB dependency of common single partitions in the mass controlled frequency range. The duct walls certainly are mass-controlled in the low and middle frequency range but the difference is attributed to the radiation factor of the duct, as it is a line-source of finite length.

Ducts with a circular cross section have in general a much higher reduction index than ducts with rectangular cross section. Figure 9.9 shows measured and predicted results applying to a "long-seam" circular duct of 1.22 mm galvanized steel having a diameter of 356 mm. Measured results are again given in one-third-octave bands and two different predictions are given. As pointed out above, not only the shape of the duct will be important but also details in its shape, and the duct in question had a single axial seam making the duct a little flat on both sides of it. The dashed curve applies to an ideal circular duct transporting a plane wave. The duct may then only move as a monopole source, making the walls subjected only to membrane stress, which results in a very high impedance for the internal plane wave and thereby a high reduction index.

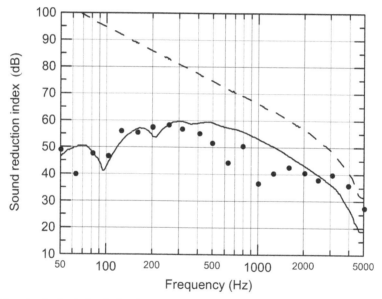

Figure 9.9 Sound reduction index for break-out. "Long-seam" circular duct, galvanized steel of wall thickness 1.22 mm and diameter 356 mm. Dashed curve – predicted for duct of ideal shape. Solid curve – predicted from "distorted" circular duct model. Points – measurement results in one-third-octave bands. Reproduced from Cummings (2001).

In the predicted result given by the solid curve, allowance has been made for a non-ideal circular shape, a shape which probably applies to all duct used in building practice, e.g. ducts with spiral seams. The internal acoustic wave then excites other vibration modes in the duct walls in addition to the pure membrane mode, resulting in an increase in the radiated sound and thereby a substantially lower reduction index.

In this case as well, plane wave propagation in the duct is assumed thereby giving a discrepancy between measured and predicted results when higher order modes may propagate. This effect may clearly be seen in the frequency range around 1000 Hz. In addition, one will always get high radiation at frequencies around the so-called *ring frequency* f_R, the frequency where the circumference of the duct is equal to the longitudinal wavelength of the duct material. Hence, the ring frequency of a circular duct of diameter D is given by

$$f_R = \frac{c_L}{\pi D}. \tag{9.10}$$

As the material of the duct is steel, for which the longitudinal wave speed c_L is approximately 5100 m/s, the ring frequency f_R will in this case be approximately equal to 4600 Hz. It should be noted than one may treat the duct walls as ordinary plane surfaces when the frequency exceeds the ring frequency.

9.2.3 Sound transmission involving suspended ceilings

A phenomenon related to the transmission by way of duct systems is transmission by way of suspended ceilings. In many cases, the partition between two rooms just extends to the ceiling (see Figure 9.10). The plenum chamber above the ceiling will then be suitable for installing building service equipment such as cables and duct systems. However, without taking proper precautions, the solution may give unsatisfactory sound insulation. Despite the fact that this is a common construction both in schools and office buildings, design tools for predicting the sound insulation are not satisfactory.

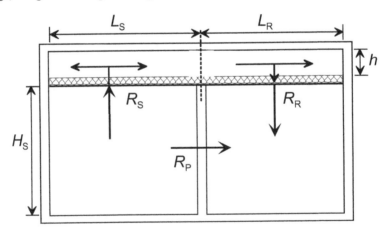

Figure 9.10 Sound transmission between rooms involving a suspended ceiling and a common plenum. The ceiling may as indicated include an absorber.

Probably, the first attempt to give a quantitative description was presented by Mariner (1959), applying diffuse field models both for the rooms and the plenum. In view of the fact that such plenums may have a height in the range 30–100 cm, maybe including an absorber, the diffuse field assumption is normally not valid. Several attempts have been made, partly by proposing simple corrections to the model of Mariner, partly by developing quite different analytical models. Mechel (1995) has presented a comprehensive theory, which composes the sound field by a forced wave solution and a modal expansion of the field in the plenum. This is analogous to the treatment of the duct transmission problem by Cummings (2001) (see the preceding section). We shall, as above, not reproduce Mechel's theory, just present one or two examples.

In addition, it might be interesting to compare these results with predictions based on a simpler model based on the diffuse field approach combined with a treatment of the plenum as a lined duct. In reality, this concerns two different models developed by Mechel (1980), of which we shall just present the one-dimensional variant.

The transmission factor for the transmission path by way of the ceiling and the plenum may be defined as

$$\tau_{cl} = \frac{W_{R,d}}{W_{S,d}}, \tag{9.11}$$

where $W_{S,d}$ and $W_{R,d}$ are the diffuse sound power incident on the ceiling in the sending (source) room and the radiated sound power from the ceiling in the receiving room, respectively. An alternative definition could also be used by referring to the power incident on the partition. This may be a more suitable definition if the task is to add contributions from several transmission paths. In this case; referring all transmission factor to a common surface area, e.g. the surface area of the partition, we may directly add the different transmission factors. The relationship between these alternative definitions of the transmission factor is

$$\tau_{cl,p} = \tau_{cl} \cdot \frac{L_S}{H_S}, \tag{9.12}$$

where the extra index p indicates reference to the partition. The quantities L_S and H_S are the length and height of the sending room, respectively (see Figure 9.10). Correspondingly, the relationship between the reduction indexes will be

$$R_{cl,p} = R_{cl} + 10 \lg \frac{H_S}{L_S}. \tag{9.13}$$

9.2.3.1 Undamped plenum (cavity)

We shall start by showing a comparison between measured and predicted results with the aid of Mechel's (1995) modal theory, where the suspended ceiling is made of 9.5 mm plasterboard and no absorbers in the cavity. Mechel does point to the fact that it is often difficult to find measured results containing sufficient specifications for materials and dimensions. He is therefore using data collected from several sources but even so, some data have to be estimated as well. Another problem concerning the sound reduction indexes is the reference surface chosen for the incident power. As discussed above, different definitions may be used. We shall, in each case, point to the definition used.

Another important assumption made in the prediction models is that there is a structural break both in the ceiling (and in a prospective absorber) above the partition. This implies that there is no direct coupling between the ceilings in the two rooms; there is no flanking transmission according to our strict definition. Whether this assumption is valid in practical cases is open to discussion but suspended ceilings are often an assembly of smaller units that may result in a less stiff structural coupling.

Measured and predicted results for the sound reduction index R_{cl} are shown in Figure 9.11, where the length of the ceiling is the same in both rooms, i.e. L_S and L_R are equal to 4.75 metre (see Figure 9.10). As stated above, the ceiling is made of 9.5 mm plasterboard and there is no absorber in the plenum. The height h of the plenum is 0.43 metre. In addition to the measured results and the results according to the model theory predicted results are given using the aforementioned one-dimensional model. The latter model is outlined in the following section.

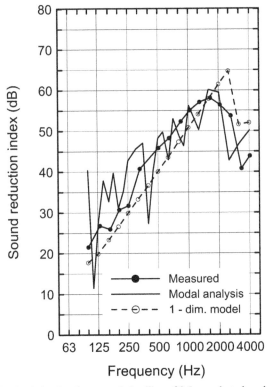

Figure 9.11 Sound reduction index R_{cl} of a suspended ceiling of 9.5 mm plasterboard without plenum absorber. Measured results and predicted results according to a modal theory, reproduced from Mechel (1995). Predicted results by a one-dimensional model by Mechel (1980). See Equation (9.20) (ε=2).

Looking at the results predicted by the modal theory these exhibit quite large excursions, presumably due to single frequency calculations, two for each one-third-octave band. Averaging over a larger number of single frequencies for each band would probably make the curve smoother. Apart from giving systematic lower results, the one-dimensional model is also quite good in this case.

9.2.3.2 One-dimensional model

In the model by Mechel (1980) we envisage that the ceiling in the sending room, having length L_S, is divided into elements of area $\Delta L_S \cdot b$ where b is the width of the room. Since the model is one-dimensional this width is certainly of no importance. The transmission through these elements is assumed to be uncorrelated, and the power transmitted into the plenum on the sending side is determined by the transmission factor of the ceiling, made up of the transmission factor $\tau_{S,pl}$ of the ceilings plates and the transmission factor $\tau_{S,a}$ of the prospective absorber. Similar conditions are imposed on the receiving side, hence we may write

$$\tau_S = \tau_{S,pl} \cdot \tau_{S,a} \quad \text{and} \quad \tau_R = \tau_{R,pl} \cdot \tau_{R,a}. \tag{9.14}$$

There is a major problem using transmission factor data, normally determined for diffuse field conditions, in this situation where the sound field on the receiving and sending side, respectively, are far from diffuse. Another problem is how the power transmitted into the plenum spreads out. One part will be transported in the direction of the partition, another goes in the direction of the backing wall, if there is one, and in turn is reflected back. The ratio of these two parts is characterized by the quantity s_S, a ratio that for lack of better alternatives is put equal to 0.5. Both parts will be attenuated during propagation in the direction of the partition, attenuation is assumed to take place exponentially as seen in the expression

$$W_{S,h}(x) \propto e^{-m_S \cdot x}, \tag{9.15}$$

where m_S is the power attenuation coefficient (m^{-1}) in the plenum of height h. With an absorber the plenum could be considered as a rectangular duct lined on one side with an absorber, and we may use routines for finding the complex propagation coefficient Γ or the complex wavenumber k' in such a duct (see e.g. Mechel (1976)). The attenuation coefficient is then found from

$$m = 2 \cdot \text{Re}\{\Gamma\} = -2 \cdot \text{Im}\{k'\}, \tag{9.16}$$

where Re and Im denote the "the real part of" and "the imaginary part of", respectively.

All power contributions are integrated over the length L_S to arrive at an expression for the total power passing over to the plenum on the receiver side. A similar derivation is carried out for the receiving side except for taking account of the attenuation partly caused by transmission through the ceiling and into the receiver room. Hence, the attenuation coefficient here is expressed as

$$m'_R = m_R + \frac{s_R \tau_R}{h}, \tag{9.17}$$

where s_R is the ratio mentioned above, applied to the receiver side of the plenum. Now assuming that all sidewalls in the plenum are totally reflecting, i.e. the reflection factor is equal to 1.0, we get

$$\tau_{cl} = \frac{s_S \, s_R \tau_S \tau_R}{m_S L_S \cdot m_R L_R} \cdot \frac{L_R}{h} \left(1 - e^{-2 m_S L_S}\right)\left(1 - e^{-2 m'_R L_R}\right). \tag{9.18}$$

The other extreme situation, assuming the sidewalls are totally absorbing, gives the same expression, however without the factor 2 in the exponential terms. In the case of minor attenuation in the plenum ($m_S L_S$, $m_R L_R \ll 1$), also putting $s_S = s_R = 0.5$, we arrive at the following very simple expression for the transmission factor

$$\tau_{cl} = \varepsilon^2 \tau_S \tau_R \frac{L_R}{4h}. \tag{9.19}$$

The constant ε is equal to 1.0 in the case of totally absorbing sidewalls and equal to 2.0 for totally reflecting ones. The corresponding sound reduction index for the transmission path through the suspended ceiling, assuming minor attenuation in the plenum, is then given by

$$R_{cl} = R_S + R_R - 10 \cdot \lg\left[\frac{\varepsilon^2 L_R}{4h}\right],$$ (9.20)

where $R_S = -10 \cdot \lg \tau_S$ and $R_R = -10 \cdot \lg \tau_R$.

The transmission factors τ_S and τ_R are calculated from Equation (9.14).

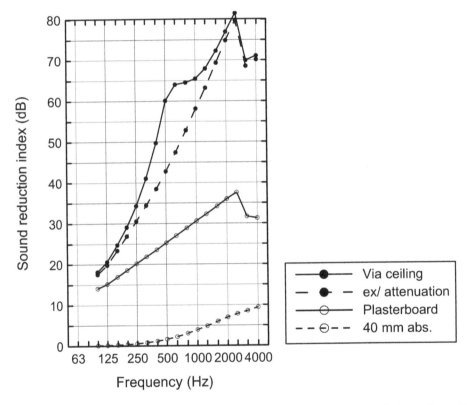

Figure 9.12 Sound reduction index R_{cl} for transmission by way of a suspended ceiling and plenum. Suspending ceiling of 9.5 mm plasterboard with 40 mm thick porous absorber. Calculated sound reduction index of plasterboard and absorber is presented by separate curves. The sound reduction index R_{cl} is also shown for the case of no attenuation in the plenum, (Equation (9.20)).

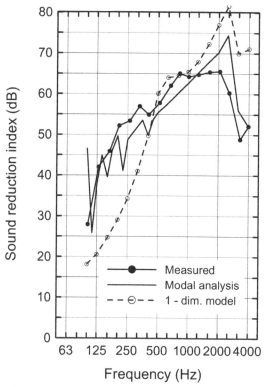

Figure 9.13 Sound reduction index R_{cl} of suspended ceiling of 9.5 mm plasterboard with 40 mm thick porous absorber. Measured results and predicted results according to a modal theory, reproduced from Mechel (1995). Predicted results by a one-dimensional model by Mechel (1980): Equation (9.18) with s_S and s_R equal to 0.5.

9.2.3.3 Damped plenum (cavity)

In the paper by Mechel (1995) an example is also given for a case where a 40 mm thick porous absorber is placed above the suspended ceiling, comparing measured and predicted result using his modal theory. However, before showing these results we shall use the full one-dimensional model, expressed by Equation (9.18), to predict the total sound reduction index and the single components included in the model as well.

Figure 9.12 shows the predicted sound reduction index R_{cl} using the full one-dimensional model, setting both s_S and s_R equal to 0.5, together with the results according to Equation (9.20). The latter result includes the reduction index of the absorber but the attenuation inside the plenum is not included. The purpose for doing so is to show the effect of the attenuation. As expected, the difference between these two curves exhibits a maximum when the height of the plenum, which here is 39 mm, is approximately equal to one half wavelength. A comment to be added here is that the attenuation is calculated for the fundamental mode only, i.e. plane wave propagation in the plenum.

In addition, the sound reduction index of the plasterboard and the porous absorber are shown separately, using material data given by Mechel (1995) and models presented in Chapter 6 (see sections 6.5.2.1 and 6.5.4). The plasterboard had a mass per unit area of 9.5 kg/m², a critical frequency of 3250 Hz and a loss factor of 0.1. The flow resistivity of the porous absorber was 10 kPa·s/m² (density 20 kg/m³).

In the same way as shown in Figure 9.11, where the plenum is without an absorber, measured and predicted results for the case when an absorber is added, is shown in Figure 9.13. We have also included predicted results using the one-dimensional model as shown in Figure 9.12. Comparing with the results for the case without any absorber, both measured and predicted results using the modal theory show that the absorber has a greater influence in the low frequency range than in the middle and high frequency ranges. This is in fact surprising and an explanation is not readily at hand. As also seen, the one-dimensional model completely fails at frequencies below approximately 400 Hz, giving a reasonable fit with the measured data just over a couple of octaves.

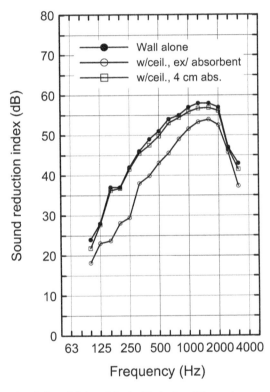

Figure 9.14 Sound reduction index of lightweight double leaf partition and apparent sound reduction index including suspended ceiling transmission path. Sound reduction indexes for suspended ceiling path are measured data from Figure 9.11 and Figure 9.13.

9.2.3.4 Apparent sound reduction index with suspended ceiling

Apart from being able to predict the sound reduction index for the transmission path across a suspended ceiling, our main interest will be the overall sound insulation between two rooms; i.e. the resulting apparent sound reduction index. We shall use the case of the suspended ceiling as an introduction to the general case where a number of different transmission paths are participating, including flanking transmission in a strict sense.

We shall then choose a partition that is reasonably "matched to" the suspended ceiling, which means that there is no point in choosing a very good partition if the expected transmission by way of the ceiling will be considerably larger than the transmission directly through the partition. As we wish to apply the measured results presented above, we choose a partition with sound reduction index as shown in Figure 9.14. This is a lightweight double leaf construction, two layers of 13 mm plasterboard on common steel studs, the cavity of 70 mm filled with mineral wool.

To calculate the apparent sound reduction index including the transmission path across the ceiling using measured data from Figures 9.11 and 9.13, we have to decide on the height of the room, i.e. the height of the partition. We may then refer the sound reduction index R_{cl} of the ceiling transmission path to the partition by applying Equation (9.13), thereafter summing up the pertinent transmission factors. This procedure is followed, giving the results as shown in Figure 9.14, where we can see a substantial decrease in sound insulation where there is no absorber present in the plenum. With the absorber, however, the sound insulation is only slightly poorer than the one offered by the partition alone.

9.3 FLANKING TRANSMISSION. APPARENT SOUND REDUCTION INDEX

The prediction of the effective sound insulation between rooms in a building, either airborne sound or impact sound; presupposes a model that includes all types of transmission path. An important type of transmission involves the flanking constructions of the wall or floor in question, and as pointed out several times, we shall reserve the notion of flanking transmission for this type of energy transport. In the model used here, we shall confine ourselves to neighbouring rooms; two rooms separated by a wall or floor. We shall also assume that the transmissions involving the different paths are independent and that all wave fields are diffuse.

We shall furthermore, primarily be treating airborne sound insulation due to the fact that it will normally represent a greater problem for prediction than the impact sound insulation. In addition, data used when calculating the apparent sound reduction index, e.g. vibration reduction index of junctions, may directly be applied to impact sound problems. For each transmission path of airborne sound we shall, as before, allocate a transmission factor. Referring these factors to the partition, we may express the apparent sound transmission index by

$$R' = 10 \cdot \lg\left(\frac{1}{\tau'}\right)$$

$$\text{with} \quad \tau' = \tau_d + \tau_f + \sum \tau_{dt} + \sum \tau_{it}. \tag{9.21}$$

The indices "d" and "f" indicate the transmission directly through the partition and by way of the flanking constructions, respectively. To complete the picture, two terms are added representing the sum of other direct or indirect transmission. The index "dt" indicates direct transmission through parts of the partition "added to" the basic wall or floor, such as doors, air vents or leaks (apertures, slits etc.). This theme has already been treated (see section 9.2) and we may include these factors directly in the transmission factor τ_d if appropriate. The last term includes all types of indirect transmission between the rooms, e.g. transmission by way of a duct system or a suspended ceiling, by way of windows in an outer wall etc. A note of warning: Reduction index product data for small units such as grills, diffusers etc. is normally referred to a standard area of 1 or 10 m², not to their actual area.

In the following we shall go into some details on the transmission expressed by the first two terms in Equation (9.21). Figure 9.15 gives an indication on the different transmission paths that we shall have to take into account. It should be noted, however, that the sketch only indicate what we may denote first order flanking paths; paths involving *one* element in the sending room, *one* junction or connection and *one* element in the receiving room. Applying the notions from the figure, the transmission factors τ_d and τ_f may be expressed as

$$\tau_d = \tau_{Dd} + \sum_n \tau_{Fd}$$

$$\text{and} \quad \tau_f = \sum_m \tau_{Df} + \sum_k \tau_{Ff}, \tag{9.22}$$

where the number of elements n, m and k will normally be four. The main contribution from the flanking transmission will normally be by paths indicated by "Ff". For multi leaf constructions, i.e. a double wall, a floating floor construction etc. other flanking paths are possible. The same applies for flanking constructions of such types.

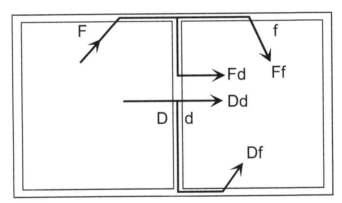

Figure 9.15 Sound transmission paths between two rooms. Letters "d" and "f" indicate direct and flanking transmission, respectively. Capital letters indicate sending room, small letters receiving room. Adapted from EN 12354–1.

The most obvious effect of flanking transmission is seen in cases where the flanking constructions are lightweight and have no structural breaks at the main partition; a partition which have good sound insulation properties as a stand-alone element. The requirement "not lightweight" is constantly underrated by builders. A classical example is given in Figure 9.16, where the curve shows the measured standardized level difference D_{nT} between two rooms for music practice. As shown in the insert to the figure, the 50 mm thick floating concrete slab passes unbroken below the partition. The situation is not helped by the fact that the partition should be good enough, having a total of six layers of 13 mm plasterboard. The sound insulation is a total failure at frequencies above 300 Hz. The result here represents approximately the sound insulation predicted for direct transmission through the 50 mm concrete slab, which has a critical frequency of about 400 Hz.

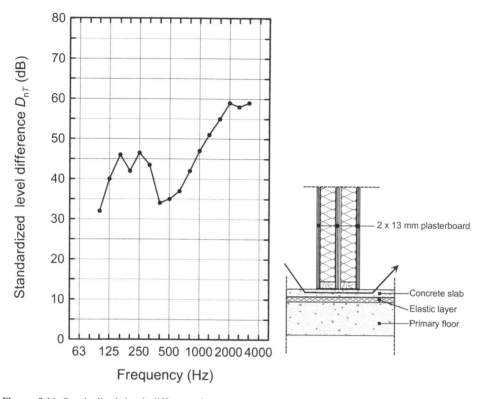

Figure 9.16 Standardized level difference between two rooms for music practice. Effect of flanking transmission by way of a floating concrete slab.

9.3.1 Flanking sound reduction index

As an example on the calculation of the apparent sound reduction index, where the flanking transmission is included, we shall assume that only flanking paths of the type indicated by "Ff" on Figure 9.15 are contributing. To make use of the flanking reduction index we shall have to express it by relevant and measurable quantities applied to the flanking elements involved. We then have a situation as sketched in Figure 9.17, where there is a direct transmission through the partition of area S_S and reduction index R_d together with a transmission path of the type mentioned above.

The apparent sound reduction index may then be written

$$R' = -10 \cdot \lg \left[10^{-\frac{R_d}{10}} + \sum 10^{-\frac{R_f}{10}} \right]. \tag{9.23}$$

The flanking sound reduction index R_f, involving the flanking element of area S_i in the sending (source) room and the corresponding one in the receiver room of surface area S_j, we shall define as

$$\left(R_\mathrm{f}\right)_{ij} = R_{ij} = 10\cdot\lg\left(\frac{1}{\tau_{ij}}\right) = 10\cdot\lg\left(\frac{W_\mathrm{S}}{W_{ij}}\right) = 10\cdot\lg\left(\frac{I_i\cdot S_\mathrm{S}}{I_j\cdot S_j}\right). \tag{9.24}$$

The quantity W_S is the sound power incident on the partition, and W_{ij} is the radiated power from element j in the receiving room caused by vibration transmission from the element i in the sending room. The sound intensity I_i is the intensity at the walls, assumed to be the same at all surfaces in the sending room. The intensity I_j, however, is the one radiated from the element j.

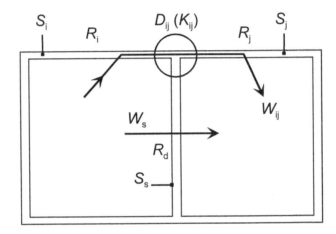

Figure 9.17 Sound transmission between two rooms, direct and flanking transmission paths.

Another choice, may be a more natural one, could be to use the actual area S_i in the definition given by Equation (9.24) in place of the area S_S of the partition. The advantage of using the latter, as pointed out in section 9.2.3, is that having a common reference area for all transmission paths one may directly sum up the accompanying transmission factors. The sound powers W_S and W_{ij} may as before be expressed as

$$W_\mathrm{S} = \frac{\left\langle \tilde{p}_\mathrm{S}^2 \right\rangle}{4\rho_0 c_0}\cdot S_\mathrm{S} \quad \text{and} \quad W_{ij} = \frac{\left\langle \tilde{p}_\mathrm{R}^2 \right\rangle_{ij}}{4\rho_0 c_0}\cdot A_\mathrm{R}, \tag{9.25}$$

where the sound pressure in the receiving room, having the total absorbing area A_R, is caused by flanking transmission only. The brackets indicate, as usual, a space averaging. Inserting into Equation (9.24), we get

$$R_{ij} = L_\mathrm{S} - \left(L_\mathrm{R}\right)_{ij} + 10\cdot\lg\frac{S_\mathrm{S}}{A_\mathrm{R}}. \tag{9.26}$$

To be able to represent the reduction index R_{ij} by the properties of the flanking elements we shall make use of the fact that the power W_{ij} may be expressed by the radiation factor of the pertinent element on the receiving side. We shall write

$$W_{ij} = \rho_0 c_0 \, S_j \left\langle \tilde{u}_j^2 \right\rangle \sigma_j, \tag{9.27}$$

which together with the second Equation (9.25) gives us

$$\left\langle \tilde{p}_R^2 \right\rangle_{ij} = \frac{4\rho_0^2 c_0^2 \, S_j \left\langle \tilde{u}_j^2 \right\rangle \sigma_j}{A_R}. \tag{9.28}$$

A corresponding equation may be found for the sending room, linking the sound pressure level and the velocity u_i of the flanking element there, by using the transmission factor τ_i of the flanking element. Hence, we shall write

$$\tau_i = \frac{W_t}{W_i} = \frac{\rho_0 c_0 \, S_i \left\langle \tilde{u}_i^2 \right\rangle \sigma_i}{W_i} = \frac{\rho_0 c_0 \, S_S \left\langle \tilde{u}_i^2 \right\rangle \sigma_i}{W_S}, \tag{9.29}$$

where W_t and W_i denote the transmitted and incident power on the flanking element, respectively. In the last expression, we have made use of the fact the sound intensity everywhere is the same at all surfaces in the sending room. Using the expression for W_S (see Equation (9.25)), we get

$$\left\langle \tilde{p}_S^2 \right\rangle = \frac{4\rho_0^2 c_0^2 \left\langle \tilde{u}_i^2 \right\rangle \sigma_i}{\tau_i}. \tag{9.30}$$

Equations (9.24), (9.25), (9.28) and (9.30) then give

$$\tau_{ij} = \tau_i \cdot \frac{\left\langle \tilde{u}_j^2 \right\rangle}{\left\langle \tilde{u}_i^2 \right\rangle} \cdot \frac{\sigma_j}{\sigma_i} \cdot \frac{S_j}{S_S} \tag{9.31}$$

or

$$R_{ij} = R_i + 10 \cdot \lg \frac{\left\langle \tilde{u}_i^2 \right\rangle}{\left\langle \tilde{u}_j^2 \right\rangle} + 10 \cdot \lg \frac{\sigma_i}{\sigma_j} + 10 \cdot \lg \frac{S_S}{S_j}, \tag{9.32}$$

where R_i is the sound reduction index of the flanking element (wall or floor) in the sending room. Whereas R_i tells us how easily the flanking element in the sending room is excited into vibrations, the second term gives us the velocity level difference of the respective elements when element i in the sending room is excited. We shall denote this term by the symbol $D_{v,ij}$.

Following the standard EN 12354–1, we may instead define the flanking sound reduction index as a mean value from measurements in two directions exchanging the sending and receiving rooms. We shall then write

$$\overline{R}_{ij} = \frac{R_i + R_j}{2} + \overline{D}_{v,ij} + 10 \cdot \lg\left(\frac{S_S}{\sqrt{S_i S_j}}\right), \qquad (9.33)$$

where we have introduced the direction averaged velocity level difference

$$\overline{D}_{v,ij} = \frac{1}{2}(D_{v,ij} + D_{v,ji}). \qquad (9.34)$$

9.3.2   Vibration reduction index

Determining data for flanking sound transmission is complicated, both by prediction and by measurement. In practice, one is normally compelled to use less accurate data than e.g. reduction indexes for walls and floors. Recently, a series of international standards has been developed for laboratory measurements of flanking sound transmission, both for airborne and impact sound transmission (see ISO 10848). This should contribute to a greater understanding of the problem and make more accurate data available.

The velocity level difference across a junction, which was introduced above is, as opposed to a sound reduction index, not an invariant quantity as it depends on the actual energy losses in the receiving element. This is quite analogous to the difference in sound pressure level between two rooms which is dependent on the absorption area in the receiving room. An invariant quantity for transmission across a junction is defined in EN 12354–1, being called *vibration reduction index* having the symbol K_{ij}. From this quantity, we may find the velocity level difference between elements i and j by correcting for the actual energy losses. We shall present examples below but for a complete picture we shall start with a presentation of the "classical" calculations concerning bending wave transmission across plate intersections involving three (T-junction) and four plates (see Cremer et al. (1988)).

9.3.2.1 Bending wave transmission across plate intersections

In Cremer's pioneering work one assumes that a plane bending wave is incident on an intersection involving three plates or four plates (see Figure 9.18), showing cross sections. All plates are assumed to be of infinite extent and a bending wave in the plate of thickness h_1 is assumed to be incident normally to the axis of the intersection, which is normal to the plane of the paper.

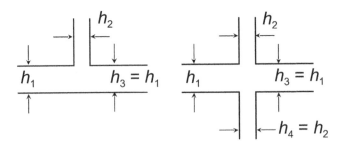

Figure 9.18 Cross sections through junctions involving three and four plates.

The reduction indexes calculated are defined by the ratio of the power in the bending wave transmitted to the power of the incident wave. These reduction indexes need not be identical to the velocity level difference as defined above. It should also be noted that the calculations do not account for longitudinal or transversal waves being generated at the intersection.

Two auxiliary quantities were introduced as follows:

$$\psi = \sqrt{\frac{m_2 \, B_2}{m_1 \, B_1}} \quad \text{and} \quad \chi = \frac{k_{B2}}{k_{B1}} = \sqrt[4]{\frac{m_2 \, B_1}{m_1 \, B_2}}, \tag{9.35}$$

which is the ratio of the impedances and the ratio of the wavenumbers of the actual plates, respectively. In practice, we shall normally find that the plates in line have identical material and thickness. With this assumption one may express the reduction indexes using a single parameter, which is the ratio of these auxiliary quantities. We then get

$$\frac{\psi}{\chi} = \frac{B_2 \, c_{B1}}{B_1 \, c_{B2}}, \tag{9.36}$$

where c_B is the phase speed. If the plates involved have identical material properties, we will further find that this ratio is given by the thickness ratio of the plates:

$$\left(\frac{\psi}{\chi} \right)_{\text{Same material}} = \left(\frac{h_2}{h_1} \right)^{\frac{5}{2}}. \tag{9.37}$$

The reduction indexes R_{12} and R_{13} for a T-junction are then given by

$$R_{12} = 20 \cdot \lg \left[\sqrt{\frac{2\chi}{\psi}} + \sqrt{\frac{\psi}{2\chi}} \right],$$

$$R_{13} = 10 \cdot \lg \left[2 + 2 \left(\frac{\psi}{\chi} \right) + \frac{1}{2} \cdot \left(\frac{\psi}{\chi} \right)^2 \right]. \tag{9.38}$$

The corresponding expressions applied to an intersection involving four plates are

$$R_{12} = 20 \cdot \lg \left[\sqrt{\frac{\chi}{\psi}} + \sqrt{\frac{\psi}{\chi}} \right] + 3 \, \text{dB},$$

$$R_{13} = 20 \cdot \lg \left[1 + \left(\frac{\psi}{\chi} \right) \right] + 3 \, \text{dB}. \tag{9.39}$$

The above equations are depicted in Figure 9.19 assuming that all plates have identical material properties. If this is not the case one has to substitute the dimension ratio h_2/h_1 by the quantity $(\psi/\chi)^{2/5}$, the latter with (ψ/χ) expressed by Equation (9.36). As pointed out above, these reduction indexes are defined by the bending wave power, i.e.

not by the velocities. The consequences are that the reduction index R_{13} will be equal to $D_{v,13}$ but we shall have to correct R_{12} to get $D_{v,12}$:

$$D_{v,12} = R_{12} + 10 \cdot \lg(\psi\chi). \tag{9.40}$$

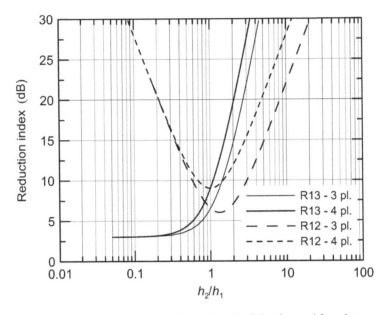

Figure 9.19 Reduction index of bending waves at intersections involving three and four plates, respectively, as a function of plate thicknesses. See Figure 9.18.

The idealized reduction indexes as shown in Figure 9.19 have limitations as to their practical applications due to the fact that no account is taken of the generation of other wave types at the intersection. Furthermore, the plates are of infinite extent and only normal incidence is treated. However, the data have been useful, serving as a first estimate in monolithic concrete building constructions. Comparing with the estimates given in EN 12354 (see below), we shall find that they are not too different from Cremer's data. Certainly, the former is based on later work by Kihlman (1967), who treated the case of random incidence on an intersection of four plates and Gerretsen (1979, 1986), who also compared with measurement data.

9.3.2.2 Vibration reduction index K_{ij}

This vibration reduction index is an attempt to establish a general invariant quantity characterizing the transmission across a joint between finite size elements under diffuse field conditions. Determining the index by measurements (see ISO 10848 series) implies measuring space time averaged velocities and structural reverberation time of the actual elements. The damping of the element, given by the reverberation time, is expressed by the *equivalent absorption length* a_i of the element i. The definition is then

$$K_{ij} = \overline{D_{v,ij}} + 10 \cdot \lg\left[\frac{l_{ij}}{\sqrt{a_i a_j}}\right], \tag{9.41}$$

where l_{ij} is the length of the junction between elements i and j. The relationship between reverberation time T and absorption length a is

$$a_i = \frac{2.2\pi^2 S_i}{c_0 T_i} \cdot \sqrt{\frac{f_{ref}}{f}}. \tag{9.42}$$

The reference frequency f_{ref} is chosen to be 1000 Hz.

It should be mentioned that one may establish a relationship between K_{ij} and a transmission factor (or reduction index) based on bending wave power in conformity with Cremer's definition. However, maybe of more interest due to recent work by Nightingale and Bosmans (2003) (see below) a similar relationship between the vibration reduction index K_{ij} and the coupling loss factor η_{ij} according to a SEA model may be established.

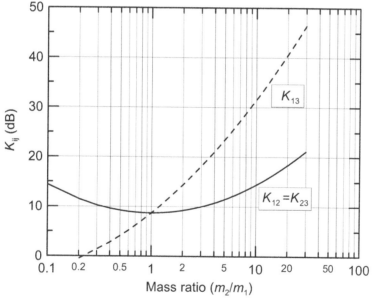

Figure 9.20 Vibration reduction index K_{ij} as a function of the mass ratio of elements at an intersection involving four plates. After EN 12354–1.

Altogether seven types of structural joints are considered in EN 12354 Part 1, for which we shall reproduce data for one type only: an intersection involving four homogeneous slabs. These data are shown in Figure 9.20, calculated using the following equations

$$K_{12} = 8.7 + 5.7 \cdot M^2 \qquad \text{(dB)},$$
$$K_{13} = 8.7 + 17.1 \cdot M + 5.7 \cdot M^2 \qquad \text{(dB)}, \qquad (9.43)$$

where M is the logarithmic ratio

$$M = \lg\left[\frac{\text{mass/area of element } j \ (\perp i)}{\text{mass/area of element } i}\right]. \qquad (9.44)$$

It should be noted that these expressions are assumed to be frequency independent.

9.3.2.3 Some examples of $D_{v,ij}$ and K_{ij}

Expressions given in the standard EN 12354 are based on a number of measurement results, both from laboratory experiments and from in situ observations. As expected, data for massive, heavy constructions are more commonly available than for lightweight, multilayered type of constructions. Estimating transmission along lightweight façades may be a real challenge. One should expect that the amount of data will increase with a prolonged use of prediction models such as given in EN 12354.

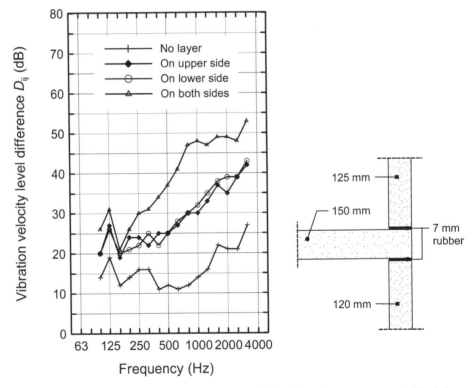

Figure 9.21 Measured vibration level differences in a test building of aerated aggregate concrete. Elastic layers (foam rubber) between wall and floor slabs. After Huse (1972).

Here we shall present results from a couple of special cases where the task is to increase the vibration reduction index by adding elastic layers to the joints. This will also give a very good illustration as to the effect of flanking transmission on the apparent sound reduction index.

The first example is taken from experiments on a small test "building", two rooms with floor area of 10 m^2, built upon one another. Walls and floors were aerated concrete slabs of density 600 kg/m^3, of thickness 125 and 150 mm, respectively. The vibration velocity differences between wall elements were measured both with direct contact between wall and floor elements and with elastic layers placed in the joints (see results and insert in Figure 9.21). The elastic layers were foam rubber bands, 7 mm thick and of 120 mm width.

Simultaneous results on K_{ij} cannot be presented as the structural reverberation times were not measured in these early experiments. Using the expression for a T-junction in EN 12354–1 we arrive at the value 6.9 dB for K_{ij}. What we see here is that $D_{v,ij}$ is frequency dependent having a minimum value of 11 dB. The elastic layers give quite a dramatic increase in the velocity level difference, which, as presented in the next section, nearly completely offset the effect of the flanking transmission on the sound insulation between the rooms.

In a NORDTEST project (see Brøsted Pedersen (1993)), aiming to work out a standard method for in situ determination of transmission properties of structural joints, similar results were obtained. The work included a number of laboratory measurements but measurements performed in a two-storey dwelling are maybe of special interest, where elastic layers were introduced on both sides of the floor separating two apartments (see results with a sketch of the situation in Figure 9.22). The floor was concrete of density 1750 kg/m^3, the walls being lightweight concrete of density 650 kg/m^3. The elastic layers are 4 mm thick polyurethane with cement (Sylomer P).

Concerning the addition of elastic layers to a joint, a point worth mentioning is that it could affect the energy losses from the floor to the connected structures. The total loss factor will decrease, which may affect e.g. the impact sound pressure level. This adverse effect may probably be of less importance than that which is gained by the decreased flanking transmission but the effect should certainly be considered.

9.3.3 Complete model for calculating the sound reduction index

Being now in the position to calculate the vibration reduction index, we shall return to the model for predicting the sound insulation between two rooms given by Equation (9.22). Computer software based on this model is commercially available, e.g. Bastian®, which include prediction of airborne and impact sound insulation as well as airborne sound insulation against outdoor noise. For airborne and impact sound insulation the prediction models found in EN 12354 Parts 1 and 2 are implemented. We shall conclude this chapter by presenting a few results of the airborne sound insulation based on this software, intended to take into account all transmission paths as sketched in Figure 9.15.

For a simple illustration of the principles behind these calculations, we shall set out to find the apparent sound reduction index R' in a case where only transmission paths of type Ff are contributing in addition to the direct transmission path. We may then use Equation (9.23), which we repeat here:

$$R' = -10 \cdot \lg \left[10^{-\frac{R_d}{10}} + \sum 10^{-\frac{R_f}{10}} \right]. \tag{9.45}$$

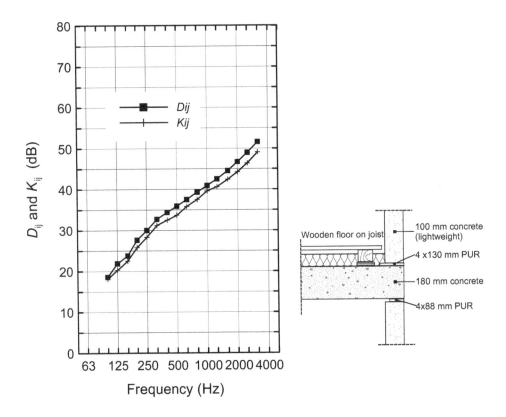

Figure 9.22 Vibration reduction index K_{ij} and velocity level differences D_{ij} with elastic layers on both sides of joint. Measurement data from a dwelling by Brøsted Pedersen (1993).

Each of these flanking reduction indexes R_f may be expressed by Equation (9.33), which combined with the appropriate vibration reduction index K_{ij} using Equation (9.41) gives

$$R_{ij} = \frac{R_i + R_j}{2} + K_{ij} + 10 \cdot \lg \left[\frac{S_S \sqrt{a_i a_j}}{l_{ij} \sqrt{S_i S_j}} \right]. \tag{9.46}$$

This expression found in EN 12354–1 was derived in a similar way as outlined in section 9.3.1. As mentioned in section 9.3.2.2 a relationship between the vibration reduction index K_{ij} and the coupling loss factor η_{ij} according to a SEA model may be established. Indeed, using the framework of SEA, Nightingale and Bosmans (2003) arrive at an identical expression for R', under the condition that the flanking reduction indexes apply to resonant transmission only. The added advantage, by using SEA, is that this enables them to formulate criteria to assess the suitability of the expressions in particular situations.

Besides finding proper estimates for K_{ij} relevant for the actual in situ situation, there remains the problem of finding corresponding data for the structural damping given by the absorption length *a*. As a first approximation, the standard suggests setting these equal to the areas *S*, which gives

$$R_{ij} \approx \frac{R_i + R_j}{2} + K_{ij} + 10 \cdot \lg\left[\frac{S_S}{l_0 l_{ij}}\right], \tag{9.47}$$

where l_0 is a reference length equal to 1.0 metre. In case of additional linings on the flanking elements, the resulting improvements ΔR_i and/or ΔR_j has to be added to the right-hand side of the equation. Before showing some predicted results based on this model, we shall return to the experiments which attempt to improve the vibration reduction index by applying elastic layers. The reason is that these measured results clearly illustrate the fine line between the dimensioning of the partition versus the flanking elements.

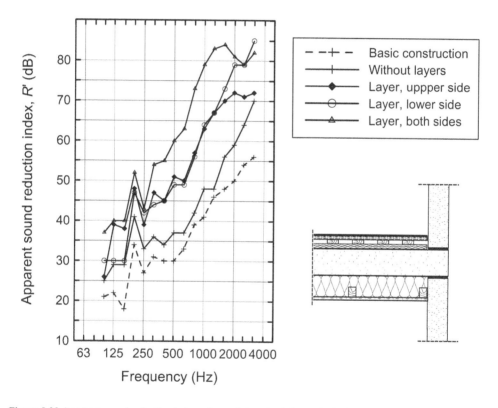

Figure 9.23 Apparent sound reduction indexes by applying elastic layers in the joints. The dashed curve applies to the primary construction; 150 mm lightweight concrete without elastic layers. The other curves apply to the case of a primary floor with additional floating floor and suspended ceiling. After Huse (1972).

In Figure 9.23 the results from a total of five separate measurements of the apparent sound reduction index are shown, which apply to the floor slab in the "building" having the measured velocity level difference depicted in Figure 9.21. The lowest curve (dashed) applies to the primary constructions without any elastic layers in the joints. The other curves show the result when the floor is improved by adding a floating floor as well as a freely suspended ceiling; see the inserted sketch. Without the elastic layers, we may, by taking the actual vibration level difference into account, easily estimate that these improvements, even when reducing the transmission through the floor to zero, will only add around 5–6 dB to our R'. In this case, the transmission between the rooms takes place solely by way of the flanking walls. Adding the elastic layers, however, we get, as expected from the measured vibration level differences, a greatly improved insulation.

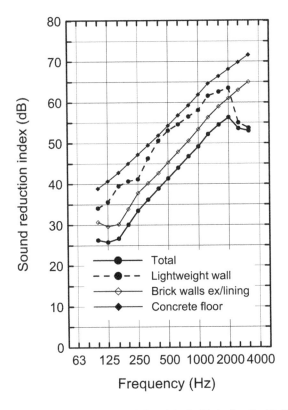

Figure 9.24 Apparent sound reduction index of a lightweight double leaf wall with flanking brick walls and concrete floors. Other curves indicate reduction index based on radiated power from the pertinent elements in the receiving room. Predictions after Bastian®.

Finally, as an example on the use of the complete model given in EN 12354 Part 1, we shall calculate R' for a conventional lightweight double leaf wall; two 13 mm plasterboard layers with a cavity of 75 mm thickness filled with mineral wool. The area of the wall is 15 m², partitioning two rooms with dimensions 8 x 5 x 3 metres (length x width x height) and 6 x 5 x 3 metres, respectively. The floors are 180 mm thick concrete, whereas the flanking walls are ½ stone brick, plastered on both sides.

Calculations are performed using Bastian®, giving results as shown in Figure 9.24. The lowest curve gives the reduction index based on the total transmitted power to the receiving room by way of all transmission paths, giving a weighted apparent sound transmission index R'_w of 44.8 dB. The other curves shows the predicted reduction index based on the power radiated from the various surfaces in the receiving room. The brick walls here seems to be the weakest part, accounting for approximately 80% of the power transmitted to the receiving room. To improve on the situation, measures have to be applied on these walls.

As an example, we have added simple linings to these walls; 13 mm plasterboard layers with a cavity depth of 50 mm, the cavity being filled with mineral wool. Linings are added in both rooms and the results are shown in Figure 9.25. The weighted apparent reduction index R'_w has now increased to 50.3 dB, and the power transmitted through the partition and via the flanking elements is approximately equal.

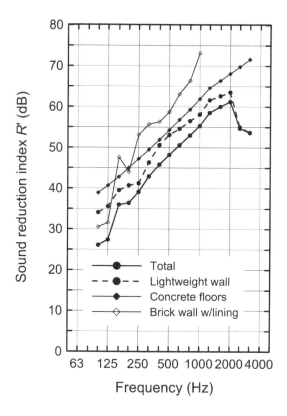

Figure 9.25 Apparent sound reduction index of a lightweight double leaf wall with flanking brick walls with linings. 180 mm concrete floors. Other curves indicate reduction index based on radiated power from the pertinent elements in the receiving room. Predictions after Bastian®.

9.4 REFERENCES

EN 12354–1: 2000, Building acoustics – Estimation of acoustic performance of buildings from the performance of elements. Part 1: Airborne sound insulation between rooms. [Parts 1–3 of this standard are adopted by ISO with number 15712 (from 2003).]

EN 12354–2: 2000, Building acoustics – Estimation of acoustic performance of buildings from the performance of elements. Part 2: Impact sound insulation between rooms.

ISO 10848: 2006, Acoustics – Laboratory measurement of the flanking sound transmission of airborne and impact sound between adjoining rooms. Part 1: Frame document. Part 2: Application to light elements when the junction has small influence. Part 3: Application to light elements when the junction has a substantial influence. [Part 4 was at draft stage in 2007.]

Abramowitz, M. and Stegun, I.A. (1970) *Handbook of mathematical functions.* Dover Publications Inc., New York.

Alvestad, S. and Cappelen, P. (1982) Sealing methods – Sound (in Norwegian). Internal Report No. 245, Norwegian Building Research Institute, Oslo.

ASHRAE (1999) *Handbook – HVAC applications.* American Society of Heating, Refrigerating and Air Conditioning Engineers, Atlanta, GA.

Brøsted Pedersen, D. (1993) Measurement of vibration attenuation through junctions of building structures. Report DTI 260 2 8011, Danish Technological Institute. (Nordtest Project No. 967-91.)

Cremer, L., Heckl, M. and Ungar, E. (1988) *Structure-borne sound*, 2nd edn. Springer-Verlag, Berlin.

Cummings, A. (2001) Sound transmission through duct walls. *J. Sound and Vibration*, 239, 731–765.

Gerretsen, E. (1979) Calculation of the sound transmission between dwellings by partitions and flanking structures. *Applied Acoustics*, 12, 413–433.

Gerretsen, E. (1986) Calculation of airborne and impact sound insulation between dwellings. *Applied Acoustics*, 19, 245–264.

Gomperts, M. C. and Kihlman, T. (1967) The sound transmission loss of circular and slit shaped apertures in walls. *Acustica*, 18, 144–150.

Huse, S. T. (1972) Flanking sound transmission in an experimental house of aerated concrete (in Norwegian). Project report. Akustisk Laboratorium, NTH, Trondheim.

Kihlman, T. (1967) Transmission of structureborne sound in buildings. Report no. 9, National Swedish Institute of Building Research, Stockholm.

Lindemann, O. A. (1974) Transient fluid reaction on a baffled plane piston of arbitrary shape. *J. Acoust. Soc. Am.*, 55, 708–717.

Mariner, T. (1959) Theory of sound transmission through suspended ceilings over partitions. *Noise Control*, 5, 13–18.

Mechel, F. P. (1976) Explizite Nährungsformel für die Schalldämpfung in rechteckigen Absorberkanälen. *Acustica*, 34, 289–305.

Mechel, F. P. (1980) Schall-Längsdämmung von Unterdecken. *wksb (Sonderausgabe)*, 16–29.

Mechel, F. P. (1986) The acoustic sealing of apertures and slits in walls. *J. Sound and Vibration*, 111, 297–336.

Mechel, F. P. (1995) Theory of suspended ceilings. *Acustica*, 81, 491–511.

Nightingale, T. R. T. and Bosmans, I. (2003) Expressions for first-order flanking paths in homogeneous isotropic and lightly damped buildings. *Acta Acustica/Acustica*, 89, 110–122.

Vigran, T. E. (2004) Conical apertures in panels: Sound transmission and enhanced absorption in resonator systems. *Acta Acustica/Acustica*, 90, 1170–1177.

Wilson, G. P. and Soroka, W.W. (1965) Approximation to the diffraction of sound by a circular aperture in a rigid wall of finite thickness. *J. Acoust. Soc. Am.*, 37, 286–297.

Subject index

Printed and bound by CPI Group (UK) Ltd, Croydon, CR0 4YY

01/11/2024

01782605-0005